# BASIC PRINCIPLES AND CLINICAL SIGNIFICANCE OF OXIDATIVE STRESS

Edited by **Sivakumar Joghi Thatha Gowder**

**Basic Principles and Clinical Significance of Oxidative Stress**
http://dx.doi.org/10.5772/59293
Edited by Sivakumar Joghi Thatha Gowder

**Contributors**

Olawale Ajuwon, Jeanine Marnewick, Lester Davids, Snezana Pejic, Ana Todorovic, Vesna Stojiljković, Ivan Pavlovic, Ljubica Gavrilović, Natasa Popovic, Snezana Pajovic, Jose Luis Silencio-Barrita, Pilar Hernandez-Sanchez, Juana Morillas-Ruiz, Ayla Ozcan, Metin Öğün, Rocio Ortiz-Butron, Vanessa Blas-Valdivia, Edgar Cano-Europa, Margarita Franco-Colin, Rodica-Mariana Ion, Baitullah Alipoor, Aziz Homayouni Rad, Leila Khalili, Davood Maleki, Dominika Boguszewska, Sivakumar Joghi Thatha Gowder

**Published by InTech**

Janeza Trdine 9, 51000 Rijeka, Croatia



**Notice**

Statements and opinions expressed in the chapters are these of the individual contributors and not necessarily those of the editors or publisher. No responsibility is accepted for the accuracy of information contained in the published chapters. The publisher assumes no responsibility for any damage or injury to persons or property arising out of the use of any materials, instructions, methods or ideas contained in the book.

**Publishing Process Manager** Iva Lipovic

**Technical Editor** InTech DTP team

**Cover** InTech Design team

First published November, 2015

Printed in Croatia

Additional hard copies can be obtained from orders@intechopen.com

Basic Principles and Clinical Significance of Oxidative Stress, Edited by Sivakumar Joghi Thatha Gowder

p. cm.

ISBN 978-953-51-2200-5

For EU product safety concerns:
IN TECH d.o.o., Prolaz Marije Krucifikse Kozulić 3, 51000 Rijeka, Croatia,
info@intechopen.com or visit our website at intechopen.com.

# Meet the editor

Dr Sivakumar Joghi Thatha Gowder received his academic training and carried out his research in institutions of high academic ranking in India and the USA (University of Madras, Chennai, India; All India Institute of Medical Sciences, New Delhi, India; UT Southwestern Medical Center, Dallas, TX, USA; LSH Health Sciences Center, Shreveport, LA, USA; and University of Pittsburgh School of Medicine, Pittsburgh, PA, USA). Currently, he is working as an Associate Professor at the College of Applied Medical Sciences, Qassim University, KSA. Dr Gowder has won prizes and awards at different levels of his academic career. He has developed his own research methods and techniques relevant to his research disciplines and has published several journal articles and book chapters. Currently, he serves as an author and editor of books; editor in chief for an international journal; editorial member and reviewer for journals; fellow and advisory board member of international organizations; and external examiner of doctoral thesis work for international universities. Dr Gowder has also served as an invited speaker and chairperson for international conferences.

# Contents

# Preface

"Stress" refers to the inadequate physiological response of an organism to certain demands (physical, mental, or emotional), whether actual or imagined. It is a natural phenomenon for all living organisms in the world to undergo different kinds of stress during their life span. "Stress" has become a common problem for human beings in this materialistic world. In this period, a publication of any material on stress would be helpful for human society. I believe it is an extremely ample opportunity for me to present the book *Basic Principles and Clinical Significance of Oxidative Stress* to the audience. This book targets all aspects of oxidative stress, including principles, mechanisms, and clinical significance. It covers four sections which are titled: Free Radicals and Oxidative Stress, Natural Compounds as Antioxidants, Antioxidants – Health and Disease, and Oxidative Stress and Therapy. Each of these sections is interwoven with the theoretical aspects and experimental techniques of basic and clinical sciences. This book will be a significant source to scientists, physicians, healthcare professionals, and students who are interested in exploring the effects of stress on human life.

In the first section, the authors have disclosed the physiological significance of the redox environment, involvement of proteolysis in the turnover of proteins oxidized by endogenously generated reactive oxygen species, mechanisms involved in the generation of reactive oxygen and nitrogen species, and the biological significance of antioxidants. In the second section, the authors have revealed the antioxidant potential of certain food items in order to be used as functional foods, the inadequate consumption of foods rich in antioxidants in a particular population, and the importance of rooibos and its flavonoids in the treatment of diseases. In the third section, we can find information on the impact of physical exercise on oxidative stress and antioxidant status in endometrial hyperplasia. The last section focuses on the application of photodynamic therapy in the treatment of cancer through the induction of oxidative stress.

I extend my gratitude toward my late mother and late father and my brothers for introducing me to higher education. While I was at home in India for a vacation, I had received blessings from my mother when I started this book project on "Oxidative Stress." After having completed it, I was under terrible stress since my affectionate mother was no longer with us.

My thanks go to Dr Yousef Aldebasi (Dean of the College of Applied Medical Sciences, Qassim University) for encouraging me to carry out this project. I should also thank Dr Sulaiman Al-Yahya (Vice President, Qassim University), Dr Naser Alwabel, Dr Mohammad Alorainy, and Dr Ahmed Aljarbou (members of Qassim University) for their encouragement. I am greatly indebted to my wife Anitha for her encouragement throughout this project as well as for her technical support. I must acknowledge the interest of and commitment from the Publishing Processing Manager of InTech, Ms Iva Lipović, whose patience and focus were a fantastic support in this project. Finally, I express deep and sincere appreciation to all the authors for their valuable contributions and scholarly cooperation for timely completion of this book.

**Dr Sivakumar Joghi Thatha Gowder**
Qassim University, College of Applied Medical Sciences
Kingdom of Saudi Arabia

Section 1

# Free Radicals and Oxidative Stress

Chapter 1

# Regulation of the Redox Environment

Edgar Cano-Europa, Vanessa Blas-Valdivia, Margarita Franco-Colin and Rocio Ortiz-Butron

Additional information is available at the end of the chapter

http://dx.doi.org/10.5772/61515

**Abstract**

All organisms maintain a strict redox environment, crucial for cell physiology, by preserving the pro-oxidant compounds generated during cell metabolism and from antioxidant system elements. In pathophysiological conditions, the redox environment is altered, causing oxidative stress, cell damage, and eventually cell death. In this chapter, we review the elements involved in the redox environment, including the oxidant, antioxidant, and glutathione systems. In addition, we summarize the physicochemical bases of the redox environment and the biological functions of the glutathione cycle. Finally, we propose a redox environment regulation model that considers some regulated variables that are actively involved in maintaining the redox environment: reactive oxygen species, reactive nitrogen species, and the redox couple $GSH^2/GSSG$.

**Keywords:** Redox environment, oxidant system, ROS, antioxidant system, glutathione

## 1. Introduction

All organisms maintain a strict redox environment, crucial for cell physiology, by preserving the pro-oxidant compounds generated during cell metabolism and from antioxidant system elements. In pathophysiological conditions, the redox environment can be altered, causing oxidative stress, cell damage, and eventually cell death. Two regulated variables are actively involved in maintaining the redox environment: the concentration of reactive oxygen species (ROS) and reactive nitrogen species (RNS), and the redox couple $GSH^2/GSSG$.

## 2. Main body

### 2.1. The redox system

Oxidizing system elements are free radicals and reactive species of various atoms or compounds such as oxygen, nitrogen, iron, copper, and glutathione (GSH). With respect to free radicals, they are molecules or molecular fragments containing one or more unpaired electrons in their atomic or molecular orbitals, which cause the molecule to be very reactive [1]. However, not all reactive species are free radicals: at a pH of 7.4 ± 0.1 they may be electroneutral molecules, able to donate electrons to free radicals, and oxidize transition metals present in cells. Although several groups of compounds are considered oxidants, those considered to be the most important from the physiological point of view are those compounds derived from oxygen and nitrogen: ROS and RNS. Under physiological conditions, the presence of ROS and RNS is required for diverse signaling pathways [2,3]. Among the most important elements of the oxidizing system in living organisms are superoxide anion radicals ($O_2^{\cdot-}$), hydroxyl groups ($^{\cdot}OH$), hydrogen peroxide ($H_2O_2$), nitric oxide (NO), and peroxynitrite ($ONOO^{\cdot}$) groups.

The presence of an electron in molecular oxygen ($O_2$) forms the free radical superoxide ($O_2^{\cdot-}$), which is considered to be a primary ROS that can interact with other molecules to generate secondary ROS or RNS [2,3]. Various metabolic pathways within the cell generate $O_2^{\cdot-}$, but the principal incomplete reduction route occurs in the mitochondrial respiratory chain: between 1 and 4% of $O_2^{\cdot-}$ is formed by the incomplete reduction of the total $O_2$ consumed in complex I (NADH: ubiquinone oxide-reductase) and III (cytochrome C oxide-reductase). The production of $O_2^{\cdot-}$ is also promoted by enzymatic complexes like xanthine oxidase (EC 1.1.3.22), cytochrome P450, nitric oxide synthase (NOS), or monoamine oxidase (EC 1.4.3.10).

Figure 1 shows how $O_2^{\cdot-}$ promotes the formation of $H_2O_2$, $^{\cdot}OH$, and $ONOO^-$ by various chemical pathways.

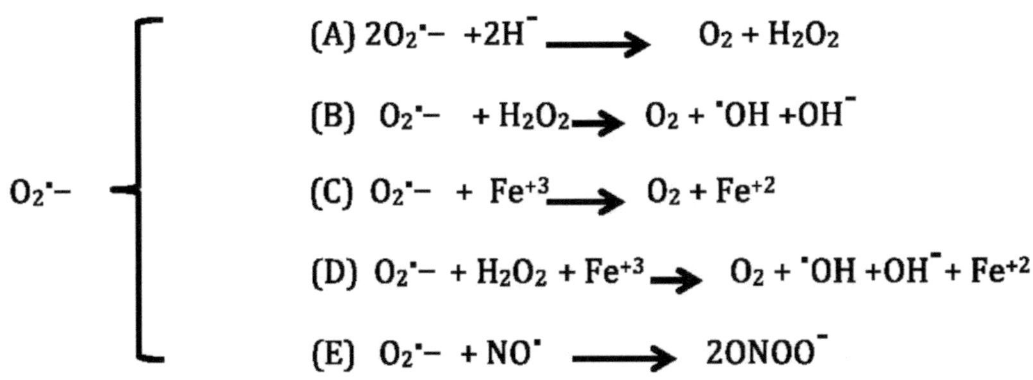


**Figure 1.** Reactive oxygen and nitrogen species formation from the superoxide anion radical. A) $O_2^{\cdot-}$ dismutation, which can be formed spontaneously or can be catalyzed by SOD. B) Haber-Weiss reaction. C) $Fe^{+3}$ to $Fe^{+2}$ reduction. D) Fenton reaction. E) Peroxynitrite ($ONOO^-$) formation.

In addition to forming ROS and RNS, $O_2^{\cdot-}$ can also inactivate enzymes involved in the antioxidant system, or in metabolic and signaling pathways such as catalase (EC 1.11.1.6), glutathione peroxidase (EC 1.11. 1.19), glyceraldehyde-3-phosphate dehydrogenase (EC 1.2.1.12), ornithine decarboxylase (EC 2.1.3.3), and adenylyl cyclase (EC 4.6.1.1) [1,4]. For example, $H_2O_2$ is formed by spontaneous dismutation or catalyzation by superoxide dismutase (SOD, EC 1.15.1.1). Although this compound has practically no oxidative effect on biomolecules, it plays a major role in the oxidative stress process by generating $^{\cdot}OH$ groups. When $H_2O_2$ molecules pass through cell membranes, they reach compartments containing transition metals, such as $Fe^{+2}$ or $Cu^{+}$, which can oxidize them and induce the formation of $^{\cdot}OH$ radicals. Thus, $^{\cdot}OH$ radicals may be formed through pathways involving $O2^{\cdot-}$ radicals or $H_2O_2$ [1,2,3,5].

The neutral form of the hydroxide ion ($OH^-$) is an $^{\cdot}OH$ group with high reactivity ($10^7$-$10^{10}$ $M^{-1}$ $s^{-1}$) and a very short half-life ($10^{-9}$ s) [6]. Thus, when $^{\cdot}OH$ groups are produced *in vivo*, they react near their site of formation and are very toxic to biomolecules such as DNA, proteins, and membrane phospholipids.

Many investigators consider NO to be radical because it contains an unpaired electron in its $2\pi_y{}^*$ orbital. When NOS (EC 1.14.13.39) metabolizes the conversion of L-arginine to L-citrulline, NO radicals are formed [7]. Under physiological conditions, NO is involved in processes such as neurotransmission, blood pressure regulation, defense mechanisms against pathogens, and immune response regulation [2]. In aqueous media, NO has a short half-life but in a hypoxic environment it presents greater stability with a half-life of more than 15 s [8]. The toxic effect of NO is in fact closely related to the formation of the secondary RNS $ONOO^-$, which occurs when NO is overproduced. $ONOO^-$ has a very high reaction constant ($7 \times 10^9$ $M^{-1}s^{-1}$), making it a powerful membrane-oxidizing agent that produces lipid peroxidation [2]. In fact, the reaction between $ONOO^-$ and carbon radicals ($^{\cdot}CO_2$) produces the secondary RNS, a nitrite radical ($^{\cdot}NO_2$). These radicals mediate protein nitrotyrosilation by reacting with the hydroxyl group of the tyrosine amino acid (TyrOH) to produce the tyrosyl free radical ($Tyro^{\cdot}$), which can then be neutralized by another $^{\cdot}NO_2$ [9].

### 2.2. Antioxidant system

In aerobic organisms, various processes such as growth, cell differentiation, apoptosis, and immune response require low concentrations of oxidants such that an increase generates a state of oxidative stress that causes cellular malfunction and even death. Therefore, an antioxidant system capable of neutralizing ROS and RNS is essential to the optimal function of many cellular operations.

The antioxidant system is divided into two subsystems: enzymatic and non-enzymatic. Within the non-enzymatic antioxidant system are organic compounds such as ascorbic acid, $\alpha$-tocopherol, carotenoids, flavonoids, and reduced GSH. While enzymes such as catalase (EC 1.11.1.6), glutathione peroxidase (GPX, EC 1.11.1.19), and SOD are considered the first line of enzymatic antioxidant defense, other enzymes, such as glutathione reductase (GR, EC 1.6.4.2), glutathione S-transferase (GST, EC 2.5.1.18), and thioredoxin reductase (EC 1.6.4.5) [1,2,5,10], also contribute to defense. Thus, maintaining the oxidation concentration of an organism at homeostasis is complex and must be executed within narrow limits. Figure 2 shows an overview of the interactions between the oxidant and the antioxidant system.

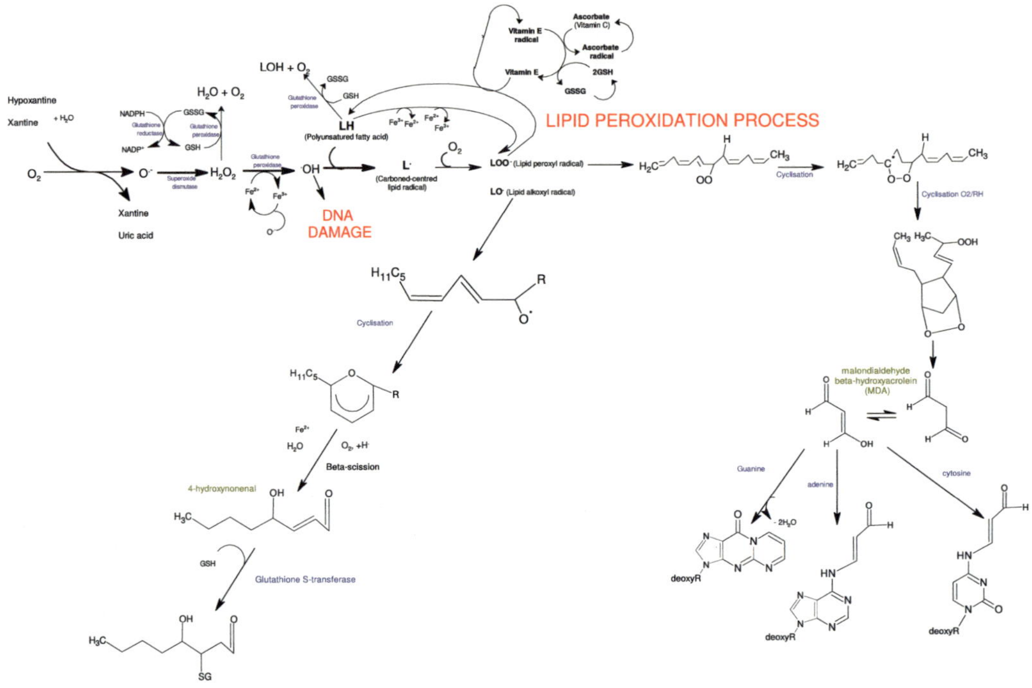


**Figure 2.** Reduced glutathione (GSH) role and the participation of other antioxidants (lipoic acid, vitamins C and E) in the ROS formation and lipid peroxidation. Reaction 1: $O_2^{\cdot-}$ is formatted from the $O_2$ reduction process mediated by the NAD(P)H oxidase and xanthine oxidase complex or produced by no enzymatic pathways such as those involving semi-ubiquinone in the mitochondrial respiratory chain. Reaction 2: $O_2^{\cdot-}$ is dismutated by superoxide dismutase (SOD) to $H_2O_2$. Reaction 3: $H_2O_2$ is neutralized by glutathione peroxidase (GPX) using GSH as cofactor. Reaction 4: Oxidized glutathione (GSSG) is reduced to GSH by reductase glutathione (GR) using NADPH.

### 2.3. Non-enzymatic antioxidant system

Inside the cell, several nucleophilic organic compounds, such as vitamin A, vitamin E, ascorbic acid, and dihydro lipoic acid, function to neutralize reactive species. For example, $\alpha$-tocopherol, commonly known as vitamin E, is the most active of tocopherols [11]. It has a chroman ring, which is responsible for its antioxidant activity, while the carbon phytyl side chain of vitamin E remains anchored to the cell membrane (Figure 3). Furthermore, $\alpha$-tocopherol is the lipid antioxidant that most potently inhibits *in vitro* lipid peroxidation propagation, and it is the dominant tocopherol found in the bloodstream.

Vitamin C or ascorbic acid is essential for the synthesis of various proteins such as collagen, oxytocin, and vasopressin. Its antioxidant capacity lies in the tocopheryl radical reduction that is anchored to cell membranes (reactions 8 to 12 of Figure 2). In addition, ascorbic acid can reduce nitrites and inhibit the formation of nitrosamines [12].

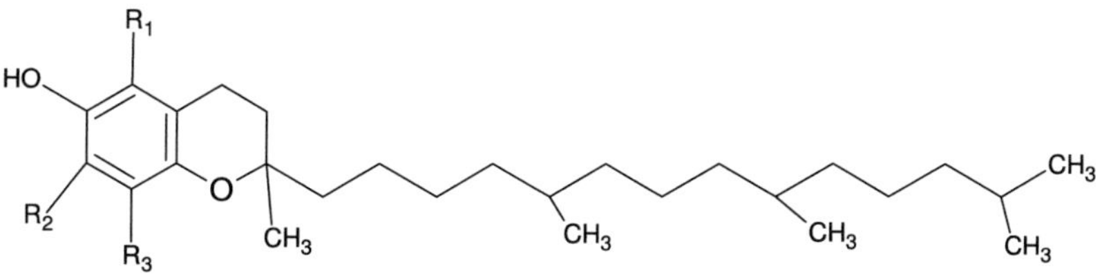


**Figure 3.** α-tocopherol structure (2R, 4'R, 8'R-tocopherol).

Dihydrolipoic acid (DHLA) or 6,8-dimercapto-octanoic acid is an organic compound that acts as a cofactor for some enzymes. The high electron density in the two SH groups gives the molecule characteristics of a nucleophile, which favors the 1,2-dithiolane-ring formation of the α-lipoic acid (LA) (chemical name: dithiolane-3-pentanoic acid), when it interacts with reactive species. LA and DHLA exhibit direct free radical scavenging properties, and as a redox couple, with a low redox potential of −0.32 V, is a strong reductant [13]. At a concentration of 0.05–1 nM, DHLA neutralizes the species $^{\cdot}OH$, $LOO^{\cdot}$, $ONOO^{-}$, and hypochlorous acid; furthermore it can form stable complexes with $Mn^{+2}$, $Cu^{+2}$, $Zn^{+2}$, and $Fe^{+2}$, preventing oxidative interactions between these transition metals and biomolecules. Finally, LA/DHLA also functions as an antioxidant by regenerating ascorbic acid when it reduces dihydroascorbate and the semidihydroascorbic radicals [12].

Now we move on to GSH, one of the most important antioxidant organic compounds because of its dependence on the cell redox environment.

## 2.4. The reduced glutathione-oxidized glutathione ($GSH^2$/GSSG) ratio and the reduction–oxidation (redox) environment

### *2.4.1. The physicochemical basis of the redox environment*

Cellular processes in aerobic organisms depend on oxidation processes, which promote the mobilization of electrons from organic molecules to oxygen; this produces the energy required to maintain cellular processes. In general, the presence of redox couples controls electron flow, but a reducing environment is also necessary: the redox environment is the sum of all the redox states of the reduction–oxidation couple that are inside the cell (intracellular redox environment) or in the extracellular fluid (extracellular redox environment).

Historically, the term redox state has been used to define the redox environment; however, based on physicochemical studies, Schafer and Buettner defined the redox state as the reduction potential of a redox couple [14]. Figure 4 shows the half-cell reduction potential ($E_{PC1/2}$) change of the redox pair involved in maintaining the redox environment, which involves the transfer of two electrons.

The next redox couple is involved in maintaining the redox environment because the curves generated show a third-order model. The p$K_a$ group is over the physiological pH and the reduced oxidized ratio is 1:100, 1:1000, or greater:

1. NADPH/$NADP^+$ system considering a 0.1-mM concentration.
2. Reduced thioredoxin/oxidized thioredoxin ($TrxSH_2$/TrxSS) system considering a 15-μM concentration.
3. $GSH^2$/GSSG system considering a 3-mM concentration.

$$\% \text{ de GSH oxidado} = -2255 - 32.59E_{hc} - 0.1450E_{hc}^2 - 0.00020488E_{hc}^3$$

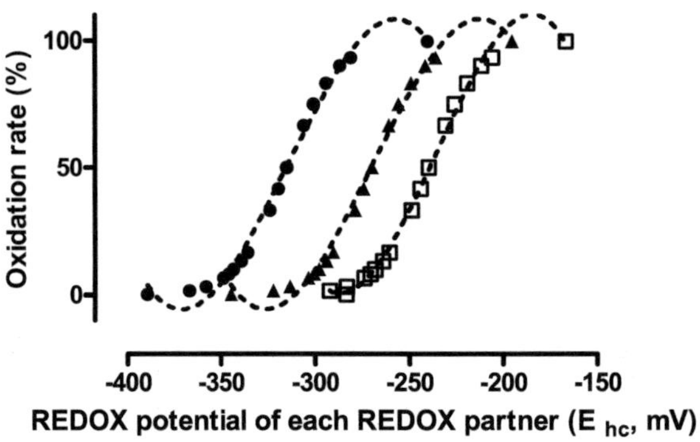


**Figure 4.** Theoretical reduction potentials of redox couple involved in maintaining the redox environment in response to the oxidized pair proportion increase. Symbol (·) represents the pair NADPH/$NADP^+$, symbol (▲) represents the pair $TrxSH_2$/TrxSS, and symbol (□) represents the pair $GSH^2$/GSSG. Modified from [14].

Furthermore, although all three intracellular buffer systems contribute to maintaining the redox environment, the most influential couple appears to be $GSH^2$/GSSG: it had the highest concentration, indicating that it better buffered the potential changes in the range between −300 and −100 mV. Interestingly, varying the reduced pair concentration caused changes in the $E_{PC1/2}$ that are perfectly associated with various processes like cell proliferation, differentiation, apoptosis, and necrosis [14–19].

## 2.5. The GSH cycle and biological functions

The synthesis of reduced GSH (γ-L-glutamyl-cysteinyl-glycine) occurs in the cell cytoplasm and involves two ATP-dependent enzymatic steps (Figure 5).

GSH synthesis begins when amino acids (e.g., glutamate, cysteine, and glycine) enter the cell. While glutamate and glycine may enter the cell by secondary active transport, cysteine enters by a neutral amino acid transport mechanism. (Some glutamate transporters can also transport cysteine.) In fact, cysteine is considered the limiting amino acid for GSH synthesis because it is present at a lower concentration in the plasma and has a lower Km [20].

Once amino acids have entered the cell, γ-glutamylcysteine synthetase (γ-GCL, EC 6.3.2.2) produces γ-glutamylcysteine. This formation process involves two steps: the interaction

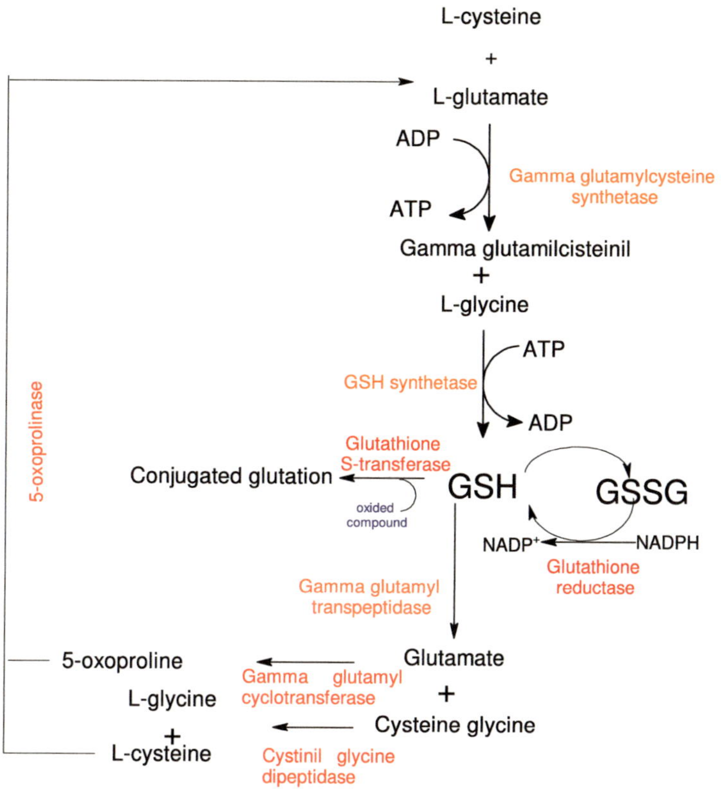


**Figure 5.** Reduced glutathione (GSH) cycle. Glutathione reductase (GR); glutathione S-transferase (GST), and oxidized glutathione (GSSG).

between glutamate and ATP in the presence of $Mg^{2+}$ to form γ-glutamylphosphate (intermediate) and the intermediate interaction with cysteine and ADP release [21]. The first step is the most important in the formation of GSH because γ-GCL is the GSH synthesis-limiting enzyme. γ-GCL is a heterodimeric enzyme composed of a catalytic subunit called a heavy subunit (γ-$GCL_H$ Mr ≈ 73 kDa) and a regulatory subunit or light subunit (γ-$GCL_L$ Mr ≈ 31 kDa). The activity of γ-GCL depends primarily on the substrates and is inhibited by GSH. Specifically, the activity of γ-$GCL_L$ is controlled by kinases such as protein kinase A (PKA) and protein kinase C (PKC) [22].

Thermodynamically, two processes can occur once the γ-GCL is formed: it may form GSH when it combines with glycine, acting as GSH synthetase (GS, EC 6.3.2.3), or it may interact with the γ-glutamyl cycle transferase to form 5-oxo-L-proline and L-cysteine. The prevailing way depends on the $K_m$ of each enzyme; under physiological conditions, the $K_m$ of GS is 12 times higher than that of γ-glutamyl cycle transferase. Thus, more than 95% of the time, GSH formation is promoted [23].

Once GSH is synthesized, there are several processes in which it participates:

1. Hydrolyse plasma GSH hydrolysis for cell GSH *in novo* synthesis; for example, when hepatocytes secrete GSH, another cell can hydrolyze it in its precursors (cysteinyl-glycine and glutamate) by γ-glutamyl-transpeptidase (γ-GT, EC 2.3.2.2) that are expressed on the outer plasma membrane. Compounds, such as cysteinyl-glycine or its S-conjugates, may be subject to peptidases, causing them to form free amino acids that can be introduced into the cell and initiate GSH formation [23,24].

2. Detoxify electrophiles to conjugate electrophiles with α and β unsaturated carbonyls by glutathione S-transferase (GST, EC. 2.5.1.18). This reaction results in electrophile removal and glutathione S-conjugate metabolism by the γ-GT enzyme and cysteinyl-glycine peptidase [24]. However, this process is not always beneficial for the cell because sometimes more toxic species can be produced, as we will discuss later.

3. Detoxify hydrogen peroxide by glutathione peroxidase (GPX, EC 1.11.1.19) [25].

4. Maintain ascorbic acid and vitamin E levels [11,12].

5. Communicate intracellular processes as internal modulator of several signaling pathways [26].

6. Modulate NMDA receptors in the central nervous system [27].

7. Transport metals (e.g., $Cu^{+2}$, $Hg^{+2}$, $Pb^{+2}$, and $Zn^{+2}$) [28].

GSH mitochondrial concentration is approximately 11–15 mM. Entry of GSH into the mitochondria is dependent on electroneutrality conveyors such as tricarboxylic or dicarboxylic acids [28]. In general, the GSH/GSSG ratio is greater than 10 for cells and organelles like mitochondria and nuclei, whereas the endoplasmic reticulum has a lower GSH/GSSG ratio between 1 and 3 [14].

### 2.6. Redox system regulation model

For several years, physiologists have sought to establish a simple model for studying physiological variables in comparison with cybernetic models [30].

The generalities of the model are presented in Figure 6. To be regulated, any physiological variable requires the following characteristics:

1. A value that fluctuates over a narrow range.

2. A sensor that reports to a comparator, which compares the variable value to a set point.

3. An integrator center that sends information loops to the input (feedback) or output (feed forward) elements so that these loops will activate variables that are generally controlled to keep the regulated variable within a narrow range.

Although this model is generalized for all physiological variables, it may be adapted for use with intracellular variables.

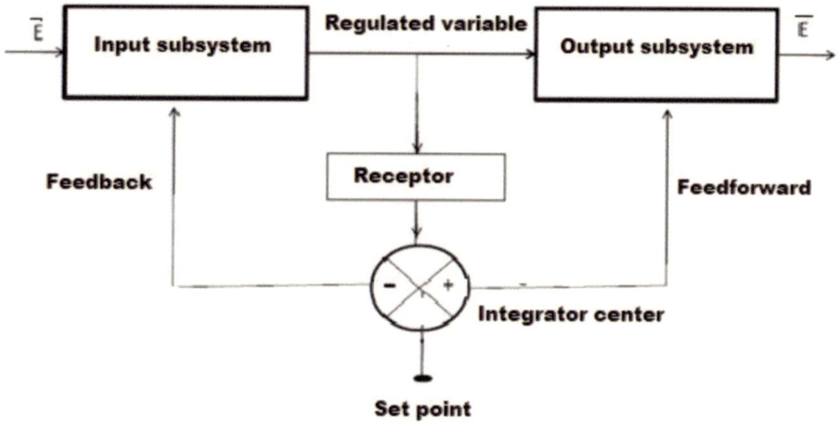


**Figure 6.** Control systems general model. Modified from [30].

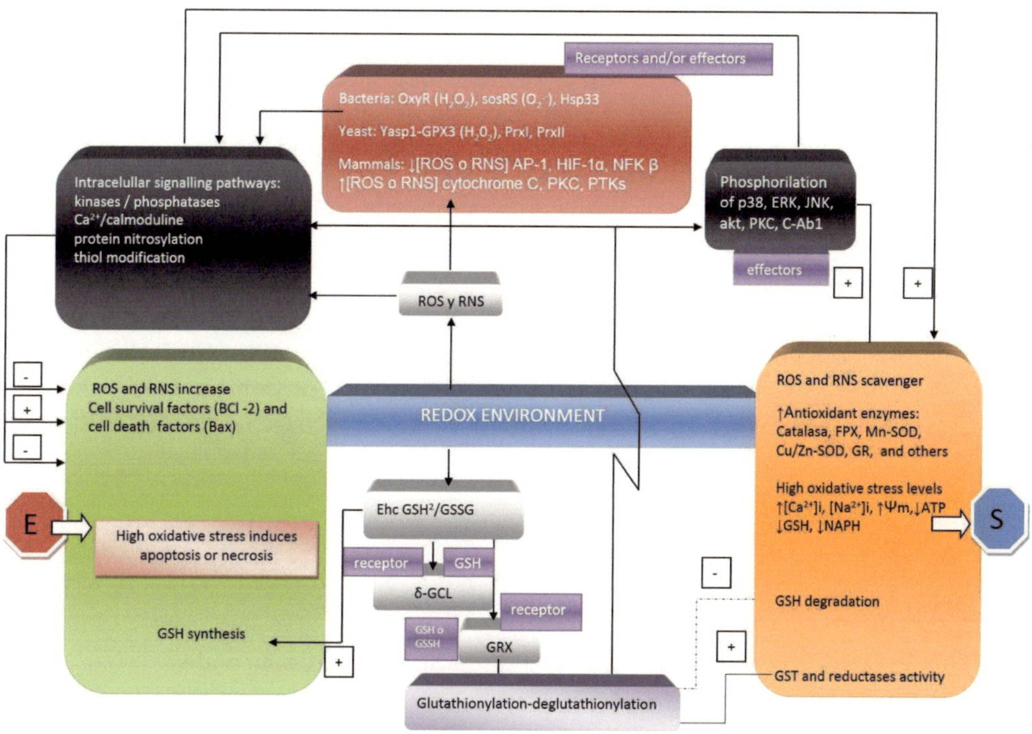


**Figure 7.** Redox environment regulation model proposed. Our model considers the GSH and ROS as regulated variables, which are sensed by intracellular proteins. Both act as receptors and integrators. In turn is produced a response to compensate the $E_{PC1/2}$ reduction or to increase ROS production to maintain the redox environment.

In particular, our research group considers the intracellular redox environment as a regulated variable since it has the elements described above. In Figure 7, we present the proposed model.

In our model we consider two actively regulated variables involved in maintaining the redox environment: ROS and RNS concentrations, and the couple $GSH^2/GSSG$ $E_{PC1/2}$ redox.

ROS and RNS are sensed in microorganisms predominantly by proteins like oxyR, sosRS, Hsp33, PRXI, PrXII, and the Yap1-GPX3 complex. These proteins not only act as receptors, they also modulate signaling pathways. The final function of intracellular communications is to reduce the reactive species generation and/or neutralize them by stimulating the expression of antioxidant enzymes. Furthermore, it has been proposed that these routes activate cell survival pathways to prevent death, although in extreme oxidative stress apoptosis is favored even above necrosis.

This representation using the basis of a cybernetic model is new, and it allows the correlation of the effects of all systems, especially the GSH cycle to maintain homeostasis of the redox environment.

Finally, we can conclude that the redox environment is an essential variable to the entire system: maintaining a good immune response [31] and suitable neurogenic events [32], controlling growth, behavior, propagation, and differentiation of tumor cells [33], and most of all ensuring the correct function of organelles like the endoplasmic reticulum [34]; however, the redox environment has also been related with some undesirable events such as drug resistance in certain tumor cells [35]. Therefore, continued study of the redox environment is important to uncovering its role in signal transduction, disease, and health.

## Acknowledgements

This study was partially supported by SIP 20140748 and 20150798. E.C-E, M.F-C., and R.O-B are fellows of EDI, COFAA, and SNI. V.B-V is a fellow of SNI. Carolyn Unck edited the language of this manuscript. We thank Instituto Politecnico Nacional for supporting this manuscript.

## Author details

Edgar Cano-Europa[1], Vanessa Blas-Valdivia[2], Margarita Franco-Colin[1] and Rocio Ortiz-Butron[2*]

*Address all correspondence to: rocipn@yahoo.com.mx

1 Metabolism Laboratory of Physiology Department, National School of Biological Sciences, National Polytechnic Institute, Mexico City, Mexico

2 Neurobiology Laboratory of Physiology Department, National School of Biological Sciences, National Polytechnic Institute, Mexico City, Mexico

## References

[1] Halliwell B, Gutteridge JMC. Free radicals in biology and medicine, Oxford University Press, 1999.

[2] Valko M, Leibfritz D, Moncol J, Cronin MT, Mazur M, Telser J. Free radicals and antioxidants in normal physiological functions and human disease. Int J Biochem Cell Biol 2007; 39: 44–84. DOI: 10.1016/j.biocel.2006.07.001

[3] Buettner GR, Wagner BA, Rodgers VG. Quantitative redox biology: an approach to understand the role of reactive species in defining the cellular redox environment. Cell Biochem Biophys 2013; 67(2): 477–483.

[4] Gutteridge J, Halliwell B. Free radicals and antioxidants in the year 2000: A historical look to the future. Ann New York Acad Sci 2000; 899: 136–147. DOI: 10.1111/j.1749-6632.2000.tb06182.x

[5] Wlodek L. Beneficial and harmful effects of thiols. Pol J Pharmacol 2002; 54: 215–223.

[6] Pastor N, Weinstein H, Jamison E, Brenowitz M. A detailed interpretation of OH radical footprints in a TBP-DNA complex reveals the role of dynamics in the mechanism of sequence-specific binding. J Mol Biol 2000; 304: 55–68. DOI: 10.1006/jmbi.2000.4173

[7] Brett DS, Snyder SH. Isolation of nitric oxide synthetase, a calmodulin-requiring enzyme. Proc Nat Acad Sci 1990; 87: 682–685.

[8] Chiueh CC. Neuroprotective properties of nitric oxide. Ann New York Acad Sci 1999; 890: 301–311. DOI: 10.1111/j.1749-6632.1999.tb08007.x

[9] Winterbourn CC, Kettle A J. Radical-radical reactions of superoxide: a potential route to toxicity. Biochem Biophys Res Commun 2003; 305: 729–736. DOI: 10.1016/S0006-291X(03)00810-6

[10] Dorado-Martínez C, Rugerio-Vargas C, Rivas-Arancibа S. Estrés oxidativo y neurodegeneración. Rev Fac Med UNAM 2003; 46: 229–237.

[11] Van Acker SABE, Koymans LMC, Bast A. Molecular pharmacology of vitamin E: structural aspects of antioxidant activity. Free Rad Biol Med 1993; 15: 311–328.

[12] Fang YZ, Yang S, Wu G. Free radicals, antioxidants, and nutrition. Nutrition 2002; 18: 872–879. DOI: 10.1016/S0899-9007(02)00916-4

[13] Moini H, Packer L, Saris NE. Antioxidant and prooxidant activities of alpha-lipoic acid and dihydrolipoic acid. Toxicol Appl Pharmacol 2002; 182: 84–90. DOI:10.1006/taap.2002.9437

[14] Schafer FQ, Buettner GR. Redox environment of the cell as viewed through the redox state of the glutathione disulfide/glutathione couple. Free Rad Biol Med 2001; 30: 1191–1212. doi:10.1016/S0891-5849(01)00480-4

[15] Cai J, Jones DP. Superoxide in Apoptosis. Mitocondrial generation triggered by cytocrome c loss. J Biol Chem 1998; 273: 11401–11404. DOI: 10.1074/jbc.273.19.11401

[16] Cai J, Wallace DC, Zhivotovsky B, Jones DP. Separation of cytochrome c-dependent caspase activation from thiol-disulfide redox change in cells lacking mitochondrial DNA. Free Rad Biol Med 2000; 29: 334–342. DOI: 10.1016/S0891-5849(00)00312-9

[17] Hwang C, Sinskey AJ, Lodish HF. Oxidized redox state of glutathione in the endoplasmic reticulum. Science 1992; 257: 1496–1502. DOI: 10.1126/science.1523409

[18] Jones DP, Maellaro E, Jiang S, Slater AF, Orrenius S. Effects of N-acetyl-L-cysteine on T-cell apoptosis are not mediated by increased cellular glutathione. Immunol Lett 1995; 45: 205–209. DOI: 10.1016/0165-2478(95)00004-O

[19] Kirlin WG, Cai J, Thompson SA, Diaz D, Kavanagh TJ, Jones DP. Glutathione redox potential in response to differentiation and enzyme inducers. Free Rad Biol Med 1999; 27: 1208–1218. DOI: 10.1016/S0891-5849(99)00145-8

[20] Aoyama K, Watabe M, Nakaki T. Regulation of neuronal glutathione synthesis. J Pharmacol Sci 2008; 108: 227–238. DOI: 10.1254/jphs.08R01CR

[21] Griffith OW, Mulcahy RT. The enzymes of glutathione synthesis: gamma-glutamylcysteine synthetase. Adv Enzymol Relat Areas Mol Biol 1999; 73: 209–267.

[22] Griffith OW. Biologic and pharmacologic regulation of mammalian glutathione synthesis. Free Rad Biol Med 1999; 27: 922–935. doi:10.1016/S0891-5849(99)00176-8

[23] Weber GF. Final common pathways in neurodegenerative diseases: regulatory role of the glutathione cycle. Neurosci Biobehav Rev 1999; 23: 1079–1086. DOI: 10.1016/S0149-7634(99)00041-X

[24] Konigsberg-Fainstein M, Aguilar-Maldonado B. Radicales libres y estrés oxidativo. Aplicaciones médicas, Manual Moderno, México, 2008. ISBN 9707293217, 9789707293212.

[25] Beckett GJ, Arthur J R. Selenium and endocrine systems. J Endocrinol 2005; 184: 455–465. DOI: 10.1677/joe.1.05971

[26] Kenneth H, Kent AR, Prasad Gabbita S, Scott S, Floyd RA. Reactive oxygen species, cell signaling, and cell injury. Free Rad Biol Med 2000; 28: 1456–1462. DOI: 10.1016/S0891-5849(00)00252-5

[27] Oja SS, Janaky R, Varga V, Saransaari P. Modulation of glutamate receptor functions by glutathione. Neurochem Int 2000; 37: 299–306. DOI: 10.1016/S0197-0186(00)00031-0

[28] Filomeni G, Rotilio G, Ciriolo MR. Cell signaling and the glutathione redox system. Biochem Pharmacol 2002; 64: 1057–1064. DOI: 10.1016/S0006-2952(02)01176-0

[29] Lash LH. Mitochondrial glutathione transport: physiological, pathological and toxicological implications. Chem Biol Interact 2006; 163: 54–67. DOI: 10.1016/j.cbi.2006.03.001

[30] Russek M, Cabanac M. Regulación y control en Biología, CECSA, México DF, 1983.

[31] Trujillo JA, Croft NP, Dudek NL, Channappanavar R, Theodossis A, Webb AI, Purcell AW. The cellular redox environment alters antigen presentation. J Biol Chem 2014; 289(40): 27979–27991.

[32] Ostrakhovitch EA, Semenikhin OA. The role of redox environment in neurogenic development. Arch Biochem Biophys 2013; 534(1), 44–54.

[33] Ostrakhovitch EA. Redox environment and its meaning for breast cancer cells fate. Curr Cancer Drug Targets 2011; 11(4): 479–495.

[34] Margittai É, Enyedi B, Csala M, Geiszt M, Bánhegyi G. Composition of the redox environment of the endoplasmic reticulum and sources of hydrogen peroxide. Free Rad Biol Med 2015; 83: 331-340.

[35] Espey MG. Redox physical oncology: intersections between redox and the physical environment in cancer. Free Rad Biol Med 2015; 79: 251–252.

Chapter 2

# Protein Oxidation and Redox Regulation of Proteolysis

Dominika Boguszewska-Mańkowska, Małgorzata Nykiel and Barbara Zagdańska

Additional information is available at the end of the chapter

http://dx.doi.org/10.5772/61182

**Abstract**

Reactive oxygen species (ROS), beyond the role of toxic by-products of aerobic metabolism, contribute to cell redox homeostasis and are signalling molecules in pathogen defence and abiotic stress tolerance. The putative mechanism of cell responses to ROS is thiol modifications of cysteine residues, which cause changes in protein conformation and activity. These post-translational modifications include generation of disulphide bridges and formation of sulphenic, sulphinic, and sulphonic acids, as well as S-glutathionylation and S-nitrosylation. S-nitrosylation or reversible modification may change the activity of enzymes related to the metabolism of nitric oxide, ROS, and cellular metabolism, whereas S-glutathionylation regulates the activity of proteins that contain in their structure the active cysteine residue, regulates the oxidoreductive pathway of signal transduction, and participates in the regeneration of antioxidant enzymes. Carbonylation, an irreversible, non-enzymatic modification of proteins is the most commonly occurring oxidative protein modification. The formation of carbonyl groups can be linked to abnormal translation, altered chaperone system and responses to stress factors. Carbonylated proteins are marked for proteolysis mediated by different pathways in different cell compartments to counteract the formation of high molecular weight aggregates and accumulation of inactive proteins. However, products of proteolysis of carbonylated proteins could function as secondary ROS messengers that target the cell nucleus.

**Keywords:** ROS, carbonylation, S-glutathionylation, protein hydrolysis, plant abiotic stresses

## 1. Introduction

The current literature reveals that the extreme changes in water content experienced by dehydration tolerant plants are accompanied by equally extreme fluctuation in cellular redox state. Regulation of redox potential is essential not only for proper plant development but is required for plant survival under extreme environmental conditions. It is widely accepted that drought induces oxidative stress. This implies that water scarcity in environment disturbs a delicate balance between endogenous generation and elimination of reactive oxygen species (ROS). Reactive oxygen species are produced by a large number of physiological processes, mainly by the electron transport chains in chloroplasts and mitochondria. Under stress conditions, proteins may be modified by ROS in large number of reactions. The putative mechanism of cell responses to ROS is the modifications of protein cysteine residues, which can be oxidized to varying degrees. These post-translational modifications of proteins include formation of disulphide bridges, sulphenic, sulphinic and sulphonic acids as well as S-glutathionylation and S-nitrosylation. The above modification cause changes in protein conformation and activity, although these modifications seem to be potentially reversible (with the exception of sulphonic acid). Additionally, oxidation of protein by direct oxidative attack on Lys, Arg, Pro, and Thr or by secondary reactions on Cys, His, and Lys residues leads to the formation of protein carbonyl derivatives. Carbonyl (CO) groups (aldehydes and ketones) are chemically stable and thus, carbonylation is irreversible and unrepairable. Protein carbonyl derivatives can also be generated through oxidative cleavage of proteins by either the a-amidation pathway or by oxidation of glutamyl side chains. Carbonylation of proteins seems to lead to inhibition or alteration of the activities of the proteins and to increase of their susceptibility to proteolytic attack. Protein hydrolysis can be mediated by different pathways in different cell compartments. In nucleus it engages mainly ubiquitination and proteasomes activity, whereas in mitochondrion and chloroplast specific enzymes like ATP-depended protease La, ClpAP and others has been described. In cytosol hydrolysis once again is maintained by ubiquitination and proteasomes, and probably other means. One of the most effective ones is the vacuolar hydrolysis based on specific for these compatment proteases like cysteine, serine, aspartate proteases and metalloproteases. Therefore, in this chapter the changes in cellular redox state and their impact on modifications of protein functions and efficient removal of abnormal and/or non-functional proteins are discussed.

## 2. Functional activities of ROS in plants under abiotic stresses

Plants, as sessile organisms, are commonly exposed to diverse unfavorable environmental factors including drought, cold, and heat that may induce stress conditions. Throughout the life cycle of plants, reactive oxygen species (ROS) are continuously produced as a consequence of aerobic metabolism and they play a role in cellular redox regulation [1]. However, stress of any kind, biotic or abiotic, leads to an increased level of ROS production and/or to inactivation of antioxidants, particularly enzymes. The oxidative stress that accompanies unfavorable environmental conditions has been suggested to be associated with harmful effects on cellular components including metabolic activities and integrity of organelles. However, recent findings indicate that transient oxidative stress may have beneficial effects in the response of plants to unfavorable environmental conditions. Oxidative stress acting transiently seems to be a signal-activating pathway that enables plants to acclimate to unfavorable environmental conditions [2–4]. Therefore, ROS seem to play a dual role because they may either be involved in the stress-induced oxidative damage of plants or may also activate genes facilitating the development of plant tolerance to abiotic and biotic stresses. Transient oxidative stress seems to participate in the control of seed dormancy, germination and seedling growth [5–6] and other processes involved in physiological transitions [7]. It should be underlined that whether ROS acts as a damaging, protective, or signaling factor depends on sensible equilibrium between ROS production and scavenging at the proper site and time. This equilibrium depends largely on proper cooperation between photosynthesis, respiration, and photorespiration linking their redox state and energy balance in the cell. Antioxidants found in almost all cellular compartments demonstrating the importance of ROS detoxification for cellular survival [8].

Aerobic organisms are evolved mechanisms that minimize the levels of oxygen or delay the generation of ROS [9]. Mobile organisms are able to avoid oxygen by move from high $O_2$ regions, but it is not the case of plants. Comparing the redox regulation in plants and animals on the cellular level, it seems that they evolved common mechanisms. The most important source of ROS in vivo is the mitochondrial electron-transport chain in aerobic animal cells and chloroplast electron transport chain in plant cells. Plant and animal cytochrome oxidase does not release ROS, but other components of the mitochondrial electron-transport chain can leak electrons directly to $O_2$, thus generating ROS [10]. However, plants are able to minimize mitochondrial ROS formation by alternative oxidase [11]. Uncoupling proteins located in the inner mitochondrial membranes of both plants and animals allow protons to leak, preventing the escape of electrons to oxygen. Other mechanisms involved in delayed ROS formation in both animals and plants are suppression of Fenton reaction and lipid peroxidation by metal binding proteins such as transferrin, ferritins, and metallothionins [12].

It is widely accepted that proteins are the most abundant cellular targets of the ROS, constituting about 70% of the oxidized biological molecules in the cell [13]. The extent of protein damage depends on the rate at which any particular reaction occurs and on the intensity of oxidant-scavenging and repair reactions [14]. The rate constants for the reaction of reactive species with proteins vary, but they are useful in computational models to predict sites and selectivity of damage between different oxidants [14]. On the basis of kinetic data, one can predict that the least selective are the most reactive oxidants, whereas the more selective are less reactive oxidants. Thus, sulphur-containing and aromatic side chains will be depleted rapidly [14].

ROS-induced protein modifications can result in the unfolding or alteration of the protein structure, but some of them may be harmless. Proteins can be oxidized in reversible or irreversible ways. Protein oxidation chemistry, based on 20 different types of amino acid side chains, results in many potential reaction sites and products that may be grouped into four categories, i.e., aliphatic, aromatic, cysteine and cystine residues, and methionine residues [14]. Amino acids containing sulphur in the side chain are the most oxidation-susceptible and therefore are the most commonly modified [15]. The oxidation of cysteine residues can result in a number of redox-based post-translational protein modifications such as glutathionylation or S-glutathionylation, S-nitrosylation and sulphenic acid, sulphinic acid and disulfide formation [16]. Side chains of proline, threonine, lysine and arginine can be oxidized to aldehyde or ketone derivatives in the process known as carbonylation of proteins. Protein carbonylation, irreversible post-translational protein modifications, is commonly regarded as a good measure of intensity of protein oxidation [17].

## 3. Oxidation of sulphur-containing amino acids

The most frequently modified amino acids are cysteine and methionine containing sulfur in the side chain. The oxidation of protein cysteine thiol groups can generate thiyl radicals (-S.), disulfide bonds (-S-S-), as well as sulfenic (-SOH), sulfinic (-$SO_2$H), and sulfonic (-$O_3$H) acid derivatives. Methionine residues can be oxidized to methionine sulfoxides and less frequently to methionine sulphones. An increase in free cysteine levels in response to various abiotic stress factors has been reported [18]. In most studies, this increase was reported together with increased reduced glutathione (GSH) concentrations, leading to the conclusion that cysteine is mainly needed for the biosynthesis of Sulphur-rich compounds with anti-stress activity, such as GSH and stress-related proteins. Cysteine plays an important role as an extracellular reducing agent and is also a critical substrate for protein synthesis being the precursor of taurine. Thiol-containing substances are normally considered as antioxidants. However, metabolites containing -SH group exhibit a double effect of antioxidant and prooxidant properties. Cysteine is a potent chelator of heavy metals ions, but cysteine-metal ion complexes can trigger the Fenton reaction, thereby producing the highly toxic ˙OH radical. Furthermore, free cysteine is often irreversibly oxidized to different by-products [19] and reversibly to cystine (cysteine disulphide). However, it remains unclear whether cystine reductase (EC 1.8.1.6), the enzyme that catalyzes the reduction of cystine to cysteine, is active in plants [20].

Methionine, similar to cysteine, can undergo ROS-mediated oxidation to methionine sulfoxide and this can lead to changes in protein conformation and activity [21]. The conversion of methionine sulfoxide to methionine is mediated by methionine sulfoxide reductases (MSRs; EC 1.8.4.11), a class of cytosolic and plastidic enzymes that are involved in ameliorating oxidative damage [22]. Methionine is also a substrate for the synthesis of various polyamines with important roles in stress tolerance, the most prominent being putrescine, spermidine, and spermine [23]. The oxidized methionine residue is readily reduced back to methionine by methionine sulfoxide reductase. Methionine residue is proposed to be a "last chance" antioxidant defense system to protect proteins from oxidation under oxidative stress. Surface exposed methionine residues effectively scavenge oxidizing agents while generally preserving the biological function of the molecule.

## 4. Protein glutathionylation as a mechanism of antioxidant defense

Glutathionylation of proteins is the modification of a reactive protein cysteine thiol by the reduced glutathione (GSH) and occurs under normal or oxidative stress conditions [24]. It was assumed as a by-product of oxidative or nitrosative stress (S-glutathionylation). Under oxidative stress conditions, GSH may interact in a reversible manner with protein cysteinyl thiols [25]. Non-enzymatic glutathionylation reactions are non-specific and seem to be associated with oxidation stress, whereas enzymatic glutathionylation reactions catalyzed by glutaredoxin in both directions are highly specific, reversible [26], and also protect sensitive SH groups from oxidation to sulphinic and sulphonic acid [27]. Thus, various mechanisms may be involved in the formation of glutathionylated proteins and it is not clear which one is prevalent in vivo [24]. The thiol-disulfide exchange seems to be a major one, although evidences against a role of GSSG in the formation of protein-SSG have been given [26, 28].

There is still some controversy about the difference between oxidative stress and redox regulation and signalling. It seems that low levels of ROS play a regulatory role, e.g., the reversible oxidation of cysteines to mixed disulfides represents redox regulation, whereas high levels of ROS lead to oxidative stress and toxicity, i.e., irreversible oxidation to sulfonic acids is toxic [29].

Glutathionylation seems to be involved in such processes as glycolysis, signal transduction, protein degradation, intracellular trafficking, and protein folding [30]. It can protect protein thiols from irreversible inactivation but can also alter the activity of many proteins. In *Arabidopsis thaliana* glutathionylation of glycolytic enzyme, cytosolic triose-phosphate isomerase and fructose-1,6-bisphosphate (FBP) aldolase, a Calvin cycle enzyme, inhibited their activity [31]. The inactivation of a soybean protein tyrosine phosphatase and inhibition of transcription factors (such as Jun and NF-kB) by glutathionylation was noticed [32]. However, glutathionylation activated the transcription factor/cyclic AMP-responsive element binding protein [33].

Glutathionylation results in the regulation and redox signaling in plants. It has been shown that glutatione-S-transferase from *Arabidopsis thaliana* undergoing glutathionylation possess a dehydroascorbate reductase activity and/or a glutathione-dependent thiol transferase activity [34]. These enzymes, which contain a catalytically essential cysteine, are glutathionylated in vitro in the presence of GSSG with a concomitant loss of enzymatic activity would constitute an intermediary step of the catalytic mechanism, allowing glutathione-dependent reduction of dehydroascorbic acid [30].

It has also been proposed that glutathionylation of some photosynthetic enzymes regulates photosynthetic metabolism and thus allows for rearrangement of NADPH and ATP within chloroplasts under oxidative stress [35]. Glutathionylation of the f-type thioredoxin (TRX f) prevents its reduction by ferredoxin-thioredoxin reductase and these thioredoxins are involved mainly in the light-dependent regulation of carbon metabolism enzymes, including several Calvin cycle enzymes [36]. Therefore, it seems that the glutathionylation of TRX f, glyceraldehydes-3-phosphate dehydrogenase and fructose-1,6-bisphosphate aldolase suggest that this posttranslational modification could constitute a new mechanism of regulation of Calvin cycle enzymes under oxidative stress [30]. The in vivo proteomic study on photosynthesizing cells performed in *Chlamydomonas* supported such a possibility [35]. This experiment has shown that more than 20 glutathionylated proteins, mostly located in the chloroplast, are involved in diverse processes such as photosynthesis, chloroplast translation, amino acid and ATP metabolism, protein folding, acetate metabolism, and oxidative stress. Two new Calvin cycle enzymes, phosphoglycerate kinase and ribose-5-phosphate isomerase distinct from those known to undergo glutathionylation, were also identified [30].

Glutathionylation of glycine decarboxylase, a key enzyme in the process of photorespiration [37], resulted in the inactivation of plant mitochondria. This points to the role of glutathionylation in the temporary protection of enzymes against oxidative stress. Glutathionylation can also regulate the activity and function of nuclear proteins, including transcription factors, as well as influence the chromatin structure and dynamics of the process of condensation [20]. Some of the best-characterized plant proteins regulated by glutathionylation are annexins. Annexins are $Ca^{2+}$ and phospholipid binding proteins forming an evolutionary conserved multigene family with a presumed function of the enzyme associated with signal transduction. Different stress conditions cause different changes in gene expression of annexin [38]. Glutathionylation of annexin A1 in *Arabidopsis thaliana* modifies its ability to bind $Ca^{2+}$ and leads to the reduction or inhibition of enzyme activity.

## 5. Protein carbonylation and its role in plant development and stress response

Carbonylation is an irreversible, non-enzymatic modification of proteins and the most commonly occurring oxidative protein modification. Carbonyl groups are introduced into proteins by a variety of oxidative pathways. Lysine, arginine, proline, and histidine side chains can be oxidized to reactive aldehyde or ketone groups via carbonylation causing inactivation, crosslinking, or breakdown of proteins [39–40]. Carbonyl groups can be generated through oxidative cleavage of proteins by $\alpha$-amidation pathway or by formation of peptides with $\alpha$-keto derivatives at the N-terminus [41]. In living organisms, the common mechanism of carbonylation is metal-catalyzed oxidation (MCO) occurring during the interaction of reduced metal ions, such as $Fe^{+2}$ or $Cu^{+}$ with $H_2O_2$, in the Fenton reaction producing extremely reactive hydroxyl radicals oxidizing amino acid side chains or causing protein backbone cleavage [41]. In many cases, oxidative modification of proteins is a detrimental process in which irreversibly inactivated proteins lead to cellular dysfunction. On the other hand, it is now clear that oxidative modification of proteins can be specific and reversible, playing a key role in metabolic regulation and normal plant physiology. However, little is known about the involvement of protein carbonylation in developmental processes and in the response of plants to adverse environmental conditions. Oxidatively modified proteins were identified at all stages of the plant life cycle and their involvement in the control of common biological functions such as dormancy, germination, and aging has been suggested [42–43]. The intracellular level of protein carbonyl groups increases with age, reaching typical values of 1–4 nmol $mg^{-1}$protein but under oxidative stress may increase to 8 nmol $mg^{-1}$ protein, i.e., about 40% of all protein molecules have one carbonyl group [41]. The increased level of oxidized proteins in aged organisms seems to be the result of higher susceptibility of proteins to ROS due to alterations in the protein structure as well as the diminished ability for protein removal [44]. Protein carbonylation seems to be the consequence of age-induced transcriptional and translational errors and increased levels of aberrant misfolded proteins resulting in the level of oxidized proteins [44, 45].

Carbonylated proteins have been found in all plant cell cellular compartments: cytosol, chloroplasts, peroxisomes, nucleus, and mitochondria [42, 43, 46-48]. Interestingly, it has been found that in wheat leaves, the concentration of carbonylated proteins per mg protein was higher in the mitochondria than in chloroplasts and peroxisomes [46]. It may suggest that mitochondrial proteins are more susceptible to oxidative damage or are more tolerant to oxidation or mitochondrial proteases are less efficient in protein removal.

Carbonylated proteins have been identified mainly in *Arabidopsis thaliana* leaves [42], germinating seeds [43], and whole shoots [49]. However, aging and stress-induced carbonylation does not affect proteome in the same way [50]. Specific plant proteins such as ribulose-1, 5-bisphospate carboxylase/oxygenase, enzymes of the cycle of Krebs, and electron transport chains may be classified as sensitive to carbonylation [42, 45]. The findings that such carbonylated proteins as glycolytic enzymes, aldose reductase, methionine synthase, and molecular chaperons characterized in mammals, yeast, and bacteria found in plants point out the participation of carbonylation in the control of common biological functions [51]. Carbonylation of plant proteins seems to also be involved in such physiological transitions such as dormancy, germination, and aging and can act as a signal in physiological transitions [42, 43, 48, 52].

Abiotic stresses such as dehydration, heat, salinity, heavy metals gamma irradiation, and others, as well as biotic stresses, result in protein carbonylation but the extent of carbonylation correlates to the exposure time and the severity of stress [51, 53]. The inactivation of the Calvin cycle, such as ribulose-1, 5-bisphosphate carboxylase oxygenase (RuBisCO), by protein carbonylation would allow plant adaptation to stress conditions. Furthermore, several components of the glycine decarboxylase complex (GDC), a key enzyme involved in photorespiration in photosynthetic tissues, are highly prone to carbonylation, especially under stress conditions [42, 54]. The peroxisome fraction isolated from castor bean endosperm and subjected to harsh metal catalyzed oxidation, a sharp increase in carbonylated malate synthase, and isocitrate lyase were observed. Also, antioxidants proteins such as catalase, manganese superoxide dismutase (MnSOD) and peroxiredoxin (Prx), which are associated with the defense against oxidative stress, are themselves sensitive to oxidative attack [55]. Molecular chaperones such as Hsp60 can be also inactivated by carbonylation. It seems that these proteins are fairly susceptible to carbonylation and under prolonged oxidative stress they could lose structural integrity and become dysfunctional. The loss of function of the chaperones and MnSOD combined with the ongoing oxidative stress would aggravate the damaging effect to the cell, eventually resulting in cell death.

## 6. Degradation of oxidized proteins

Enhanced modification of proteins has been reported in plants under various stresses [56]. To maintain the cellular metabolism, it is highly required to possess the efficient degradation and removal of oxidized proteins. The fate of carbonylated proteins in plant cells is of paramount importance. If not degraded these proteins tend to form high molecular weight aggregates due to covalent cross-linking and/or increased surface hydrophobocity. What is more, the accumulation of such inactive proteins could functionally compete with their active counterparts. To circumvent this, several distinct proteolysis pathways balance the extent of protein carbonylation [57].

Different pathways can mediate protein hydrolysis in different cell compartments. In the nucleus, it engages mainly in ubiquitination and proteasomes activity. In mitochondrion- and chloroplast-specific enzymes such as ATP-depended protease La, ClpAP and others have been described [58–60]. In cytosol, hydrolysis once again is maintained by ubiquitination and proteasomes [61]. One of the most effective ones is the vacuolar hydrolysis based on proteases such as cysteine, serine, aspartate proteases, and metalloproteases specific for this compartment. These different proteolytic pathways do not exclusively target carbonylated proteins. However, some of them as 20S proteasome pathway seem to preferentially degrade oxidized rather than non-modified proteins [62]. In this way, protein carbonylation could promote degradation of mistranslated, damaged, and aberrant or even no longer required proteins in plant cells.

## 7. Removal of oxidized proteins by proteasome

The ubiquitin 26S proteasome system is the major mechanism of intracellular protein degradation in eukaryotes, playing a key role in basic cellular processes. Plant cells contain a mixture of 26S and 20S proteasomes that are responsible for proteolysis [63]. The 26S proteasome is proteolytic component of the ubiquitin (Ub)-dependent proteolytic system, essential in plants and also in animals. It degrades functional proteins, negatively controls the abundance of regulatory proteins involved in signaling and metabolic pathways, and degrades abnormal and denatured proteins produced under biotic and abiotic stress. This illustrates why proteasome is essential for protein quality control [64]. The 26S proteasome is an ATP-dependent, multi-subunit protease complex composed of two subcomplexes, the 20S core protease (or multi-catalytic protease) and the 19S regulatory particles. The 20S core protease is ATP- and ubiquitin-independent protease that consists of a cylindrical stack created by the assembly of four heptameric rings [61]. It has been reported that reduction in 26S proteasome abundance is connected with the increasing cell expansion and with the decreasing cell division rates [65]. It is known that abiotic stresses lead to inhibit 26S activity directly by inhibiting 26S proteasome function or indirectly by increasing the substrate load. Heat shock and other stresses can cause protein misfolding which leads to a substrate overload. Oxidative stress directly leads to 26S proteasome causing a decrease in cell division. These mechanisms seem to be relevant when the intensity of the stressor is high [63].

A strong candidate for the removal of damaged proteins is the 20S proteasome, a 700 kDa proteolytic complex responsible for degradation of short-lived and abnormal intracellular proteins. It is composed of a barrel-shaped protein comprising of four stacked rings with the catalytic side in the lumen. Two outer and two inner rings are made up of seven subunits ranging from 20 to 35 kDa each. The first and third rings contain alpha subunits; the second and fourth rings contain beta subunits.

The 26S proteasome is responsible for the degradation of misfolded proteins and the free 20S proteasome is needed for the removal of oxidized proteins [66]. Plant cells contain a mixture of the 26S and free 20S proteasome and the adequate proteasomes ratio seems to be important for plant development and stress responses. The comparative analyses of mutants showed that optimal 26S proteasome levels are needed to maintain tolerance to stresses, such as heat shock, that cause protein misfolding, while increased levels of the free 20S proteasome lead to better tolerance to oxidative stress [63].

For the first time, the importance of developmental regulation of the 26S to 20S proteasome was studied for *Drosophila*. It was proved that during aging, the 26S proteasome decreases while 20S proteasome increases [67]. Similarly, a decline in 26S proteasome levels and increase in 20S proteasome levels was presented in a proteomic study of potato tuber aging [68].

An increased production of free radicals and increased oxidative damage to cellular components are obtained during aging. Under these conditions, a shift in the proteasomes ratio in favor of the free 20S proteasome would be beneficial because it increases the cellular capacity to remove oxidized proteins that are potentially cytotoxic when allowed to accumulate. The regulation of 26S proteasome to 20S proteasome levels during aging and senescence is still poorly understood. From what we know, we can conclude that the mild loss of 26S proteasome activity strongly impacts plant growth; that is why plants need to maintain an optimal 26S proteasome activity level to ensure developmentally-programmed cell division and expansion rates. Finally, the loss of the 26S proteasome function does not disturb plant survival because it provides to the 20S proteasome biogenesis and at the same time increases oxidative stress tolerance [63]. Not all proteins are degraded after the carbonylation; some of them may form aggregates that accumulate in the cell. Inhibition of proteolysis may be a result of the creation of groups of modified proteins susceptible to degradation, effectively blocking the proteasome. Accumulation of such protein aggregates may result in cell death.

## 8. Autophagic degradation of carbonylated proteins

Autophagy, a non-specific protein degradation pathway, could degrade oxidized proteins [69, 70]. However, the potential role of autophagy in plant responses to abiotic stresses is still unknown. Although autophagy-related proteins function in plant responses to salt and osmotic stress, the relationship to autophagy is not clear [71, 72]. This process plays a critical role in the acclimation of plants to changing environmental stresses such as oxidative stress, drought, salt, and pathogen occurrence [73]. Autophagy is a process in which cytoplasmic components are taken up into the vacuole (yeast, plants) or lysosome (mammals) for degradation. There are three major forms of autophagy: macroautophagy, microautophagy, and chaperone-mediated autophagy [74–76]. In plants, chaperone-mediated autophagy has not yet been established. Microautophagy is in intravacuolar vesicles (autophagic bodies) that came from the invaginated tonoplast. The isolated body containing the cytoplasmic materials and membranes is degraded by vacuolar hydrolases. Macroautophagy (hereafter called autophagy) is a cellular mechanism of the removal of unnecessary cytosolic components in vesicles formed de novo. These double membrane-bound vesicles, called autophagosome, are fused with the tonoplast to release the internal vesicles into the lumen of the vacuole, where they are degraded by hydrolases residents in the vacuole [77]. Recently, new type of autophagy-dependent bodies specifically involved in the delivery of chloroplast (chlorophagy), mitochondria (mitophagy), and ribosome (ribophagy) components to the vacuole has been described [77].

Nutrient deprivation was one of the first common abiotic stresses shown to induce autophagy. Very common and frequent environmental stresses encountered by plants are high salinity and drought stress [78]. These stresses reveal some common features but are regulated differently. Autophagy-deprived AtAtg18a-RNAi plants show high sensitivity to these two stresses. It is commonly accepted that diverse environmental stresses may induce autophagy, but it may be regulated by different signaling pathways. It has been shown that the salinity and osmotic stress-induced autophagy in *Arabidopsis thaliana* [79] and autophagy induced by starvation and salt stress has been dependent on NADPH oxidase activity, enzyme generating superoxide by transferring electrons from NADPH to molecular oxygen to produce superoxide anion. However, it was not the case in autophagy induced by osmotic stress. To date, the isoforms of NADPH oxidase responsible for the induction of autophagy have not been found. Thus, in nutrient-starved and salinized plants, autophagy seems to be regulated by NADPH oxidase-dependent pathways and involved ROS, whereas in osmotically stressed plants, autophagy is regulated by an NADPH oxidase independent pathway in which ROS may not be involved as a signal [79]. Evidences show that autophagy is induced by high salinity and osmotic stresses and that autophagy-defective plants are more sensitive to these conditions [80].

## 9. Role of the vacuole

Vacuolar proteolysis also plays a major role in the turnover of proteins oxidized by endogenously generated ROS. These oxidized proteins, probably nonessential to survival, are selectively recognized and degraded by proteolytic enzymes [81]. Thereafter, amino acids may be released for synthesis of new proteins, aberrant proteins formed under water deficit may be degraded, and certain proteins may be activated. Controlled protein hydrolysis has therefore been recognized as essential for the adaptation of plants to environment [82]. The MEROPS database (http://merops.sanger.ac.uk/) reflects the increasing number of proteases classified based on their catalytic type, e.g., aspartic (A), cysteine (C), serine (S), threonine (T) and glutamic (G) peptidases and asparagine peptide lyases (N), all based on the amino acid residue at the active site involved directly in peptide bond hydrolysis, and into metallopeptidases (M) that require a divalent metal ion as part of the active site. Drought has been shown to induce large increases in acid protease activity in leaves of a susceptible wheat cultivar, probably associated mainly with cysteine proteinases involved in playing a house-keeping function to remove abnormal and misfolded proteins arising from unfavorable conditions [83, 84]. One may suppose that carbonylated proteins could promote degradation of misfolded, damaged, or even no longer required proteins that are nonessential to survival and should be replaced by newly formed proteins [84, 85]. However, detailed studies in this respect are needed.

## 10. The effects of vitamins present in the plants on ROS

The Food and Nutrition Board of the National Institute of Medicine in the U.S.A. has defined a dietary antioxidant as "a substance in foods that significantly decreases the adverse effects of reactive species (oxygen and nitrogen species) on normal physiological function in humans." Among the most commonly studied dietary antioxidants playing a vital role against ROS are ascorbic acid (vitamin C), α-tocopherol (vitamin E), and β-carotene. Vitamin C is a very important water-soluble antioxidant in extracellular fluids and is able to neutralize ROS in the aqueous phase before they can attack lipids. Vitamin E is a lipid-soluble antioxidant. It is important as the chain-breaking antioxidant within the cell membrane; it plays an important role in protecting membrane fatty acids from lipid peroxidation and has the highest antioxidative activity because of the presence of three methyl groups in its molecular structure [86]. Vitamin C, moreover, is able to regenerate vitamin E. β-carotene and other carotenoids also have antioxidant properties and may work in synergy with vitamin E [87]. Among the most important non-enzymatic antioxidants in plants are the glutathione (GSH), proline, carotenoids, and flavonoids. Glutathione (GSH) is aimed to be the most important intracellular defense against ROS-induced oxidative damage. It is involved in the control of $H_2O_2$ levels. Flavonoids belong to one of the most reactive secondary metabolites of plants. Flavonoids play an important role as a ROS scavenger by locating and neutralizing radicals before they damage the cell structure. Proline is considered to act as an osmoprotectant, a protein stabilizer, a metal chelator, an inhibitor of lipid peroxidation, and $OH^{\cdot}$ and $1O_2$ scavenger. Moreover, it also plays an important role as a ROS quencher as well as a signaling molecule. Carotenoids are lipid soluble antioxidants that play a multitude functions in plant metabolism including oxidative stress tolerance [88].

*Arabidopsis thaliana,* contained in leaf flavonoids is a highly diverse polyphenolic secondary metabolite and glucosinolate - another class of secondary compounds [89, 90]. Glucosinolates, due to their flavor and medicinal properties, and coumarins characterized by their antimicrobial and antifungal activities as well as pharmacological effects have attracted much interest [91]. Flavonoids appear to play a fundamental role during environmental interactions and plant coping with abiotic stresses [90].

It is possible to measure antioxidant assay value for food. There are numerous methods of assessment for dietary antioxidants. These assays can be divided into two groups depending on the reactions involved: assays based on hydrogen atom transfer (HAT) reactions and assays based on electron transfer (ET). The first group includes low-density lipoprotein autoxidation, oxygen radical absorbance capacity (ORAC), total radical trapping antioxidant parameter

(TRAP), and crocin bleaching assays. ET-based assays include the total phenols assay by Folin–Ciocalteu reagent (FCR), Trolox equivalence antioxidant capacity (TEAC), ferric ion reducing antioxidant power (FRAP), "total antioxidant potential" assay using a Cu(II) complex as an oxidant, and diphenyl-1-picrylhydrazyl radical –DPPH [87]. The antioxidant values of foods listed are expressed in ORAC (Oxygen Radical Absorbance Capacity) units, a unit of measurement for antioxidants developed by the National Institute on Aging in the National Institutes of Health [92, 93]. Numerous health food and marketers have erroneously capitalized on the ORAC rating by promoting products claimed to be "high in ORAC."

## 11. Conclusion

Redox regulatory mechanisms are a necessary part of the intracellular communication system activating the plant stress responses. Redox-based post-translational protein modification plays a fundamental role in redox regulation of protein function by disulfide bridge formation and in control of protein activity by redox-based Cys-modification such as S-glutathionylation. Another oxidative protein modification marks carbonylated proteins to proteolysis. On the other hand, carbonylation of mistranslated and aberrant proteins seems to be an important mechanism of protein quality control. It seems that plant tolerance to stress factors are mediated by both, redox modifications of proteins and their contribution to the maintenance and control of cellular homeostasis.

## Author details

Dominika Boguszewska-Mańkowska[1*], Małgorzata Nykiel[2] and Barbara Zagdańska[2]

*Address all correspondence to: dominika@boguszewska.net

1 Plant Breeding and Acclimatization Institute-National Research Institute, Division Jadwisin, Poland

2 Warsaw University of Life Sciences, Department of Biochemistry, Poland

## References

[1] Dalle-Donne I, Milzani A, Gagliano N, Colombo R, Giustarini D, Rossi R: Molecular mechanisms and potential clinical significance of Sglutathionylation. Antioxid. Redox Signal. 2008;10:445-473.

[2] Jaspers P, Kandasjarvi J: Reactive oxygen species in abiotic stress signalling. Physiol. Plant. 2010;138:405-413.

[3] Miller G, Suzuki N, Ciftci-Yilmaz S, Mittler R: Reactive oxygen species homeostasis and signalling during drought and salinity stresses. Plant Cell Environ. 2010;33:453-467.

[4] Suzuki N, Koussevitzky S, Mittler R, Miller G: ROS and redox signalling in the response of plants to abiotic stress. Plant Cell Environ. 2012;35:259-270.

[5] Job C, Rajjou L, Lovigny Y, Belghazi M, Job D: Patterns of protein oxidation in Arabidopsis seeds and during germination. Plant Physiol. 2005;138:790-802.

[6] Arc E, Galland M, Cueff G, Godin B, Lounifi I, Job D, Rajjou L: Reboot the system thanks to protein post-translational modification and proteome diversity how quiescent seeds restart their metabolism to prepare seedling establishment. Proteomics. 2011;11:1606-1618.

[7] Lounifi I, Arc E, Molassiotis A, Job D, Rajjou L, Tanou, G: Interplay between protein carbonylation and nitrosylation in plants. Proteomics. 2013;13(3-4):568-578.

[8] Mittler R, et al.: ROS signaling: The new wave? Trends in Plant Science. 2011;16(6): 300-309.

[9] Halliwell B, Gutteridge JMC. Free Radicals in Biology and Medicine, Ed 4. Clarendon Press, Oxford; 2006.

[10] Turrens JF: Mitochondrial formation of reactive oxygen species. J. Physiol. 2003;552:335-344.

[11] Möller IM: Plant mitochondria and oxidative stress: Electron transport, NADPH turnover, and metabolism of reactive oxygen species. Annu Rev Plant Physiol Plant Mol Biol. 2001;52:561-591.

[12] Halliwell B, Gutteridge JMC: The antioxidants of human extracellular fluids. Arch Biochem Biophys. 1990;280:1-8.

[13] Levine A, Tenhaken R, Dixon R, Lamb CJ: $H_2O_2$ from the oxidative burst orchestrates the plant hypersensitive disease resistance response. Cell. 1994;79:583-593.

[14] Davies, M: The oxidative environment and protein damage. Biochim. Biophys. Acta. 2005;1703:93-109.

[15] Rinalducci S, Murgiano L, Zolla L: Redox proteomics: Basic principles and future perspectives for the detection of protein oxidation in plants. J. Exp. Bot. 2008;59(14): 3781-3801.

[16] Spadaro D, Yun BW, Spoel SH, Chu C, Wang YQ, Loake GJ: The redox switch: Dynamic regulation of protein function by cysteine modifications. Physiol. Plant. 2009;138:360-371.

[17] Gill SS, Tuteja N: Reactive oxygen species and antioxidant machinery in abiotic stress tolerance in crop plants. Plant Physiol. Biochem. 2010;48:909-930.

[18] Harms K, Ballmoos P, Brunold C, Hofgen R, Holger H: Expression of a bacterial serine acetyltransferase in transgenic potato plants leads to increased levels of cysteine and glutathione. Plant J. 2000;24:335-343.

[19] Bashir H, Ahmad J, Bagheri R, Nauman M, Qureshi M I: Limited sulfur resource forces Arabidopsis thalianato shift towards non-sulfur tolerance under cadmium stress. Environ. Exp. Bot. 2012. doi:10.1016/j.envexpbot.2012.05.004.

[20] Zagorchev L, Seal CE, Kranner I, Odjakova M: A central role for thiols in plant tolerance to abiotic stress. Int. J. Mol. Sci. 2013;14:7405-7432.

[21] Vieira Dos Santos C, Rey P: Plant thioredoxins are key actors in oxidative stress response. Trends Plant Sci. 2006;11:329-334.

[22] Cabreiro F, et al: Methionine sulfoxide reductases: Relevance to aging and protection against oxidative stress. Ann N Y Acad Sci. 2006;1067:37-44.

[23] Alcázar R, Altabella T, Marco F, Bortolotti C, Reymond M, Koncz C, et al.: Polyamines: Molecules with regulatory functions in plant abiotic stress tolerance. Planta. 2010;231:1237-1249.

[24] Mailloux RJ, McBride SL, Harper M E: Unearthing the secrets of mitochondrial ROS and glutathione in bioenergetics. Trends Biochem. Sci. 2013;38:592-602. doi:10.1016/j.tibs.2013.09.001.

[25] Grimm LM, Collison MW, Fisher RA, Thomas JA: Protein mixed disulfides in cardiac cells. Biochim. Biophys. Acta. 1985;844:50-54.

[26] Meyer AJ, Hell R: Glutathione homeostasis and redox-regulation by sulfhydryl groups. Photosynth Res. 2005;86:435-457.

[27] Fiaschi T, Cozzi G, Raugei G, Formigli L, Ramponi G, Chiarugi P: Redox regulation of beta-actin during integrin-mediated cell adhesion. J Biol Chem. 2006;281:22983-22991.

[28] Dalle-Donne I, Rossi R, Milzani A, Di Simplicio P, Colombo R: The actin cytoskeleton response to oxidants: From small heat shock protein phosphorylation to changes in the redox state of actin itself. Free Rad. Biol. Med. 2001;31:1624-1632.

[29] Ghezzi P, Bonetto V, Fratelli M: Thiol-disulfide balance: From the concept of oxidative stress to that of redox regulation. Antioxidants and Redox Signaling 2005;7:964-972.

[30] Rouhier N, Lemaire SD, Jacquot JP: The role of glutathione in photosynthetic organisms: Emerging functions for glutaredoxins and glutathionylation. Annu Rev Plant Biol. 2008;59:143-166. doi:10.1146/annurev.arplant.59.032607.092811.

[31] Ito H, Iwabuchi M, Ogawa K: The sugar-metabolic enzymes aldolase and triosephosphate isomerase are targets of glutathionylation in Arabidopsis thaliana: Detection using biotinylated glutathione. Plant Cell Physiol. 2003;44:655-660.

[32] Pineda-Molina E, Klatt P, Vazquez J, Marina A, Garcia de LAcoba M, Perez-Sale D, Lamas S: Glutathionylation of the p50 subunit of NF-kB: A mechanism for redox induced inhibition of DNA binding. Biochemistry. 2001;40:14134-14142.

[33] Brar SS, Grigg C, Wilson K S, et al.: Disulfiram inhibits activating transcription factor/cyclic AMP-responsive element binding protein and human melanoma growth in a metal-dependent manner in vitro, in mice and in a patient with metastatic disease. Mol Cancer Ther. 2004;3:1049-1060.

[34] Dixon DP, Skipsey M, Grundy NM, Edwards R: Stress-induced protein sglutathionylation in arabidopsis. Plant Physiology. 2005;138(4):2233-2244.

[35] Michelet L, Zaffagnini M, Marchand C, Collin V, Decottignies P, Tsan P, Lancelin J M, Trost P, Miginiac-Maslow M, Noctor G, Lemaire DS: Glutathionylation of chloroplast thioredoxin f is a redox signaling mechanism in plants Proc. Natl Acad. Sci. U S A. 2005;102:16478-16483.

[36] Jaspers P, Kandasjarvi J: Reactive oxygen species in abiotic stress signaling. Physiol. Plant. 2010;138:405-413.

[37] Palmieri M C, Lindermayr C, Bauwe H, Steinhauser C, Durner J: Regulation of plant glycine decarboxylase by s-nitrosylation and glutathionylation. Plant Physiol. 2010;152:1514-1528.

[38] Mortimer JC, Laohavisit A, Macpherson N, Webb A, Brownlee C, Battey NH, Davies J M: Annexins: Multi-functional components of growth and adaptation. J. Exp. Bot. 2008;59:533-544.

[39] Rinalducci S, Murgiano L, Zolla L: Redox proteomics: Basic principles and future perspectives for the detection of protein oxidation in plants. J Exp Bot. 2008;59:3781-3801.

[40] Bond AE, Row PE, Dudley E: Post-translation modification of proteins: Methodologies and applications in plant sciences. Phytochemistry. 2011;72(10):975-996.

[41] Møller IM, Rogowska-Wrzesinska A, Rao RS: Protein carbonylation and metal-catalyzed protein oxidation in a cellular perspective. J Proteomics. 2011;74(11):2228-2242. doi:10.1016/j. jprot.2011.05.004.

[42] Johansson E, Olsson O, Nystrom T: Progression and specificity of protein oxidation in the life cycle of Arabidopsis thaliana. J Biol Chem, 2004;279:22204-22208.

[43] Job C, Rajjou L, Lovigny Y, Belghazi M, Job D: Patterns of protein oxidation in arabidopsis seeds and during germination. Plant Physiology. 2005;138(2):790-802.

[44] Dukan S, Farewell A, Ballesteros M, Taddei F, Radman M, Nyström T. Protein oxidation in response to increased transcriptional or translational errors. Proc Natl Acad Sci USA. 2000;97:5746-5749.

[45] Nystrom T: Role of oxidative carbonylation in protein quality control and senescence. EMBO J. 2005;24:1311-1317. doi:10.1038/sj.emboj.7600599.

[46] Bartoli CG, Gómez F, Martinez DE, Guiamet JJ, Mitochondria are the main target for oxidative damage in leaves of wheat. J. Exp. Bot. 2004;55:1663-1669.

[47] Nguyen AT, Donaldson RP: Metal-catalyzed oxidation induces carbonylation of peroxisomal proteins and loss of enzymatic activities. Arch Biochem Biophys. 2005;439:25-31.

[48] Rajjou L, Lovigny Y, Groot S P, Belghazi M, Job C, Job D: Proteome-wide characterization of seed aging in Arabidopsis: A comparison between artificial and natural aging protocols. Plant Physiol. 2008;148:620-41.

[49] Qiu QS, Huber JL, Jain V, Leakey ADB, Ort DR, Huber SC: Increased protein carbonylation in leaves of Arabidopsis and soybean in response to elevated [$CO_2$]. Photosynth Res. 2008. doi:10.1007/s11120-008-9310-5.

[50] Levine RL: Carbonyl modified proteins in cellular regulation, aging, and disease. Free Radic Biol Med. 2002;32:790-796.

[51] Lounifi I, Arc E, Molassiotis A, Job D, Rajjou L, Tanou G: Interplay between protein carbonylation and nitrosylation in plants. Proteomics. 2013;13(3-4):568-578. doi: 10.1002/pmic. 201200304.

[52] Minas IS, Tanou G, Belghazi M, Job D, Manganaris GA, Molassiotis A, Vasilakakis M: Physiological and proteomic approaches to address the active role of ozone in kiwifruit post-harvest ripening. J. Exp.Bot. 2012;63:2449-2464.

[53] Song H, Fan P, Li Y: Overexpression of organellar and cytosolic athsp90 in arabidopsis thaliana impairs plant tolerance to oxidative stress. Plant Mol Biol Rep. 2009;27:342-349.

[54] Kristensen BK, Askerlund P, Bykova NV, Egsgaard H, Moller I M: Identification of oxidised proteins in the matrix of rice leaf mitochondria by immunoprecipitation and two-dimensional liquid chromatography—tandem mass spectrometry. Phytochemistry. 2004;65:1839-1851.

[55] Smakowska E, Czarna M, Janska H: Mitochondrial ATP-dependent proteases in protection against accumulation of carbonylated proteins. Mitochondrion. 2014;Pt B: 245-51.

[56] Sharma P, Jha AB, Dubey RS, Pessarakli M: Reactive oxygen species, oxidative damage, and antioxidative defense mechanism in plants under stressful conditions. Journal of Botany. 2012;1-26. Article ID 217037.

[57] Jung T, Höhn A, Grune T: The proteasome and the degradation of oxidized proteins: Part II – protein oxidation and proteasomal degradation. Redox Biology. 2014;2:99-104.

[58] Janska H: ATP-dependent proteases in plant mitochondria: What do we know about them today? Physiologia Plantarum. 2005;123:399-405.

[59] Lupas A, Flanagan J M, Tamura T, Baumeister W: Self-compartmentalizing proteases. Trends Biochem Sci. 1997;22:399-404.

[60] Schmidt M, Lupas AN, Finley D: Structure and mechanism of ATP-dependent proteases, Curr. Opin. Chem. Biol. 1999;3:584-591.

[61] Smalle J, Vierstra RD: The ubiquitin 26S proteasome proteolytic pathway. Annu. Rev. Plant Biol. 2004; 55:555-590.

[62] Polge C, Jaquinod M, Holzer F, Bourguignon J, Walling L, Brouquisse R: Evidence for the existence in Arabidopsis thaliana of the proteasome proteolytic pathway; activation in response to cadmium. J. Biol. Chem. 2009;284,35412-35424.

[63] Kurepa J, Wang S, Li Y, Smalle J: Proteasome regulation, plant growth and stress tolerance. Plant Signaling and Behavior. 2009a;4(10):924-927.

[64] Wang S, Kurepa J, Smalle JA: The Arabidopsis 26S Proteasome Subunit RPN1a is required for optimal plant growth and stress responses. Plant Cell Physiol. 2009;50(9): 1721-1725.

[65] Kurepa J, Wang S, Li Y, Zaitlin D, Pierce AJ, Smalle JA: Loss of 26S proteasome function leads to increased cell size and decreased cell number in Arabidopsis shoot organs. Plant Physiol. 2009;150:178-89.

[66] Bader N, Grune T: Protein oxidation and proteolysis. Biol Chem. 2006;387:1351-5.

[67] Vernace VA, Arnaud L, Schmidt-Glenewinkel T, Figueiredo-Pereira ME: Aging perturbs 26S proteasome assembly in Drosophila melanogaster. FASEB J. 2007;21:2672-82.

[68] Delaplace P, Fauconnier ML, Sergeant K, Dierick JF, Oufir M, van der Wal F, America A H P, Renaut J, Hausman J F, du Jardin P: Potato (Solanum tuberosum L.) tuber ageing induces changes in the proteome and antioxidants associated with the sprouting pattern. J Exp Bot. 2009;60:1273-88.

[69] Xiong Y, Contento AL, Nguyen PQ, Bassham DC: Degradation of oxidized proteins by autophagy during oxidative stress in Arabidopsis. Plant Physiol. 2007;143:291-299.

[70] Bassham DC: Plant autophagy-more than a starvation response. Curr Opin Plant Biol. 2007;10(6):587-593.

[71] Shin JH, Yoshimoto K, Ohsumi Y, Jeon JS, An G: OsATG10b, an autophagosome component, is needed for cell survival against oxidative stresses in rice. Mol Cells. 2009;27(1):67-74.

[72] Slavikova S, Ufaz S, Avin-Wittenberg T, Levanony H, Galili G: An autophagy-associated Atg8 protein is involved in the responses of Arabidopsis seedlings to hormonal controls and abiotic stresses. J Exp Bot. 2008;59:4029-4043.

[73] Pérez-Pérez M E, Lemaire S D, Crespo JL: Reactive oxygen species and autophagy in plants and algae, Plant Physiology. 2012;160:156-164.

[74] Klionsky D J and Emr S D: Autophagy as a regulated pathway of cellular degradation. Science. 2000;290:1717-1721.

[75] Klionsky DJ: The molecular machinery of autophagy: Unanswered questions. J Cell Sci. 2005;118:7-18.

[76] Klionsky D J Autophagy: From phenomenology to molecular understanding in less than a decade. Nat Rev Mol Cell Biol. 2007;8:931-937.

[77] Li F, Vierstra RD: Autophagy: A multifaceted intracellular system for bulk and selective recycling. Trends Plant Sci. 2012;17(9):526-537.

[78] Lv X, Pu X, Quin TZ, Lin H: The roles of autophagy in development and stress responses in Arabidopsis thaliana. Apoptosis. 2014;19:905-921.

[79] Liu Y, Xiong Y, Bassham DC: Autophagy is required for tolerance of drought and salt stress in plants. Autophagy. 2009;5:954-963.

[80] Han S, Yu B, Wang Y, Liu Y: Role of plant autophagy in stress response. Protein Cell. 2011;2:784-791.

[81] Marques M, Mojzita D, Amorim MA, Almeida T, Hohmann S, Moradas-Ferreira P, Costa V: The Pep4p vacuolar proteinase contributes to the turnover of oxidized proteins but PEP4 overexpression is not sufficient to increase chronological lifespan in Saccharomyces cerevisiae. Microbiology. 2006;152:3595-3605.

[82] Vierstra RD: Proteolysis in plants: Mechanisms and functions. Plant Mol Biol. 1996;32:275-302.

[83] Simova-Stoilova L, Vaseva I, Grigorova B, Demirevska K, Feller U: Proteolytic activity and cysteine protease expression in wheat leaves under severe soil drought and recovery. Plant Physiol. and Biochem. 2010;48:200-206.

[84] Grudkowska M, Zagdańska B: Multifunctional role of plant cysteine proteinases. Acta Biochim. Polon. 2004;51(3):609-624.

[85] Nystrom T: Role of oxidative carbonylation in protein quality control and homeostasis in plants. Trends in Plant Science. 12:125-134.

[86] Kamal-Eldin A, Appelqvist LA: The chemistry and antioxidant properties of tocopherols and tocotrienols. Lipids. 1996;31(7):671-701.

[87] Prior RL: Oxygen radical absorbance capacity (ORAC): New horizons in relating dietary antioxidants/ bioactives and health benefits. Jaurnal of Functional Food. 2014;1-14.

[88] Boguszewska D, Zagdańska B: ROS as signaling molecules and enzymes of plant response to unfavorable environmental conditions, in: Oxidative Stress—Molecular Mechanisms and Biological Effects, Dr. Volodymyr Lushchak (Ed.), 2012, ISBN: 978-953-51-0554-1, InTech. doi: 10.5772/33589; 341-362.

[89] Cohen JD, Slovin JP, Hendrickson AM. Two genetically discrete pathways convert tryptophan to auxin: More redundancy in auxin biosynthesis. Trends Plant Sci. 2003;8(5):197-199.

[90] Martens S, Preuß A, Mate M: Multifunctional flavonoid dioxygenases: Flavonol and anthocyanin biosynthesis In Arabidopsis thaliana L. Phytochem. 2010;71,1040-1049.

[91] Gnonlonfin GJB, Sanni A, Brimer L: Review scopoletin—A coumarin phytoalexin with medicinal properties. Critical Reviews in Plant Sciences. 2012;31(1):47-56.

[92] Wu X, Beecher GR, Holden JM, Haytowitz DB, Gebhardt SE, Prior RL: Lipophilic and hydrophilic antioxidant capacities of common foods in the United States. Journal of Agricultural and Food Chemistry. 2004;52:4026-4037.

[93] Wu X, Gu L, Holden J, Haytowitz D, Gebhardt SE, Beecher G, Prior RL: Development of a database for total antioxidant capacity in foods: A preliminary study. Journal of Food Composition and Analysis. 2004;17:407-422.

Chapter 3

# Biochemistry of Reactive Oxygen and Nitrogen Species

Ayla Ozcan and Metin Ogun

Additional information is available at the end of the chapter

http://dx.doi.org/10.5772/61193

**Abstract**

Reactive species or free radicals include reactive oxygen and nitrogen species that are called reactive oxygen nitrogen species. Reactive oxygen species are formed as a natural by-product of the normal metabolism of oxygen and have significant roles in cell signaling and homeostasis. The reactive oxygen species are generated as a by-product of biochemical reactions, in mitochondria, peroxisomes, cytochrome P450, and other cellular components. When oxygen homeostasis is not maintained, oxidative stress is increased in the cellular environment. Superoxide, hydrogen peroxide and hydroxyl radicals are normal metabolic by-products which are generated continuously by the mitochondria in growing cells. Microsomal cytochrome P450 enzymes, flavoprotein oxidases and peroxisomal enzymes are other significant intracellular sources of reactive oxygen species.

**Keywords:** Reactive oxygen species, mitochondria, NADPH oxidase, 5-lipoxygenase

## 1. Introduction

All living aerobic multicellular organisms require molecular oxygen ($O_2$) to survive rather than oxygen, which is susceptible to radical formation due to its electronic structure. Reactive oxygen species (ROS) are small molecules derived from oxygen molecules including free oxygen radicals, such as superoxide ($O_2^{\cdot -}$), hydroxyl ($\cdot OH$), peroxyl ($RO_2 \cdot$), and alkoxyl ($RO\cdot$) as well as hypochlorous acid (HOCl), ozone ($O_3$), singlet oxygen ($^1O_2$), and hydrogen peroxide ($H_2O_2$), which are non-radicals. These non-radicals are either oxidizing agents or easily converted into radicals. Nitrogen-containing oxidants, such as nitric oxide ($NO^{\cdot}$) peroxynitrite ($ONOO^{\cdot}$), nitrogen dioxide ($NO_2$) are called reactive nitrogen species (RNS) [1, 2].

Reactive species or free radicals include reactive oxygen and nitrogen species collectively and are called reactive oxygen nitrogen species (RONS). They are released from macrophages, neutrophils and dendritic cells in response to an inflammatory stimulus. RONS are highly reactive due to the presence of unpaired valence shell electrons or non-static bonds, and their proper regulation is vital for an efficient immune response and for limiting tissue damage [3].

Reactive oxygen species, chemically reactive molecules, containing oxygen, are formed as a natural by-product of the normal metabolism of oxygen and have significant roles in cell signaling and homeostasis. However, during times of environmental stress (e.g., UV radiation, heat exposure and ionizing radiation), their levels could increase dramatically. At high concentrations, ROS reacts readily with lipids, proteins, carbohydrates, and nucleic acids. This may result in significant damage to cell structures, and cumulates into a situation known as oxidative stress [4, 5]. Oxidative stress is a condition when the balance between the production of oxidants and their removal by antioxidants gets disturbed leading to increased production and accumulation of oxidants in the body. Aging, chronic inflammatory diseases, smoking, diabetes neurodegenerative diseases, cancer, etc., lead to generation of oxidative stress [3].

## 2. Types of free radicals

Free radicals are a group of highly reactive chemical molecules with one or more unpaired electrons. These substances are capable of giving rise to chain reactions involving a number of steps, each of which forms a free radical that triggers the next step. Initiation, propagation and termination are three phases of these reactions. There are some different types of free radicals. Oxygen-centered, nitrogen-centered, carbon-centered and sulphur-centered radicals are different free radical species [6, 7].

### 2.1. Oxygen-centered radicals

In living system oxygen radicals represent the most important class of radical species. In biochemistry, the free radicals are often referred as ROS, because biologically, the most significant free radicals are oxygen-centered [8, 9].

Radical formation is realized through subsequent steps [8].

$$\begin{aligned}
&O_2 + 1e^- + H^+ \leftrightarrow HO_2^{\bullet -} \leftrightarrow H^+ + O_2^{-\bullet} \\
&HO_2^{\bullet -} + 1\,e^- + H^+ \leftrightarrow H_2O_2 \\
&H_2O_2 + 1e^- + H^+ \leftrightarrow \left[H_3O_2\right] \leftrightarrow H_2O + {}^{\bullet}OH \\
&{}^{\bullet}OH + 1e^- + H^+ \leftrightarrow H_2O
\end{aligned}$$

http://dx.doi.org/10.5772/61193

Due to oxygen reduction, three important intermediate products are generated: superoxide anion, hydrogen peroxide and hydroxyl radical.

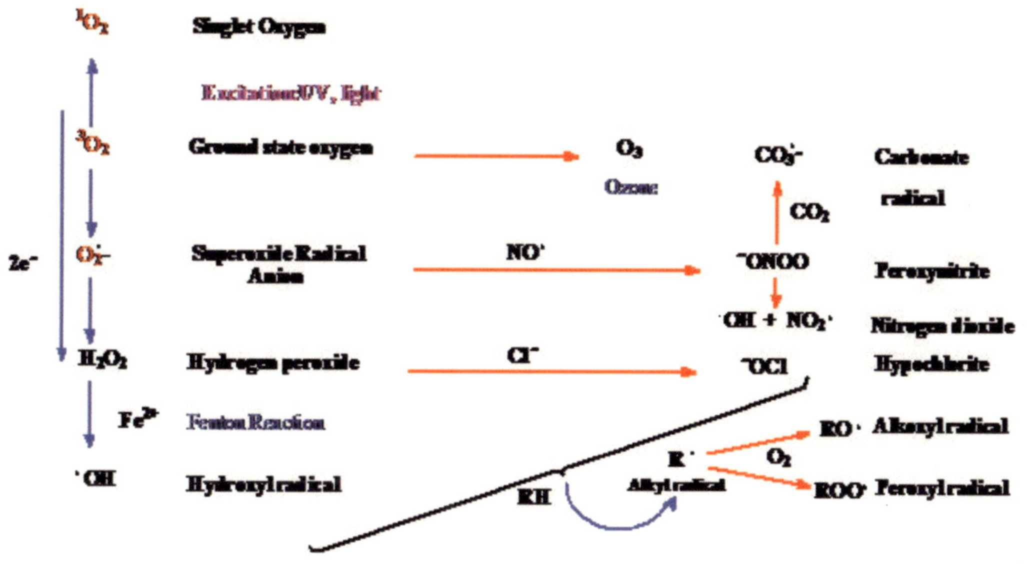

**Figure 1.** Main reactive oxygen species

### *2.1.1. Superoxide anion*

Superoxide anion ($O_2^{\bullet-}$), the most common ROS, is generated in mitochondria, in cardiovascular system and other parts of the body [3, 10]. The electron transport chain (ETC) is responsible for most of the superoxide generation through partial reduction of oxygen [11].

Radical formation is possible:

$$X + O_2^{\bullet -} + H^{+}® \rightarrow XH + O_2$$
$$YH + O_2^{\bullet -} + H^{+} \rightarrow Y^{\bullet} + H_2O_2$$

In aerobic organism most of the oxygen is reduced to water in mitochondrial respiratory chain. However, a small proportion of the oxygen molecules (1% -2%) is converted to superoxide anion radical. These reactions occur in respiratory chain by complex I (NADH: ubiquinone oxidoreductase) and complex III (ubiquinol: cytochrome *c* oxidoreductase) [12]. Another important pathway to form superoxide is represented by heme oxidation. The iron (of heme group) is reduced to ferrous (Fe II) in the deoxyhemoglobin and when it attaches to oxygen an intermediate structure is formed [8].

$$\text{Heme Fe}^{+2}\text{–O}^{2-} \rightarrow O_2^{-\bullet} + \text{Heme Fe}^{+3}$$

In addition to this, dihydrorotate dehydrogenase, aldehyde oxidase, and xanthine oxidase are oxidative enzymes that can also produce superoxide anion.

### *2.1.2. Hydrogen peroxide*

Hydrogen peroxide ($H_2O_2$) is a neutral molecule, which is the least reactive molecule among ROS and is stable under physiological pH and temperature in the absence of metal ions. It is highly diffusible and does cross the plasma membrane easily. $H_2O_2$ is formed in our body by large number of reactions and yields potent species [3]. $H_2O_2$ could be produced from superoxide anion by superoxide dismutase (SOD) through a dismutation reaction. Amino acid oxidase and xanthine oxidase can also produce $H_2O_2$ from superoxide anion. In the presence of metal ions and superoxide anion $H_2O_2$ can produce the hydroxyl radical [13, 15].

$$O_2^{\cdot -} + H_2O_2 \rightarrow \cdot OH + OH^- + O_2$$

### *2.1.3. Hydroxyl radical*

Hydroxyl radical (·OH), the most reactive and dangerous radical, can be formed from superoxide anion and $H_2O_2$ in the presence of metal ions, and it has a very short *in vivo* half-life of about $10^{-9}$s [13]. As a result, when produced *in vivo*, OH reacts close to its site of formation. Most of hydroxyl radicals are produced in the excess of superoxide anion and $H_2O_2$ by Haber-Weiss reaction [8, 14].

$$O_2^{\cdot -} + H_2O \rightarrow OH^- + {}^{\cdot}OH + O_2$$

It can also be produced by multiple pathways; such as, decomposition of water because of ionizing radiation that form hydroxyl radicals and hydrogen atoms, and photolytic decomposition of alkyl hydroperoxides. *In vivo*, primarily hydroxyl radicals emerge from the metal-catalyzed breakdown of $H_2O_2$, through *Fenton reaction* [8, 14].

$$M^{n+}\left(Cu^{+}, Fe^{2+}, Ti^{3+}, Co^{2+}\right) + H_2O_2 \rightarrow M^{(n+1)}\left(Cu^{2+}, Fe^{3+}, Ti^{4+}, Co^{3+}\right) + \cdot OH + OH^-$$

The Fenton reaction is generally considered to yield the ·OH; following as shown from spin-trapping and hydroxylation [15].

### *2.1.4. Singlet oxygen*

Singlet oxygen (${}^1O_2$), a non-radical, can be generated by an input of energy that rearranges the electrons, and it is rather mild and nontoxic for mammalian tissue. It is formed during

http://dx.doi.org/10.5772/61193

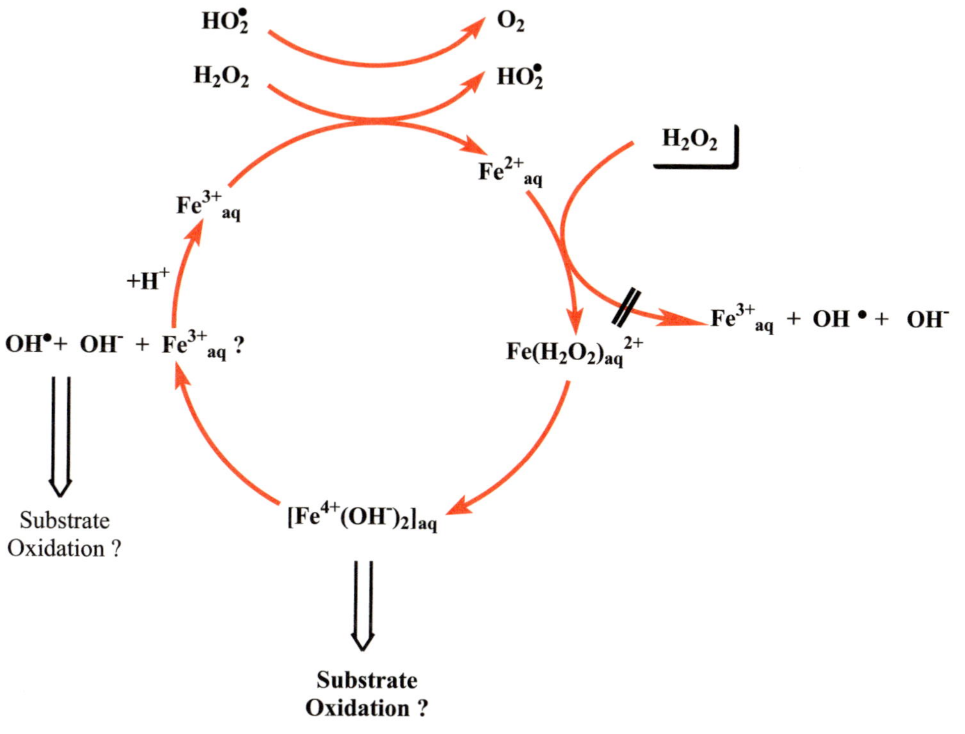


**Figure 2.** Mechanistic presentation of possible reactions involved in the thermal Fenton reaction with simplified notations used for the various iron complexes

photosensitization and chemical reactions [3]. In the human beings, $^1O_2$ is both a signal and a weapon, with therapeutic potency against various pathogens such as microbes, viruses, and cancer cells. Two different pathways in biology can produce singlet oxygen. $^1O_2$ can directly oxidize proteins, DNA, and lipids and has been known to be involved in cholesterol oxidation that can participate in *Dielse-Alder* reactions. It can be generated by chemical processes, such as spontaneous decomposition of hydrogen trioxide in water or the reaction of $H_2O_2$ with hypochlorite. $^1O_2$ reacts with an alkene (-C=C-CH-) by abstraction of the allylic proton in an ene-type reaction to the allyl hydroperoxide HO-O-R (R = alkyl), which can then be reduced to the allyl alcohol [16].

### 2.2. Carbon-centered radicals

Carbon centered radicals are described as follows.

#### 2.2.1. *Peroxyl and alkoxyl radicals*

Peroxyl (ROO ) and alkoxyl (RO ) radicals are good oxidizing agents, having more than 1000 mV of standard reduction potential. Irradiation of UV light or the presence of transition metal ions can cause hemolysis of peroxides to produce peroxyl and alkoxyl radicals.

$$ROOH \rightarrow ROO\cdot + H\cdot,\ ROOH + Fe^{3+} \rightarrow ROO\cdot + Fe^{2+} + H^+$$

Aromatic alkoxyl and peroxyl radicals are less than respective open chain radicals because of the delocalization of electrons in the ring. ROO is reactive and formed from lipids, proteins, DNA and carbohydrates during oxidative damage [3]. Peroxyl radicals are generated by a direct reaction of oxygen with alkyl radicals (R ) [13].

### 2.3. Nitrogen-centered free radicals

NO and its by-products, such as nitrate ($NO_3^-$), nitrite ($NO_2^-$), peroxynitrite ($ONOO^-$), and 3-nitrotyrosine are called RNS. More recently the role of the RNS has been shown to have a direct role in cellular signaling, vasodilatation, and immune response [10].

#### *2.3.1. Nitric oxide*

NO is generated during the breakdown of arginine to citrulline by a family of NADPH-dependent enzymes called nitric oxide synthase [10, 11]. It is an uncharged lipophilic molecule containing a single unpaired electron, which causes it to be reactive with other molecules such as oxygen, superoxide radicals and glutathion. While NO is not a very reactive free radical, it is able to form other reactive intermediates, which have an effect on protein function and on the function of the entire organisms. These reactive intermediates can trigger nitrosative damage on biomolecules [10]. Therefore; NO could function as an oxidant or an antioxidant. NO is neurotransmitter and blood pressure regulator, and it can yield potent oxidants during pathologic states [3]. Overproduction of NO is involved in ischemia reperfusion, and neuro-degenerative and chronic inflammatory diseases, such as rheumatoid arthritis and inflamma-tory bowel disease [3,13].

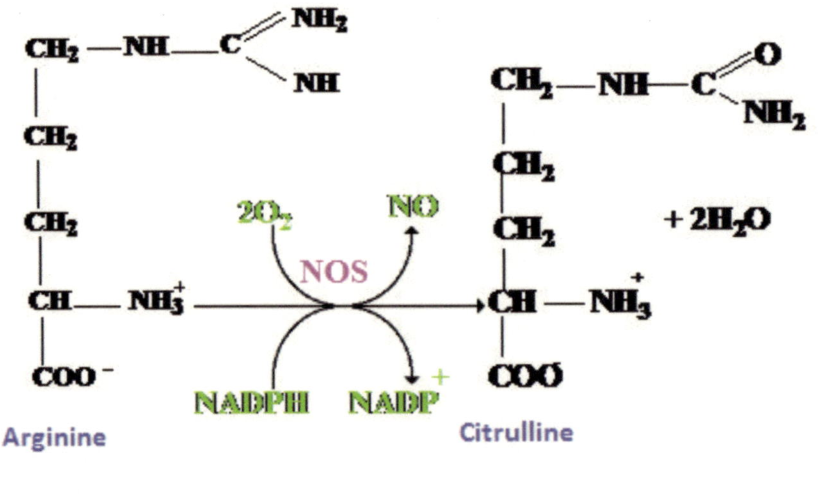


**Figure 3.** Nitric oxide formation

After NO is produced, it can inhibit cytochrome c oxidase. NO found in the mitochondria increases the production of ROS and RNS which can alter the activity of various processes such as mitochondrial biogenesis, respiration and oxidative stress [11].

NO is more stable and diffusible than hydroxyl radicals. These have shown hyperphosphorylatation and inactivate retinoblastoma protein resulting in increased proliferation of human colon cancer cells. NO modifies DNA directly and inactivates the DNA repair enzymes. NO is very reactive signaling molecule and it is an important regulator for cellular functions. Nitrosative stress also plays a critical role in inflammation-associated carcinogenesis by activating a representative redox sensitive transcription factor. NO modifies DNA directly and inactivates the DNA repair enzymes [3]. NO is exposed to human blood plasma which can deplete the concentration of ascorbic acid, uric acid, and initiate lipid peroxidation.

### 2.3.2. *Peroxynitrite*

NO reacts with $O_2^-$ with a high rate constant to give peroxynitrite ($O_2^- + NO \rightarrow OONO^-$) [15, 17], which may spontaneously decompose to yield $NO_2^-$ and hydroxyl radical (.OH) [18]. At physiological pH, $ONOO^-$ is a stronger oxidant than $O_2^-$ or NO, and it oxidizes lipids, proteins, nitrated amino acids and DNA [19, 20].

Peroxynitrite is a cytotoxic and causes tissue injury and oxidizes low-density lipoprotein. The significance of $OONO^-$ as a biological oxidant comes from its high diffusibility across cell membranes. $OONO^-$ appears to be an important tissue-damaging species generated at the sites of inflammation and has been shown to be involved in various neurodegenerative disorders and several kidney diseases. The molecule can cause direct protein oxidation and DNA base oxidation and modification acting as a "hydroxyl radical-like" oxidant. Nitrotyrosine, which can be formed from peroxynitrite-mediated reactions with amino acids, has been found in age-associated tissues [9, 13].

Peroxynitrite as a potential biomarker of inflammation-associated cancers and a product formed by a reaction between NO radical and superoxide anion, causes DNA damage by generating 8-nitroguanine. Thus, oxidative and nitrosative DNA damage products have been implicated in the initiation of inflammation-driven carcinogenesis [3].

### 2.3.3. *Nitrogen dioxide*

Nitrogen dioxide ($NO_2\cdot$) is formed from the reaction of peroxyl radical and NO. $NO_2\cdot$ initiates lipid peroxidation for production of free radicals and also oxidizes ascorbic acid [13]. $NO_2\cdot$, a major decomposition product of NO and very reactive oxidant, is able to oxidize tyrosine to 3-nitrotyrosine. Moreover, $NO_2\cdot$ is a substrate for the mammalian peroxidase and lactoperoxidase and forms $NO_2\cdot$ via peroxidase-catalyzed oxidation of $NO_2\cdot$. This provides an additional pathway contributing to cytotoxicity or host defense associated with increased NO production and an alternative pathway for the formation of 3-nitrotyrosine [10].

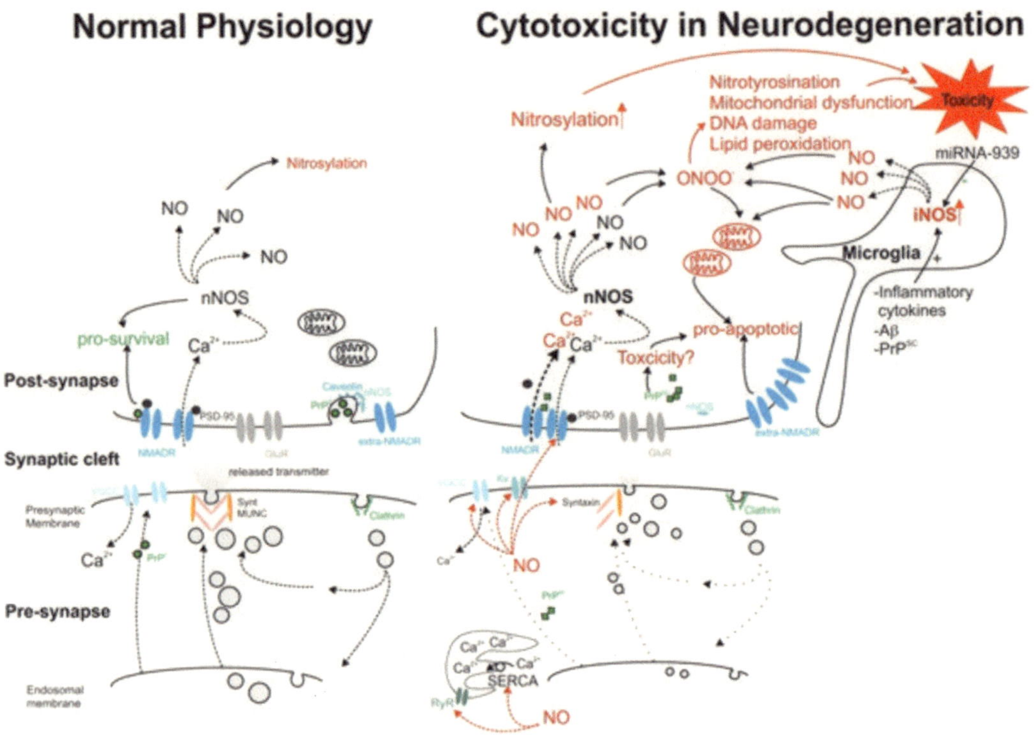


**Figure 4.** Nitric oxide and cytotoxicty

## 3. Biochemistry of ROS and RNS

The generation of ROS can occur as a product of biochemical reactions, in mitochondria, peroxisomes, cytochrome P450, and other cellular components [21, 22]. ROS are generated mainly by the mitochondrial ETC. Almost all cells, and tissues continuously convert a small proportion of molecular oxygen to ROS in ETC. ROS produced by other pathways, including the respiratory burst taking place in activated phagocytes, are ionizing the damaging effect of the radiation on components of cell membranes, and by-products of several cellular enzymes (NADPH oxidases, xanthine oxidase, nitric oxide synthase) [23]. The formation of ROS is a natural consequence of aerobic metabolism and is integral for maintaining tissue oxygen homeostasis. When oxygen homeostasis is not maintained, there is an increase in oxidative stress in the cellular environment. Superoxide, hydrogen peroxide and hydroxyl radicals are normal metabolic by-products which are generated continuously by the mitochondria in growing cells. Microsomal cytochrome P450 enzymes, flavoprotein oxidases and peroxisomal enzymes are other significant intracellular sources of ROS [24].

ROS play the key roles in both health and disease. ROS also have an important role in several physiologic processes such as normal vascular cell functioning and maintaining vascular

http://dx.doi.org/10.5772/61193

diameter regulation. ROS carry out this function by mounting effective immune response, acting as possible signaling molecules and regulating glucose uptake by skeletal muscle [3]. They have a role in response to growth factor stimulation and control of inflammatory responses. They participate in the regulation of differentiation, proliferation, growth, apoptosis, cytoskeletal regulation, migration, and contraction [4].

ROS contributes to a wide range of pathologies and many of the implicated diseases which lead to death, such as chronic inflammation and autoimmune diseases (diabetes, rheumatoid arthritis, lupus), sensory impairment (ocular disease, hearing loss), cardiovascular diseases (atherosclerosis, hypertension, ischemia/reperfusion injury), cancer (breast, renal, lung), fibrotic disease (pulmonary and liver fibrosis, diabetic nephropathy), obesity, insulin resistance, neurological disorders (Parkinson's, Alzheimer's, ALS, schizophrenia), and infectious diseases (septic shock, influenza, hepatitis, HIV) [3, 4].

ROS generation is generally a cascade of reactions which starts with the production of superoxide. Superoxide rapidly dismutates to hydrogen peroxide spontaneously, particularly at low pH or is catalyzed by SOD. The other generations include the reaction of superoxide with NO to form peroxynitrite, the peroxidase-catalyzed formation of hypochlorous acid from hydrogen peroxide, and the iron-catalyzed Fenton reaction leading to the generation of hydroxyl radical [2,25].

ROS act with a large number of biomolecules (proteins, lipids, carbohydrates, and nucleic acids). ROS may irreversibly destroys and alter the function of these molecules after interacting with them. As a result, they have been increasingly identified in biological organisms as major contributors to damage. Harman has showed the role of ROS in the aging process [26], after these ROS become cell damaging agents in aging theory. ROS also have an important role in host defense because ROS generation deficiencies reduce the killing ability of leukocytes. However, over the last decades, a second important concept of ROS have been evolving. Furthermore, ROS have a reversible regulatory process in virtually all cells and tissues [27].

### 3.1. Biological sources of ROS

In mammalian cells the biological sources of ROS are as follows; mitochondria [28], endoplasmic reticulum [29], peroxisomes [12], cytosol [30], plasma membrane [31] and extracellular space [32]. Major sources of ROS include metabolic processes and cellular respiration processes. During the metabolic processes the peroxisome catabolizes biomolecules that remove hydrogen in an oxidative reaction creating $H_2O_2$. While the cellular respiration oxygen is reduced the intermediates with odd electrons can escape the chain.

Typically the initial reaction is an electron transfer of oxygen to $O_2^-$, which then dismutases to $H_2O_2$. $O_2^-$ does not readily cross membranes, and is short-lived and local in its effects, but SOD converts $O_2$ to longer-lasting and membrane-diffusible $H_2O_2$. When $O_2$ react with NO, the highly reactive ONOO is formed. Peroxidases catalyze reactions involving $H_2O_2$, resulting in the generation of HOCl and $^1O_2$ among other species. Finally the Haber-Weiss reaction uses an iron ion catalyst to generate hydroxyl radicals from $O_2^-$ and $H_2O_2$. Major sources of ROS

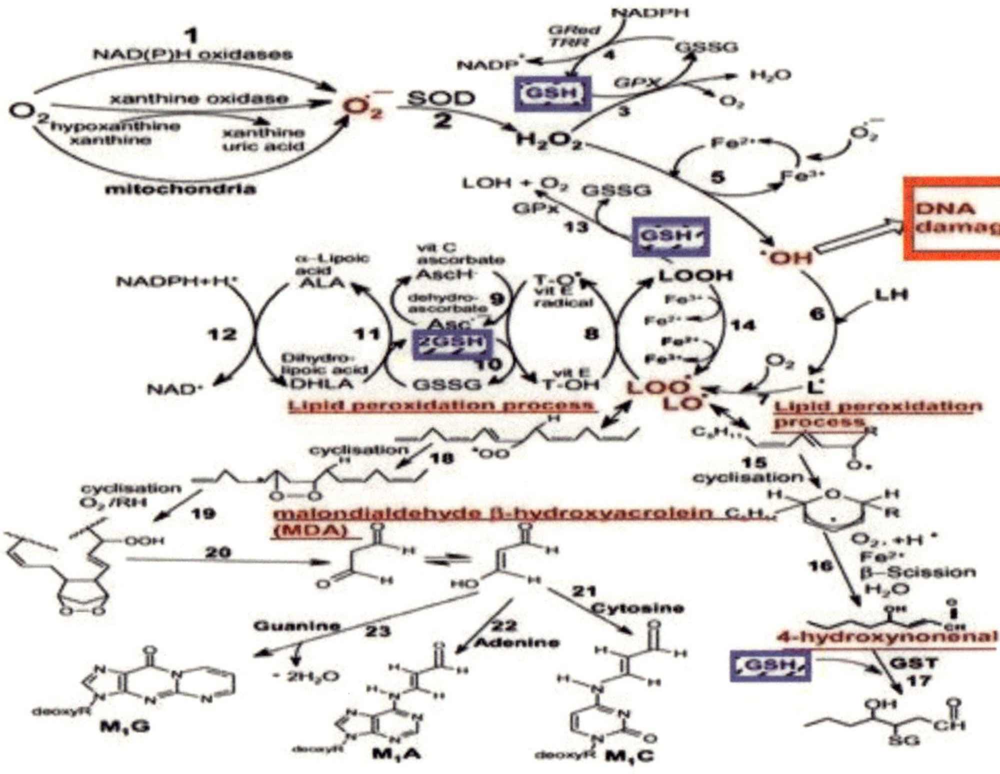

**Figure 5.** Pathways of ROS formation

includes cellular respiration and metabolic processes, though ROS may also be generated by radiation [4].

### 3.2. Biosynthesis pathway of ROS

Mitochondria, NADPH oxidase and 5-lipoxygenase (5-LOX) are three major sources of ROS formation.

#### *3.2.1. Mitochondrial ROS production*

The generation of mitochondrial ROS is the result of oxidative phosphorylation. In cytochrome chain, electrons derived from FADH or NADH can directly react with oxygen or other electron acceptors, and generate free radicals [3]. ROS, as generated by mitochondria during respiration, induce oxidative stress, which accumulates over life and is considered as the proximal mechanism of aging and a major determinant of degenerative disease, including cancer, and lifespan (the free radical or mitochondrial theory of aging). Recent data showed that specific ROS in particular $H_2O_2$ is directly implicated in the physiological regulation of different signal transduction pathways, including the insulin/$IGF_1$ pathway. Indeed mitochondrial ROS

http://dx.doi.org/10.5772/61193

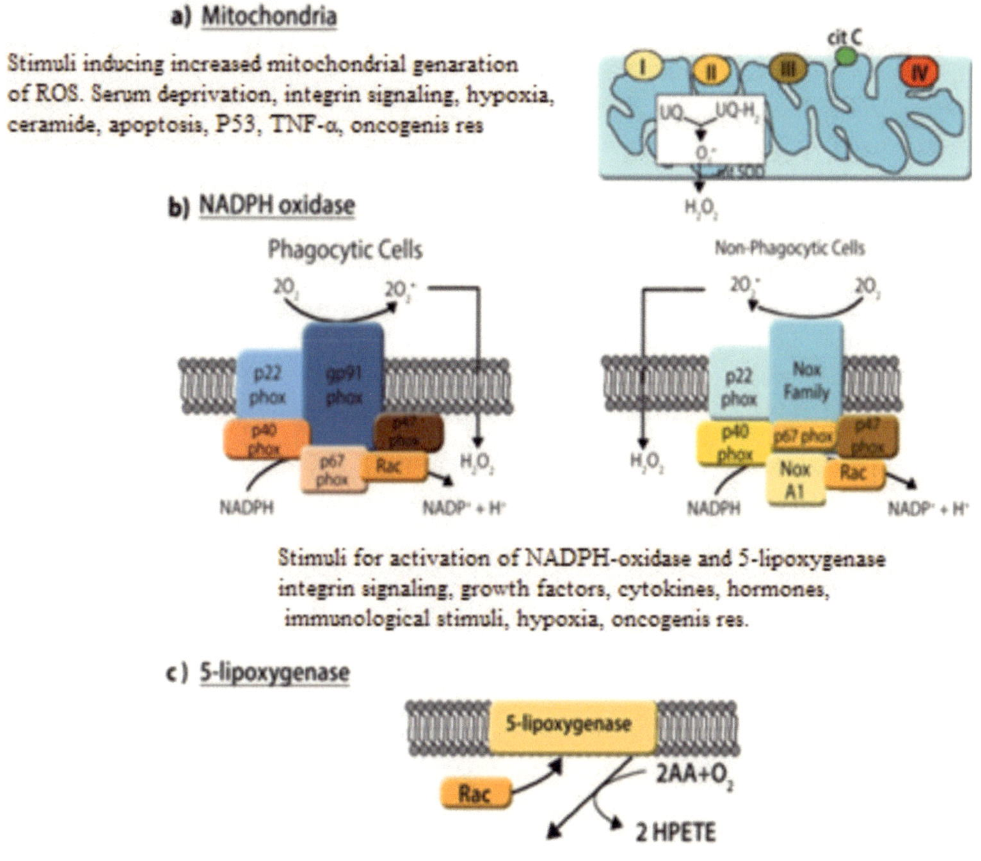


**Figure 6.** The major pathway formation of ROS

production has been found to contribute to the regulation of several cellular processes in a specific manner. A role for ROS as signaling molecules is further supported by recent findings that the generation of $H_2O_2$ by mitochondria is not only as a by-product of respiration but also as a result of specific enzymatic systems, such as p66Shc. P66Shc functions as an inducible redox enzyme, which is activated by stress and triggers apoptosis to regulate signal transduction and transcription. Regardless of the purpose p66Shc utilizes to shift the intracellular redox balance towards oxidation, it appears that mitochondrial ROS formation ability is evolved to set intracellular ROS levels. Therefore; one could hypothesize that the mitochondrial-mediated oxidative stress may have a critical oncogenic role when tumor suppressor mechanisms decreased rather than acting as primary mutagens [33-35].

Mitochondria are responsible for 90% of the energy production in cells, and thus tissues, organs and the body as at the whole needs to function. Hence, they are known as the "powerhouse of a cell": the core of cellular energy metabolism, being the site of most ATP generation through mitochondrial oxidative phosphorylation [36]. In this process, electrons liberated from reducing substrates are delivered to $O_2$ establishing an electrochemical gradient used to drive

ATP synthesis. During the oxidative phosphorylation, the reduction of oxygen by one electron at a time ($O_2 \rightarrow O_2^{\cdot} \cdot \rightarrow H_2O_2 \rightarrow \cdot OH \rightarrow H_2O$) produces ROS, relatively stable intermediates with one unpaired electron [37]. There is a lot of evidence supporting the point of view that mitochondria are not a major source of ROS in the cell. In fact, about 90% to 95% of cellular oxygen is used up in oxidative phosphorylation and 3% from that pool can be converted to superoxide which is a very strong argument to mitochondria as a main source of this oxygen radical. Consequently, in mammalian cells mitochondria are the major source of ROS [38].

The primary ROS are generated by mitochondria, as a result of monoelectronic reduction of $O_2$ which is the precursor of most ROS and a mediator in oxidative chain reactions. $O_2$ is produced in mammalian cells enzymatically by NADPH oxidase, cytochrome P450-dependent oxygenizes and xanthine oxidase as well as in course of a single electron is directly transferred to $O_2^{\cdot}$. Furthermore, $O_2$ can produce RNS which react with other radicals, such as NO.

Hydrogen peroxide is produced by dismutation of $O_2$, catalyzed by SODs, whether spontaneously or through a reaction in mitochondria. Because $H_2O_2$ is relatively stable and membrane permeable, it can diffuse within the cell and be eliminated by cytosolic or mitochondrial antioxidant systems such as catalase, glutathione peroxidase, and thioredoxin peroxidase. Mitochondrial generated $H_2O_2$ can also act as a signaling molecule in the cytosol, affecting multiple networks which control, for instance, cell cycle, stress response, energy metabolism, and redox balance [39, 40].

Hydroxyl radical ($\cdot$OH) can be formed in the presence of metal ions by Fenton reaction when $H_2O_2$ is not metabolized. As one of the strongest oxidants hydroxyl radical is highly reactive and may damage the other molecules. Therefore, mitochondria have developed an efficient $H_2O_2$ removal systems. These metal-chelating mechanisms include chaperone proteins and prevent the formation of this radical. Even if there are at least ten enzymes in mammalian mitochondrial for ROS production, their capacity to produce ROS greatly differ in a tissue-specific manner. *In vitro* experiments have demonstrated that $H_2O_2$ generation is both substrate specific and organ specific [41, 42].

Mitochondrial ROS are a product of respiration and generally occur at the ETC. As a terminal component of the ETC, cytochrome c oxidase (Complex IV) receives four electrons from cytochrome c and reduces one $O_2$ molecule to two $H_2O$. The first mitochondrial site producing ROS was identified at the Complex III located at the inner side of inner mitochondrial membrane (bc1 complex, ubiquinone: cytochrome c reductase). The primary ROS produced at this site are $O_2$, through the referred Q-cycle [43].

Unsaturated fatty acids in mitochondrial membranes, which are components of phospholipids, can be easily oxidized by the hydroxyl radical. Lipid 4-hydroxynonenal (4-HNE), malondialdehyde (MDA) and acrolein are the lipid peroxidation products. Generation of these lipid radicals lead to the production of a new radical, which causes various effects in the cell such as affecting membrane fluidity and increasing permeability [44]. Lipid radicals can covalently modify membrane proteins as well as cause "lipoxidative" damage to the mtDNA because they diffuse easily in the membranes [45]. For instance, 4-hydroxy 2-nonenal inhibits

http://dx.doi.org/10.5772/61193

adenine nucleotide translocase activity in mitochondria on account of the modification of sulfhydryl groups in adenine nucleotide translocase [46].

The nine known sources of ROS in mitochondria are shown in Fig. 6 in the context of location within a mitochondrion. ROS producing seven enzymes will be discussed in the following text.

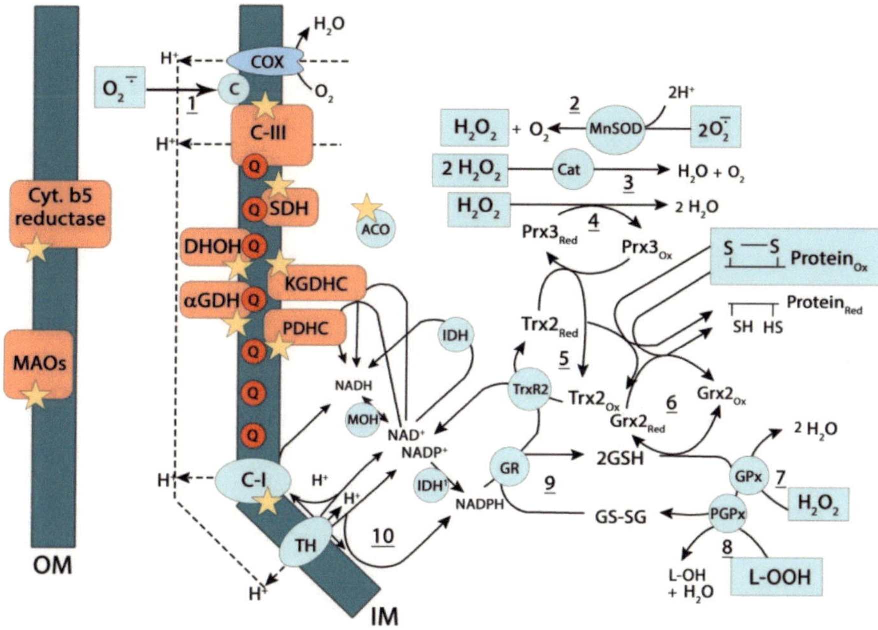


**Figure 7.** Sources of ROS in mitochondria

1. **Complex III** (cytochrome c oxidoreductase, EC 1.10.2.2) enzyme is located closer to the outer and the inner coupling membrane and widely distributed in mammalian tissues. Complex III produces superoxide at both site of membrane. The enzyme oxidizes cytoplasmic NAD(P)H and reduces cytochrome *b*5 in the outer membrane. It may also reduce ascorbyl free radical and, therefore, be involved in regeneration of ascorbate in mammalian liver. The enzyme is upregulated in the patients suffering from schizophrenia, thus implying a role in the etiology of the disease. There is a single report that mitochondrial cytochrome *b*5 reductase may produce superoxide with a high rate of ~300 nmol/ min per mg protein [36, 47].

2. **Monoamine oxidase** (MAO, EC 1.4.3.4) is located in the outer mitochondrial membrane and contain flavin adenine dinucleotide (FAD). Two isoforms MAO-A and MAO-B are identified with their specificity and sensitivity to inhibitors. Norepinephrine and serotonin are oxidized by MAO-A, whereas the substrates for MAO-B are phenylethylamine and benzylamine. This enzyme is found in all the tissues such as, lung, liver blood vessels,

heart, the nervous system, etc. The enzyme catalyze oxidation of biogenic amines and produce toxic molecules, aldehydes, ammonium hydroxide and hydrogen peroxide. MAO cause development of oxidative stress in heart and nervous system. MAO is also a major source of $H_2O_2$ in ischemia [48] and aging [42]. Furthermore MAO gives rise to peroxidation of phospholipids in mitochondria. The inhibitors of MAO decreased production of hydrogen peroxide and thus protect the heart [43].

3. **Dihydroorotate dehydrogenase** (DODH, dihydroorotate: ubiquinone oxidoreductase, EC 1.3.99.11) is a flavoprotein located in the outer surface of inner mammalian mitochondrial membrane. The enzyme catalyzes the conversion of dihydroorotate to orotate, which is a step in the *de novo* synthesis of uridine monophosphate. DODH has been found in many cells, such as tumor, mucosal, ileum, colon, crypts, kidney cortex, heart and liver. The most active enzyme is found in heart and liver mitochondria. The enzyme contains flavin mononucleotide (FMN) containing active site as a source of superoxide. Superoxide is formed during aerobic oxidation of dihydroorotate in the presence of cyanide. However, histochemical studies showed that heart and kidney cortex mitochondria produce hydrogen peroxide during dihydroorotate oxidation. Drugs specifically interacts with the hydrophobic channel of the enzyme by preventing FMN-ubiquinone oxydoreduction thus decreasing ROS formation in cancer cell cultures.

4. **Mitochondrial glycerophosphate dehydrogenase** (mGPDH, $\alpha$-glycerol-3-phosphate: ubiquinone oxidoreductase, EC 1.1.99.5) is a FAD-linked enzyme and located in the outer surface of inner mitochondrial membrane. The enzyme catalyze oxidation of glycerol-3-phosphate to dihydroxyacetone phosphate by utilizing mitochondrial coenzyme Q as electron acceptor. The mGPDH has influence on lipid metabolism and catalyze transfer of reducing equivalents from cytosolic NADH to the respiratory chain in mitochondria, which is so-called glycerol phosphate shuttle. Its activity and content substantially vary in different tissues. Placenta, brown adipose tissue, testis, skeletal muscle, langerhans and brain tissue have high mGPDH activity, whereas heart, liver and kidney mitochondria have low activity. In heart mitochondria the ratio of mGPDH to succinate dehydrogenase is approximately 1/15 [40, 49]. Thyroid hormones increase its activity. ROS production is occurred during $\alpha$-glycerophosphate oxidation. These productions are increased by the inhibitors of complex III, antimycin A and myxothiazol.

5. **Complex II** [succinate dehydrogenase complex (SDH, succinate:ubiquinone oxidoreductase, EC1.3.99.1)] is a marker enzyme located at the inner surface of inner mitochondrial membrane. The enzyme oxidizes succinate to fumarate, which is one of reactions in Krebs cycle by using coenzyme Q as an electron acceptor. Although oxidation of succinate by good-quality mitochondria from most mammalian tissues can produce ROS with a high rate, the source of ROS is Complex I, not SDH. The mechanism involves reverse electron transfer from SDH-reduced coenzyme Q to Complex I. Nevertheless, isolated SHD reconstructed in liposomes can produce ROS by itself. Authors concluded that reduced FAD of SDH generates ROS in the absence of its electron acceptor. There is also a report implying that SDH can generate ROS in submitochondrial particles. However; the conclusion was based solely on the inhibition of ROS production by carboxin, a specific

inhibitor of SDH. The same inhibitor also suppressed antimycin-induced ROS production and ROS production supported by NADH oxidation. The former is thought to originate from Complex III that is not inhibited by carboxin whereas the effect of carboxin on NADH-supported ROS production may not be readily explained either. Therefore, it is unclear whether SDH produces ROS *in situ*, in mitochondria [43].

6. **Mitochondrial aconitase** (m-aconitase, EC 4.2.1.3) is an enzyme localized to the matrix space of mitochondria; it participates in tricarboxylic acid cycle catalyzing a conversion of citrate to isocitrate. The enzyme contains an iron-sulfur cluster that can be oxidized by superoxide, inactivating m-aconitase. Recently, it was found that isolated aconitase oxidized by either superoxide or hydrogen peroxide produces hydroxyl radical. Vasquez-Vivar et al. [50] state that similar continuous hydroxyl radical production may occur upon superoxide-driven redox-cycling of aconitase in mitochondria.

7. **Ketoglutarate dehydrogenase complex** (KGDHC, 2-oxoglutarate dehydrogenase) is an integral mitochondrial enzyme tightly bound to the inner mitochondrial membrane on the matrix side. In the tricarboxylic acid cycle, it catalyzes the oxidation of $\alpha$-ketoglutarate to succinyl-CoA using $NAD^+$ as electron acceptor. Structurally, KGDHC is composed of multiple copies of three enzymes: $\alpha$-ketoglutarate dehydrogenase, dihydrolipoamide succinyltransferase, and lipoamide dehydrogenase. The E3 component of KGDHC is a flavin-containing enzyme; it is identical to the E3 component of another integral mitochondrial enzyme located in the matrix, pyruvate dehydrogenase (PDHC). The E3 component is also known as dihydrolipoamide dehydrogenase, which is ubiquitously present in mammalian mitochondria. Both PDHC and KGDHC can generate superoxide and hydrogen peroxide; ROS production was shown with isolated purified enzymes from bovine heart and in isolated mitochondria. The source of ROS in KGDHC appears to be the dihydrolipoamide dehydrogenase component. Earlier, isolated dihydrolipoamide dehydrogenase was shown to produce ROS. In mitochondria and with isolated enzyme, ROS production from KGDHC was stimulated by a decrease in availability of its natural electron acceptor, $NAD^+$ [51, 52].

### *3.2.2. NADPH oxidase*

In aerobic organisms, electrons are transferred from the reductant to the oxidant. These are always catalyze by oxidoreductases that gives rise to superoxide production the primary ROS molecule in biological systems. In mammalian cells, cyclooxygenase, lipoxygenase, cytochrome P450 enzymes, nitric oxide synthase xanthine oxidase, mitochondrial NADH:ubiquinone oxidoreductase (complex I) and NADPH oxidase have been identified as potential sources of superoxide. Contrary to the other oxidoreductase, the NADPH oxidase is the major enzymatic source of ROS generation in cells but the other enzymes produce ROS only as by-products. It catalyzes the production of superoxide by the reduction of oxygen, using NADPH as the electron donor.

$$\rightarrow 2O_2 + NADPH \rightarrow 2O_{2^-} + 2NADP^+ + H^+$$

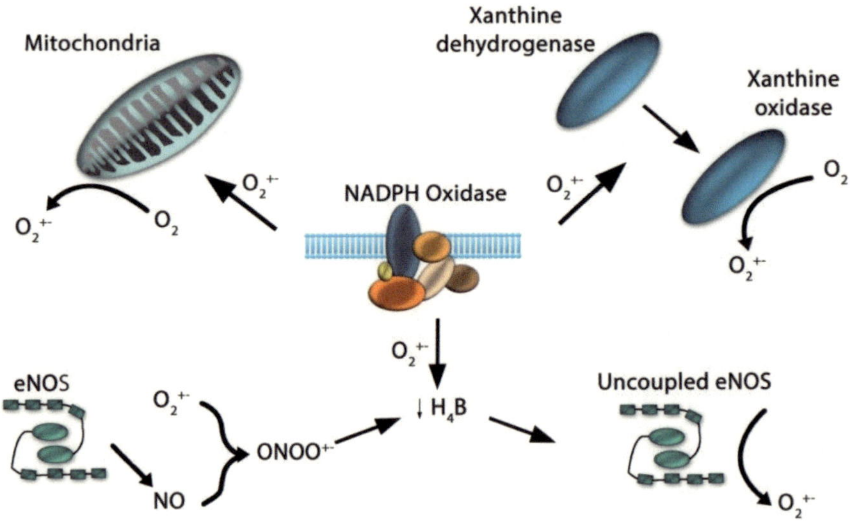


**Figure 8.** Interplay between NADPH oxidase and other sources of ROS

The enzyme is largely derived from leukocytes, such as macrophages and neutrophils and consists of two subunits gp91phox (NOX2) and p22phox. Enzyme complex generates superoxide reducing oxygen via its gp91phox subunit with reduced NADPH as the electron donor. There are lots of NADPH oxidase (NOX) isoform identified in mammalian cells (NOX1, NOX-3, NOX-4, NOX-5, etc.) [53-55]. NADPH oxidase plays a key role in mammalian cells intracellular signaling pathways by generating ROS molecules of NADPH oxidase. ROS signaling system is overwhelming in comparison with other signaling mechanisms [1].

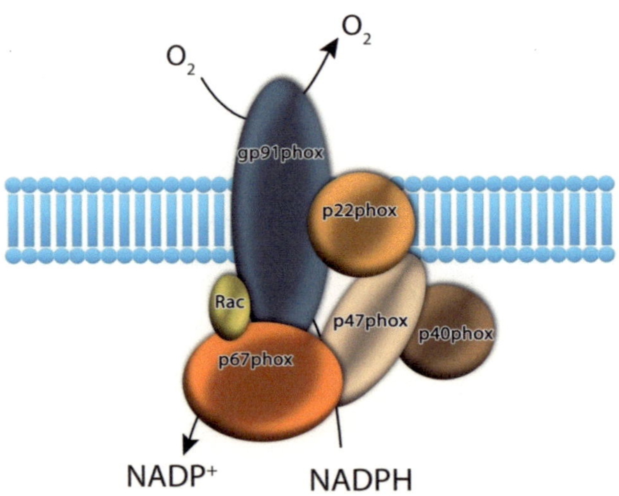


**Figure 9.** Schematic diagram of the structure of the active NADPH oxidase complex

http://dx.doi.org/10.5772/61193

### 3.2.3. *Lipoxygenases and ROS*

The 5-lipoxygenase (5-LOX) has 78 kDa molecular weight and catalyzes the biosynthesis of potent bioactive eicosanoids, such as leukotriene (LT) and hydroxy eicosatetraenoic acid (HETE). The biosynthesis of LT begins with the metabolism of arachidonic acid by 5-LOX. These soluble dioxygenases incorporate oxygen molecules at position C5 of the fatty acid, yielding 5 (S)-hydroperoxyeicosatetraenoic acid (5-HPETE), which is subsequently metabolized by 5-LOX to generate the unstable epoxide, leukotriene A4 (LTA4). LTA4 can be converted to LTB4 via LTA4 hydrolase, or to LTC4 via LTC4 synthase [56]. The most relevant pathophysiological function performed by LTs involves the regulation of inflammatory immune responses. Leukotriene B4 (LTB4) is a potent activator of neutrophil chemotaxis, whereas the cysteinyl leukotrienes (CysLTs) (i.e., LTC4, LTD4, and LTE4) are key mediators of allergic inflammation. LTB4, which was the first leukotriene to be isolated, elicits a variety of inflammatory responses, including leukocyte activation, chemotaxis, and degranulation [57]. Over the past few years, it has been reported that LTB4 treatment of fibroblasts and neutrophils results in ROS generation, and that LTB4-induced chemotaxis is mediated by a NADPH oxidase-dependent cascade. These observations suggest that LTB4-induced ROS generation occurring via NOX is crucial to cell chemotaxis [58]. LTB4 has also been previously shown to promote the phosphorylation and translocation of p47phox, thereby stimulating NADPH oxidase. Collectively, these results suggest that the 5-LOX metabolite, LTB4, appears to stimulate NOX, thereby generating ROS that mediate a variety of signaling pathways in non-phagocytic cells [59].

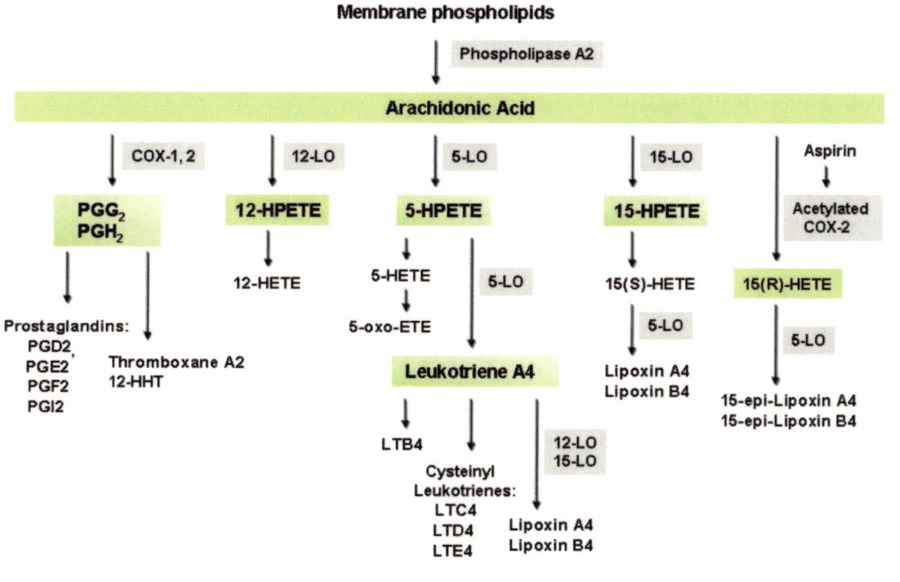


**Figure 10.** Generation of various bioactive eicosanoids via the metabolism of arachidonic acid by lipoxygenases and cyclooxygenases

## 4. Conclusion

ROS are inevitable by-products of normal metabolism. ROS are produced in ETC by electron transfer to molecular oxygen. They act as signaling molecules which mediate response in living cells. By the time ROS level increased in cell it causes cellular oxidative damage to DNA, protein and lipid. ROS lead to altered membrane properties. Membrane fluidity and ion transport are impressed. They cause loss of enzyme activities, DNA damage, and inhibition of protein synthesis which result in cell death. Today it is well known that ROS produced in the cells are involved in the redox regulation of signal transduction pathways.

## Author details

Ayla Ozcan* and Metin Ogun

*Address all correspondence to: aylabicer@hotmail.com

Faculty of Veterinary Medicine, University of Kafkas, Turkey

## References

[1] Bedard K., Krause KH. The NOX family of ROS- generating NADPH oxidases: Physiology and pathophysiology. Physiological Reviews. 2007; 87: 245–313.

[2] Klebanoff SJ. Oxygen metabolism and the toxic properties of phagocytes. Annals of Internal Medicine. 1980; 93: 480–489.

[3] Salman KA., Ashraf S. Reactive oxygen species: A link between chronic inflammation and cancer. Asia-Pacific Journal of Molecular Biology and Biotechnology. 2013; 21(2): 42-49.

[4] Brieger K., Schiavone S., Miller Jr FJ., Krause KH. Reactive oxygen species:from health to disease. Swiss Medical Weekly. 2012; 142: 13659.

[5] Conner E.M., Grisham M.B. Inflammation, free radicals, and antioxidants. Nutrition. 1996; 12(4): 274-281.

[6] Agarwal A., Prabakaran S., Allamaneni S. What an andrologist/urologist should know about free radicals and why. Urology. 2006; 67(1): 2–8.

[7] Perrone S., Tataranno M.L., Negro S., et al. Early identification of the risk for free radical-related diseases in preterm newborns. Early Human Development. 2010; 86: 241–244.

[8] Buonocore G., Perrone S., Tataranno ML. Oxygen toxicity: Chemistry and biology of reactive oxygen species. Seminars in Fetal and Neonatal Medicine. 2010;15: 186–190.

[9] Halliwell B. Biochemistry of oxidative stress. Biochemical Society Transactions. 2007; 35: 1147–1150.

[10] Drew B., Leeuwenburgh C. Aging and the role of reactive nitrogen species. Annals of the New York Academy Science., 2002; 959: 66–81.

[11] Bolisetty S., Jaimes EA. Mitochondria and reactive oxygen species:physiology and pathophysiology, International Journal of Molecular Science, 2013; 14: 6306–6344.

[12] Boveris, A., Cadenas, E., Stoppani, AO. Role of ubiquinone in the mitochondrial generation of hydrogen peroxide. Biochemical Jounal. 1976; 156:435–444.

[13] Lee J., Koo N., Min DB. Reactive oxygen species, aging, and antioxidative nutraceuticals. Comprehensive Reviews in Food Science and Food Safety. 2004;3:21–33.

[14] Liochev I., Fridovich I. The Haber-Weiss cycle 70 years later: An alternative view. Redox Report. 2002; 7: 55–57.

[15] Barbusıński, K. Fenton reaction controversy concerning the chemistry, Ecological Chemistry and Engineering Science. 2009; 16(3): 309–314.

[16] Foote N., Peterson J., Gadsby PM., Greenwood C., Thomson AJ. Redox-linked spin-state changes in the di-haem cytochrome c-551 peroxidase from Pseudomonas aeruginosa. Biochemical Journal. 1985;230: 227–237.

[17] Beckman, J.S., Beckman, T.W., Chen, J., Marshall, P.A., Freeman, B.A. Apparent hydroxyl radical production by peroxynitrite: implications for endothelial injury from nitric oxide and superoxide. Proceedings of the National Academy of Science. 1990;87: 1620–1624.

[18] Radi R, Cosgrove T., Beckman JS., Freeman B. Peroxynitrite-induced luminal chemiluminescence. Biochemical Journal. 1993; 290: 51–57.

[19] Crow, J.P., Beckman, J.S. The role of peroxynitrite in nitric oxide-mediated toxicity. Current Topics in Mcrobiology and Immunology. 1995; 196: 57–73.

[20] Pryor WA, Squadrito GL. The chemistry of peroxynitrite: A product from the reaction of nitric oxide with superoxide. American Journal of Physiology. 1995; 268: 699–722.

[21] Balaban RS., Nemoto S., Finkel T. Mitochondria, oxidants, and aging. Cell. 2005; 120: 483–495.

[22] Gonzalez FJ. Role of cytochromes P450 in chemical toxicity and oxidative stress: Studies with CYP2E1. Mutation Research. 2005; 569: 101–110.

[23] Alfadda AA., Sallam RM. Reactive oxygen species in health and disease Journal of Biomedicine and Biotechnology. 2012; 1–14.

[24] Seifried HE., Anderson DE., Fisher EI, Milner J. A review of the interaction among dietary antioxidants and reactive oxygen species. The Journal of Nutritional Biochemistry. 2007; 18,: 567–579.

[25] Thannickal VJ., Fanburg BL. Reactive oxygen species in cell signaling. American Journal of Physioloy Lung Cellular and Molecular Physiology. 2000; 279: 1005–1028.

[26] Harman D. Aging: A theory based on free radical and radiation chemistry. Journals of Gerontology. 1956; 11: 298–300.

[27] Beckman KB. Ames BN. The free radical theory of aging matures Physiological Reviews. 1998; 78: 547–581.

[28] Starkov AA. The role of mitochondria in reactive oxygen species metabolism and signaling. Annals of the New York Academy of Sciences. 2008; 1147: 37–52.

[29] Gross E, Sevier CS., Heldman N., et al. Generating disulfides enzymatically: reaction products and electron acceptors of the endoplasmic reticulum thiol oxidase Ero1p. Proceedings of the National Academy of Sciences. 2006; 103: 299–304.

[30] Kukreja RC., Kontos HA., Hess ML., Ellis EF. PGH synthase and lipoxygenase generate superoxide in the presence of NADH or NADPH. Circulation Research. 1986; 59(6): 612–619.

[31] O'Donnell VB., Azzi A. High rates of extracellular superoxide generation by cultured human fibroblasts: involvement of a lipid-metabolizing enzyme. Biochemical Journal. 1996; 318: 805–812.

[32] McNally JS.,, Davis ME., Giddens DP., et al. Role of xanthine oxidoreductase and NAD(P)H oxidase in endothelial superoxide production in response to oscillatory shear stress. American Journal of Physiology. 2003; 285: 2290–2297.

[33] Cadenas, E., Davies, K.J. Mitochondrial free radical generation, oxidative stress, and aging. Free Radical Biology and Medicine. 2000; 29: 222–230.

[34] Turrens JF. Mitochondrial formation of reactive oxygen species. The Journal of Physiology. 2003;552: 335–344.

[35] Wosniak JJ., Santos CXC., Kowaltowski AJ., Laurindo F. Cross-talk between mitochondria and NADPH oxidase: Effects of mild mitochondrial dysfunction on angiotensin II mediated increase in NOX isoform expression and activity in vascular smooth muscle cells. Antioxidants and Redox Signaling, 2009; 11: 1265–1278.

[36] Whatley SA., Curti D., Das Gupta F., et al. Superoxide, neuroleptics and the ubiquinone and cytochrome b5 reductases in brain and lymphocytes from normal and schizophrenic patients. Molecular Psychiatry, 1998; 3: 227–237.

[37] Hauptmann N., Grimsby J., Shih JC., Cadenas E. The metabolism of tyramine by monoamine oxidaseA/B causes oxidative damage to mitochondrial DNA. Archives of Biochemistry and Biophysics. 1996; 335: 295–304.

[38] Maurel A., Hernandez C., Kunduzova O., et al. Age-dependent increase in hydrogen peroxide production by cardiac monoamine oxidase A in rats. American Journal of Physiology; Heart and Circulatory Physiology. 2003; 84: 1460–1467.

[39] Forman JH., Kennedy J. Superoxide production and electron transport in mitochondrial oxidation of hydroorotic acid. The Journal of Biological Chemistry. 1975; 250: 4322–4326.

[40] Dummler K., Muller S., Seitz HJ. Regulation of adenine nucleotide translocase and glycerol 3-phosphate dehydrogenase expression by thyroid hormones in different rat tissues. Biochemical Journal. 1996; 317 (3): 913–918.

[41] Estabrook RW., Sacktor B. Alpha-glycerophosphate oxidase of flight muscle mitochondria. The Journal of Biological Chemistry. 1958; 233: 1014–1019.

[42] Kwong LK., Sohal RS. Substrate and site specificity of hydrogen peroxide generation in mouse mitochondria. Archives of Biochemistry and Biophysics. 1998; 350: 118–126.

[43] Zhang L., Yu L., Yu CA. Generation of superoxide anion by succinatecytochrome c reductase from bovine heart mitochondria. The Journal of Bioogical Chemistry. 1998; 273: 33972–33976.

[44] Loschen G., Flohe L., Chance B. Respiratory chain linked $H_2O_2$ production in pigeon heart mitochondria. FEBS Letters. 1971; 18: 261–264.

[45] Cadenas E., Boveris A, Enhancement of hydrogen peroxide formation by protophores and ionophores in antimycin-supplemented mitochondria. Biochemical Journal., 1980; 188: 31–37.

[46] Skulachev VP. Role of uncoupled and non-coupled oxidations in maintenance of safely low levels of oxygen and its one-electron reductants Quarterly Reviews of Biophysics. 1996; 29: 169–202.

[47] Nishino H., Ito A.cSubcellular distribution of OM cytochrome b-mediated NADH-semidehydroascorbate reductase activity in rat liver. Journal of Biochemistry. 1986; 100: 1523–1531.

[48] Koza RA., Kozak UC., Brown LJ., et al. Sequence and tissue-dependent RNA expression of mouse FAD-linked glycerol-3-phosphate dehydrogenase. Archives of Biochemistry and Biophysics. 2004; 336: 97–104.

[49] Lee YP, Lardy HA. Influence of thyroid hormones on L-alpha-glycerophosphate dehydrogenases and other dehydrogenases in various organs of the rat. The Journal of Biological Chemistry. 1965; 240: 1427–1436.

[50] Vasquez-Vivar J., Kalyanaraman B., Kennedy MC. Mitochondrial aconitase is a source of hydroxyl radical. An electron spin resonance investigation. The Journal of Biological Chemistry. 2000; 275: 14064–14069.

[51] Starkov AA., Chinopoulos C., Lorenzo BJ., et al. Mitochondrial alpha-ketoglutarate dehydrogenase complex generates reactive oxygen species. Journal of Neuroscience. 2004; 24(36): 7779–7788.

[52] Tretter LA-VV. Generation of reactive oxygen species in the reaction catalyzed by alpha-ketoglutarate dehydrogenase. Journal of Neuroscience. 2004; 24(36): 7771–7778.

[53] Geiszt M., Kapus A., Ligeti E. Chronic granulomatous disease: More than the lack of superoxide? Journal of Leukocyte Biology. 2001; 69: 191–196.

[54] Geiszt M., Leto TL. The NOX family of NAD(P)H oxidases: Host defense and beyond. The Journal of Biological Chemistry. 2004; 279: 51715–51718.

[55] Banfi B., Molnar G., Maturana A. A Ca(2ş)-activated NADPH oxidase in testis, spleen, and lymph nodes. The Journal of Biological Chemistry. 2001;276: 37594–37601.

[56] Werz O., Steinhilber D. Therapeutic options for 5-lipoxygenase inhibitors. Pharmacology and Therapeutics. 2006; 112: 701–718.

[57] Woo CH., You HJ., Cho SH., et al. Leukotriene B(4) stimulates Rac-ERK cascade to generate reactive oxygen species that mediates chemotaxis. The Journal of Biological Chemistry. 2002; 277: 8572–8578.

[58] Woo CH., Yoo MH., You HJ., et al. Transepithelial migration of neutrophils in response to leukotriene B4 is mediated by a reactive oxygen species-extracellular signal-regulated kinase-linked cascade. The Journal of Immunology. 2003; 170: 6273–6279.

[59] Perkins R.S., Lindsay MA., Barnes PJ., Giembycz MA. Early signalling events implicated in leukotriene B4-induced activation of the NADPH oxidase in eosinophils: Role of Ca2+, protein kinase C and phospholipases C and D. Biochemical Jounal. 1995; 310(3): 795–806.

Chapter 4

# Members of Antioxidant Machinery and Their Functions

Shalini Kapoor Mehta and Sivakumar Joghi Thatha Gowder

Additional information is available at the end of the chapter

http://dx.doi.org/10.5772/61884

**Abstract**

In this modern world, due to the rapid advancement of civilization, industrialization, and overpopulation, scientific knowledge on antioxidants is important since most of the diseases are mediated through reactive oxygen species (ROS). An antioxidant is a molecule that inhibits the oxidation of another molecule. Antioxidants may work through single or combined mechanisms, and based on their activity, they have been categorized into primary, secondary, and tertiary antioxidants. Enzymatic and non-enzymatic antioxidants are the two widely accepted categories of antioxidants. In addition to natural antioxidants, synthetic antioxidants have been extensively used in medicinal and food industries. In brief, antioxidants play a significant role in ameliorating toxicity through free radical scavenging reactions and therefore have potential therapeutic value.

**Keywords:** Antioxidants, Free Radicals, Oxidative Stress, Drugs, Therapy

## 1. Introduction

Halliwell and Gutteridge [1] defined antioxidants as "any substance that delays, prevents or removes oxidative damage to a target molecule" [1, 2]. Khlebnikov et al. [3] defined antioxidants as "any substance that directly scavenges ROS or indirectly acts to up-regulate antioxidant defenses or inhibit ROS production". In other words, we can define antioxidants as any molecule that inhibits the oxidation of another molecule. A chemical reaction involving the loss of electrons and increase in the oxidative state is termed as "oxidation." Oxidation results in the formation of free radicals that are unstable atoms and molecules deficit in electrons. They have unpaired electrons and are extremely reactive and are capable of initiating chain reactions that destabilize other molecules and generate free radicals. These free radicals are also termed as reactive oxygen species or ROS and create a homeostatic imbalance that generates oxidative stress and causes cell death and tissue injury. ROS includes: superoxide (O2 ·−), hydroxyl (OH·), peroxyl (RO2 ·), hydroperoxyl (HO2 ·), alkoxyl (RO·), peroxyl

(ROO·), nitric oxide (NO·), nitrogen dioxide (NO2 ·), and lipid peroxyl (LOO·) and the non-radicals hydrogen peroxide (H2O2), hypochlorous acid (HOCl), ozone (O3), singlet oxygen (1Δg), and lipid peroxide (LOOH) [4]. Free radicals are known to be formed as a result of environmental pollution, stress, cigarette smoke, UV Light, ionizing radiations, and xenobiotics. Toxic effect of the free radicals causes oxidative stress and results in the pathogenesis of diseases (Figure 1).

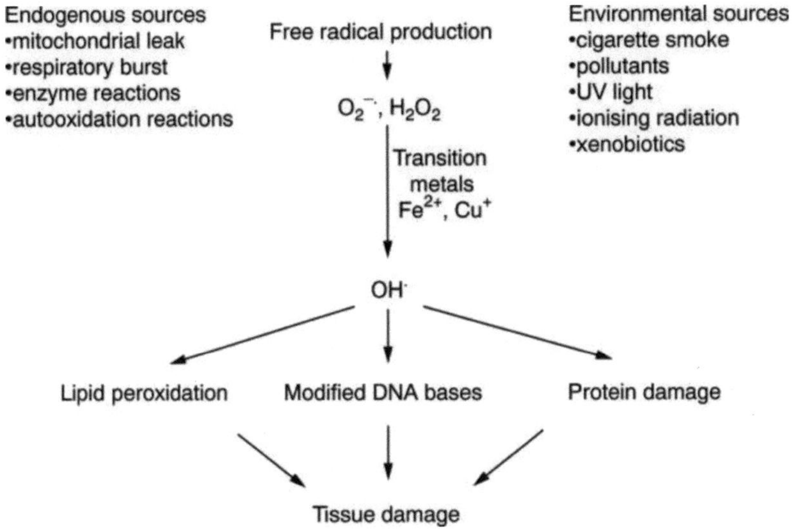


**Figure 1.** Free radicals: Production and damage (Adapted from [5]).

Involvement of ROS is implicated in neurodegenerative and other disorders such as Alzheimer's disease, Parkinson's disease, multiple sclerosis, Down syndrome, inflammation, viral infection, autoimmune pathology, and digestive ulcers. Recent developments in biomedical science emphasize the involvement of free radicals in many diseases, such as brain dysfunction, cancer, heart disease, and immune system [6]. Antioxidants normally terminate many reactions by removing free radical intermediates and inhibit other oxidation reactions. Thus, antioxidants often serve as reducing agents (examples: thiols, ascorbic acid, or polyphenols) [7]. Depending on the balance between ROS and the availability of antioxidants in the microenvironment of the cell, antioxidants can inhibit or delay the initiation or propagation of oxidative chain reaction and thus prevent or repair cell damage caused by reactive oxygen [8]. Antioxidants have been reported to work through single or combined mechanisms, namely, free radical scavenging, reducing activity, complexing of pro-oxidant, scavenging lipid peroxyl radicals, and quenching of singlet oxygen. Preventive oxidants are the antioxidants that act as inhibitors of free radical oxidation reactions. Chain-breaking antioxidants inhibit formation of free lipid radicals as follows: by obstructing the propagation of the autoxidation chain reactions; as singlet oxygen quenchers; as reducing agents which convert hydroperoxides into stable compounds; as metal chelators that convert metal pro-oxidants (iron and copper derivatives) into stable products; and finally as inhibitors of pro-oxidative enzymes (lipoxygenases) [9]. Antioxidant approach to disease management holds potential as most of the diseases are mediated through ROS, also with the rapid advancement of civilization, industrialization, and overpopulation. Epidemiological researches strongly suggest that foods containing antioxidants and scavengers have a potential protective effect against disorders caused by ROS [10].

### 1.1. Classification of antioxidants

Guttering and Halliwell classified the antioxidants into three categories: primary, secondary, and tertiary antioxidants [11]. Primary antioxidants are involved in the prevention of oxidant formation; secondary antioxidants are known to be scavengers of ROS, and tertiary antioxidants repair the oxidized molecules through sources like dietary or consecutive antioxidants.

Antioxidants may also be classified as enzymatic or non-enzymatic antioxidants (Figure 2).

#### *1.1.1. Enzymatic antioxidants*

The antioxidant enzymatic system directly/indirectly contributes to defense against the ROS. Catalase, superoxide dismutase (SOD), glutathione peroxidase, glutathione reductase, etc., are enzymatic antioxidants.

#### *1.1.2. Non-enzymatic antioxidants*

These antioxidants are quite a few, namely vitamins (A, C, E, and K), enzyme cofactors (Q10), minerals (Zn, Se, etc.), organosulfur compounds (allium and allium sulfur), nitrogen compounds (uric acid), peptides (glutathione), and polyphenols (flavonoids and phenolic acid).

#### *1.1.3. Hydrophilic antioxidants*

Antioxidants that react with oxidants in the cell cytoplasm and the blood plasma are termed as hydrophilic antioxidants (ascorbic acid, glutathione, and uric acid).

#### *1.1.4. Hydrophobic antioxidants*

These compounds are known to protect cell membranes from lipid peroxidation (ubiquinol, carotenes, and $\alpha$-tocopherol). They are obtained either from the diet or synthesized in the body [12].

#### *1.1.5. Endogenous antioxidants*

Endogenous antioxidants can be categorized into primary antioxidants and secondary antioxidants. Primary antioxidants inactivate the ROS into their intermediates. SOD, catalase, and glutathione peroxidase are the primary antioxidant enzymes [13]. They can be water soluble or lipid soluble (ascorbate, glutathione, uric acid, etc., are water soluble, and tocopherols, ubiquinols, and carotenoids, etc., are lipid soluble). Secondary antioxidant enzymes act directly to detoxify ROS. They maintain their proper functioning by decreasing the peroxides level and continuously supplying NADPH (nicotinamide adenine dinucleotide phosphate) and glutathione for primary antioxidant enzymes. Glutathione reductase, glucose-6-phosphate dehydrogenase, glutathione-s-transferase, and ubiquinone are the secondary antioxidants. Iron, copper, zinc, manganese, and selenium also increase the antioxidant enzyme activities [14, 15].

### *1.1.6. Exogenous antioxidants*

Many foods and various dietary components exhibit antioxidant activities. Several herbs, spices, vitamins, foods, vegetables, etc., are reported to be sources of exogenous antioxidants. These antioxidant drugs could be used for the treatment of various pathological diseases, and therefore gained importance in clinical as well as research areas. Many polyphenolic compounds such as flavonoids, isoflavones, flavones, anthocyanins, coumarins, lignans, catechins, isocatechins, epicatechins, and phenolic acids such as hydrocinnamic acid, hydrobenzoic acid, gallic acid, ellagic acid, etc., have gained importance as antioxidant phytochemicals. These bioactive compounds are being tested in clinical and preclinical trials. Plant-derived drugs are medicinally useful as they contain phytochemicals like terpenoid, alkaloids, glycosides, polyphenolics, and steroids and are of great significance in research area [16, 17]. Dietary nutrients, protein, and amino acids are responsible for the synthesis of antioxidant enzymes and hence play an important role in the defensive mechanism. GSH, creatine, and uric acid act as the direct scavengers of reactive metabolites [18]. Antioxidants of natural origin such as polyphenols, tannins, and flavonoids act by donating electrons to the intermediate radicals formed in oxidative stress or tissue damage that help in inhibition of the lipid peroxidation.

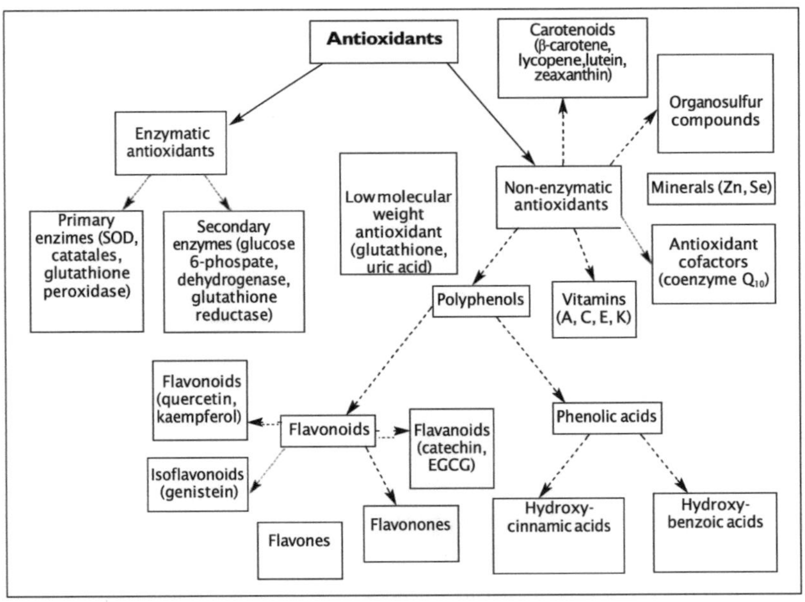


**Figure 2.** Antioxidants: Classification (Adapted from [19]).

http://dx.doi.org/10.5772/61884

## 2. Enzymatic and non-enzymatic antioxidants

When cells are exposed to oxidative stress, a defense system endorses the expression and regulation of a number of antioxidant enzymes as a defense mechanism to protect them from the damage induced by free radicals. These antioxidant defenses could be non-enzymatic or enzymatic (Table 1).

### 2.1. Enzymatic antioxidants

Enzymatic antioxidants are categorized into primary and secondary enzymatic defenses. Primary defense is composed of three important enzymes that prevent the formation or neutralize free radicals: glutathione peroxidase, which donates two electrons to reduce peroxides by forming selenols and also eliminates peroxides; catalase, which converts hydrogen peroxide into water and molecular oxygen; and SOD, which converts superoxide anions into hydrogen peroxide as a substrate for catalase [20].

Glutathione reductase and glucose-6-phosphate dehydrogenase are involved in the secondary enzymatic defense system. Glutathione reductase reduces glutathione (antioxidant) from its oxidized form to its reduced form, thereby recycling itself to continue neutralizing more free radicals. Glucose-6-phosphate regenerates NADPH (a coenzyme used in anabolic reactions) creating a reducing environment [21, 22]. These two enzymes do not neutralize free radicals directly but have supporting roles to other endogenous antioxidants.

Glutathione peroxidase, catalase, and SOD metabolize toxic oxidative intermediates and therefore form the primary antioxidant enzymes. These form the body's endogenous defense mechanism and help protect against free radical-induced cell damage. For optimum catalytic activity, these enzymes also require co-factors such as selenium, iron, copper, zinc, and manganese. It has been indicated that an inadequate dietary intake of these trace minerals may compromise the effectiveness of these antioxidant defense mechanisms. The consumption and absorption of important trace minerals may decrease with aging.

#### *2.1.1. Superoxide dismutase*

The superoxide dismutases catalyze the dismutation of superoxide to hydrogen peroxide:

$$O_2^- + O_2^- + 2H^+ \rightarrow H_2O_2 + O_2.$$

Catalase or glutathione peroxidase removes hydrogen peroxide. Catalase converts hydrogen peroxide into water and molecular oxygen.

Mammalian tissues have three forms of superoxide dismutase, each with a specific subcellular location and different tissue distribution (Figure 3).

1. Copper zinc superoxide dismutase (CuZnSOD): CuZnSOD has a molecular mass of approximately 32,000 kDa and has two protein subunits, each containing a catalytically active copper and zinc atom and is present in the cytoplasm and organelles of all mammalian cells.

2. Manganese superoxide dismutase (MnSOD): MnSOD is found to have a molecular mass of 40,000 kDa. It consists of four protein subunits, each probably containing a single manganese atom. It is present in the mitochondria of almost all cells [23]. The amino acid sequence of MnSOD is very dissimilar to that of CuZnSOD and is not inhibited by cyanide, and thereby MnSOD activity can be distinguished from that of CuZnSOD in mixtures of the two enzymes.

3. Extracellular superoxide dismutase (ECSOD): Marklund described EC-SOD in 1982 [24]. It is a secretory copper and zinc containing SOD and is different from the CuZnSOD. Only a few cell types, including fibroblasts and endothelial cells, synthesize EC-SOD and are expressed on the cell surface where it is bound to heparin sulfates. EC-SOD is the major SOD detectable in extracellular fluids and following the injection of heparin, it is released into the circulation from the surface of vascular endothelium [25]. EC-SOD might play a role in the regulation of vascular tone because endothelial derived relaxing factor (nitric oxide or a closely related compound) is neutralized in the plasma by superoxide [26].

These superoxide enzymes are present in extracellular fluids of almost all aerobic cells. SODs contain metal ion cofactors like copper, zinc, manganese, or iron depending on the isozyme. For example, in human copper/zinc SOD is present in the cytosol while manganese SOD is present in the mitochondrion. The mitochondrial SOD is the most biologically significant of these three enzymes. SOD isozymes are present in the cytosol and mitochondria in plants, and there is also an iron SOD found in chloroplasts.

### 2.1.2. *Catalase*

Catalase was the first antioxidant enzyme to be characterized and catalyzes the two-stage conversion of hydrogen peroxide to water and oxygen. Catalases are enzymes that catalyze the conversion of hydrogen peroxide to water and oxygen, using either an iron or manganese cofactor. Here, its cofactor is oxidized by one molecule of hydrogen peroxide and then regenerated by transferring the bound oxygen to a second molecule of substrate.

$$\text{Catalase–Fe}\left(\text{III}\right) + H_2O_2 \rightarrow \text{compound I}$$
$$\text{Compound I} + H_2O_2 \rightarrow \text{catalase–Fe}\left(\text{III}\right) + 2H_2O + O_2.$$

Catalase consists of four protein subunits, each containing a heme group and a molecule of NADPH [27]. Catalase is largely located within cells in peroxisomes, which also contain most of the enzymes capable of generating hydrogen peroxide. The greatest activity is present in the liver and erythrocytes, but some catalase is found in all tissues. It is a tetrameric enzyme consisting of four identical tetrahedrally arranged subunits of 60 kDa, which contains a single ferriprotoporphyrin group per subunit and has a molecular mass of about 240 kDa

### 2.1.3. *Glutathione enzymes*

The glutathione system includes glutathione, glutathione reductase, glutathione peroxidases, and glutathione "s"-transferases. Glutathione peroxidase is an enzyme containing four

http://dx.doi.org/10.5772/61884

selenium cofactors that catalyze the breakdown of hydrogen peroxide and organic hydroperoxides. Glutathione "s"-transferases show high activity with lipid peroxides. These enzymes are noticed especially in high levels in the liver. Glutathione peroxidases catalyze the oxidation of glutathione. Hydroperoxides, such as hydrogen peroxide and lipid hydroperoxides, act as substrates for these enzymes [28].

$$ROOH + 2GSH \rightarrow GSSG + H_2O + ROH.$$

Selenium is required at the active site for effective functioning of glutathione peroxidases [29]. Kidney synthesizes the plasma form of glutathione, and the highest level of glutathione peroxidases is found within liver cells, although glutathione peroxidase is widely distributed in almost all tissues. Glutathione peroxidase is the main scavenger of hydrogen peroxide in these subcellular compartments; the predominant sub-cellular distribution is in the cytosol and mitochondria. The activity of the enzyme glutathione peroxidase is dependent on the constant availability of reduced glutathione [30].

$$GSSG + NADPH^{+}H^{+} \rightarrow 2GSH + NADP^{+}.$$

The NADPH required by this enzyme to restore the supply of reduced glutathione is supplied by the pentose phosphate pathway. Glutathione reductase is a flavine nucleotide-dependent enzyme and has a similar tissue distribution to glutathione peroxidase [31].

Amino acids such as glycine, glutamate, and cysteine are utilized in the synthesis of glutathione. It is an important water-soluble antioxidant that plays a major role in xenobiotic metabolism; it can directly neutralize ROS such as lipid peroxide. When a body is exposed to xenobiotics or toxins, there is an increase in the level of detoxification enzymes (cytochrome P-450 mixed-function oxidase). Xenobiotics conjugate with glutathione, and hence a higher concentration of the enzyme is required for conjugation to make the toxin neutral and thereby making the enzyme less available as an antioxidant. Glutathione and vitamin C work interactively to neutralize the free radicals.

#### *2.1.4. Non-enzymatic endogenous antioxidants*

There are a number of non-enzymatic antioxidants: vitamins (A, C, E, and K), enzyme cofactors (Q10), minerals (Zn and Se), organosulfur compounds (allium and allium sulfur), nitrogen compounds (uric acid), peptides (glutathione), and polyphenols (flavonoids and phenolic acid).

##### *2.1.4.1. Vitamin A*

Vitamin A is produced as a result of the breakdown of β-carotene and is a carotenoid produced in the liver. It exhibits antioxidant activity due to its ability to combine with peroxyl radicals before they propagate peroxidation to lipids. Vitamin A is known to have a beneficial impact on the skin, eyes, and internal organs [32, 33].

*2.1.4.2. Coenzyme Q10*

Coenzyme Q10 has been reported to act by preventing the formation of lipid peroxyl radicals. It neutralizes the radicals even after their formation. An important role of this coenzyme is regeneration of vitamin E. Regeneration of vitamin E through this process is more likely than through ascorbate (vitamin C). This coenzyme is present in all cells and membranes and plays an important role in the respiratory chain and other cellular metabolism processes [34].

*2.1.4.3. Uric acid*

The end product of purine nucleotide metabolism in humans is uric acid. After undergoing kidney filtration, 90% of the uric acid is reabsorbed by the body, proving that it has important functions within the body. Uric acid prevents lysis of erythrocytes by peroxidation and is a potent scavenger of singlet oxygen and hydroxyl radicals. It is also known to prevent the overproduction of oxo-heme oxidants that result from the reaction of hemoglobin with peroxides [35].

*2.1.4.4. Glutathione*

Glutathione is an endogenous tripeptide that protects the cells against free radicals by donating either a hydrogen atom or an electron. It also plays an important role in the regeneration of other antioxidants like ascorbate [36]. However, the endogenous antioxidant system is not sufficient; humans depend on dietary antioxidants to reduce free radical concentrations [37].

*2.1.4.5. Vitamin C*

Ascorbic acid and tocopherols are generic names for vitamin C and vitamin E. Ascorbic acid consists of two antioxidant compounds: L-ascorbic acid and L-dehydroascorbic acid. These two compounds are absorbed through the gastrointestinal tract and can be interchanged enzymatically *in vivo*. Ascorbic acid acts by scavenging the superoxide radical anion, hydrogen peroxide, hydroxyl radical, singlet oxygen, and reactive nitrogen oxide [38].

*2.1.4.6. Vitamin E*

Vitamin E is the only major lipid-soluble, chain-breaking antioxidant found in plasma, red cells, and tissues, thus protecting the integrity of lipid structures, mainly membranes. It inhibits lipid peroxidation by donating its phenolic hydrogen to the peroxyl radicals forming tocopheroxyl radicals that, despite also being radicals, are unreactive and unable to continue the oxidative chain reaction. There are eight isoforms of vitamin E: four tocopherols ($\alpha$-tocopherol, $\beta$-tocopherol, $\gamma$-tocopherol, and $\delta$-tocopherol) and four tocotrienols ($\alpha$-tocotrienol, $\beta$-tocotrienol, $\gamma$-tocotrienol, and $\delta$-tocotrienol), $\alpha$- tocopherol being the most potent and abundant isoform in biological systems. The antioxidant activity of tocopherols is due to the chroman head, but the phytyl tail has no effect [39]. These two vitamins also display a synergistic behavior with the regeneration of vitamin E through vitamin C from the tocopheroxyl radical to an intermediate form, therefore reinstating its antioxidant potential [40].

http://dx.doi.org/10.5772/61884

*2.1.4.7. Vitamin K*

This vitamin has two natural isoforms: vitamins K1 and K2. Vitamin K is a group of fat-soluble compounds, essential for the post-translational conversion of protein-bound glutamates into γ-carboxyglutamates in various target proteins. The antioxidant activity is due to the 1, 4-naphthoquinone structure of these vitamins [41].

*2.1.4.8. Flavonoids*

Flavonoids are a group of compounds composed of diphenyl propane (C6C3C6) skeleton. It can be classified as flavonols, flavanols, anthocyanins, isoflavonoids, flavanones, and flavones. Flavanones and flavones are usually found in the same fruits and are connected by specific enzymes while flavones and flavonols do not share this phenomenon and are rarely found together. Anthocyanins are also absent in flavanone-rich plants. Flavonoids exhibit their antioxidant activity due to the phenolic hydroxyl groups attached to ring structures. They may act as reducing agents, superoxide radical scavengers, hydrogen donators, singlet oxygen quenchers, and also as metal chelators. They activate antioxidant enzymes, reduce α-tocopherol radicals (tocopheroxyls), inhibit oxidases, mitigate nitrosative stress, and increase the levels of uric acid and low-molecular-weight molecules. Some of the flavonoids of significance are quercetin, kaempferol, catechin, and catechin-gallate [42, 43].

*2.1.4.9. Phenolic acids*

Phenolic acids are composed of hydroxycinnamic and hydroxybenzoic acids. One of the most studied and promising compounds in the hydroxybenzoic group is gallic acid that is also the precursor of many types of tannin, while cinnamic acid is the precursor of all the hydroxycinnamic acids. They are present in plant material and sometimes present as esters and glycosides. They have antioxidant activity as chelators and free radical scavengers with special impact over hydroxyl and peroxyl radicals, superoxide anions, and peroxynitrites [44, 45].

*2.1.4.10. Carotenoids*

Carotenoids are a group of natural pigments and are synthesized by plants and microorganisms. They can be classified into two different groups: the carotenoid hydrocarbons known as the carotenes containing distinct end groups like lycopene and β-carotene; and the oxygenated carotenoids known as xanthophylls, like zeaxanthin and lutein. Carotenoids display their antioxidant activity due to singlet oxygen quenching which culminates in excited carotenoids that dispel the newly acquired energy through a series of rotational and vibrational interactions with the solvent, thus returning to the unexcited state and allowing them to quench more radical species. The only free radicals that completely damage these pigments are peroxyl radicals. Carotenoids are relatively unreactive, but they may also decay and form non-radical compounds and result in terminating free radical attacks by binding to these radicals [46].

*2.1.4.11. Minerals*

Minerals are found in trace quantities in animals and are a small part of dietary antioxidants, but play significant roles in their metabolism. The most important minerals exhibiting antioxidant activity are selenium and zinc. Selenium can be found in both organic (selenocysteine and selenomethionine) and inorganic (selenite and selenate) forms in the human body. It does not act directly on free radicals but is a vital part of most antioxidant enzymes (metalloenzymes, glutathione peroxidase, and thioredoxin reductase) that would have no effect without it [47].

Various pathways in metabolism require zinc. Zinc is essential in the prevention of free radical formation and does not directly attack free radicals. Zinc is also an inhibitor of NADPH oxidases that catalyze the production of the singlet oxygen radical from oxygen by using NADPH as an electron donor. It is present in SOD, a vital antioxidant enzyme that converts the singlet oxygen radical into hydrogen peroxide. Zinc brings about the production of metallothionein that is a scavenger of the hydroxyl radical. Finally, zinc also competes with copper for binding to the cell wall, thus decreasing the production of hydroxyl radicals [48].

*2.1.4.12. Lipoic acid*

Lipoic acid and its reduced form, dihydrolipoic acid (DHLA), neutralize the free radicals in both lipid and aqueous domains and are called "universal antioxidants." It is categorized as "thiol" or "biothiol."

They are sulfur-containing molecules that catalyze the oxidative decarboxylation of alpha-keto acids, such as pyruvate and alpha-ketoglutarate, in the Krebs cycle.

*2.1.4.13. Peroxiredoxins*

These may be of three basic types: typical 2-cysteine peroxiredoxins; atypical 2-cysteine peroxiredoxins; and 1-cysteine peroxiredoxins. Peroxiredoxins are important in antioxidant metabolism as they catalyze the reduction of hydrogen peroxide, organic hydroperoxides, as well as peroxynitrite.

*2.1.4.14. Synthetic antioxidants*

Synthetic antioxidants have been developed to have a standard antioxidant activity measurement system and to compare with natural antioxidants that are incorporated into food. Synthetic antioxidants are added to food so that it can withstand various treatments and conditions to prolong shelf life and prevention of food oxidation, especially fatty acids. It has been reported that synthetic antioxidants are added to almost all processed foods, which are reported to be safe, although some studies oppose this fact. The important synthetic antioxidants are BHT (butylated hydroxytoluene) and BHA (butylated hydroxyanisole). The European Food Safety Authority (EFSA) between 2011 and 2012 classified an NOAEL (No Observable Adverse Effect Level) of 0.25 mg/kg BW/day for BHT and 1.0 mg/kg BW/day for BHA in terms of daily intake and admitted that the exposure of adults and children was

http://dx.doi.org/10.5772/61884

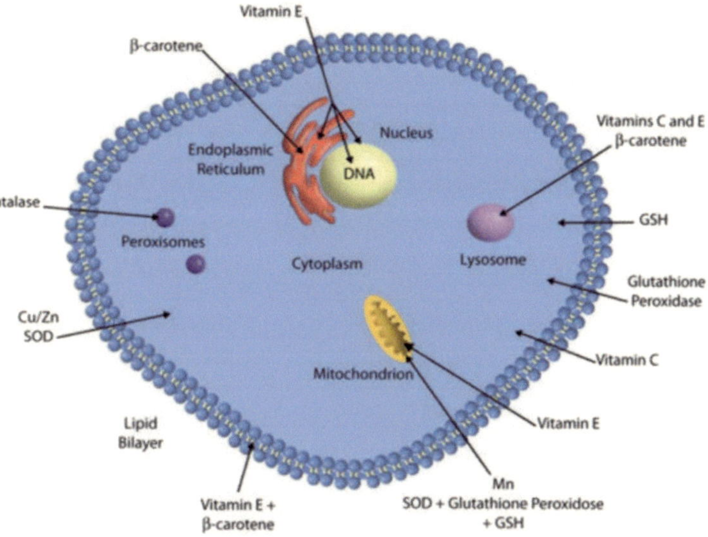


**Figure 3.** Diagrammatic representation of the site of enzymatic and non-enzymatic antioxidant defense system (Adapted from [49]).

unlikely to exceed these doses. TBHQ (*tert*-butylhydroquinone) stabilizes and preserves the freshness, nutritive value, flavor, and color of animal food products. Octyl gallate is considered as safe to use as a food additive because, after consumption, it is hydrolyzed into gallic acid and octanol, which are found in many plants and do not pose a threat to human health [50]. NDGA (nordihydroguaiaretic acid) despite being a food antioxidant is known to cause renal cystic disease in rodents [51].

### *2.1.4.15. Pro-oxidants*

Pro-oxidants are defined as chemicals that induce oxidative stress, usually through the formation of reactive species or by inhibiting antioxidant systems. Free radicals are considered pro-oxidants, but sometimes, antioxidants can also have pro-oxidant behavior. Vitamin C is a potent antioxidant, but it can also become a pro-oxidant when it combines with iron and copper reducing Fe3+ to Fe2+ (or $Cu3^+$ to $Cu2^+$), which in turn reduces hydrogen peroxide to hydroxyl radicals [52].

α-Tocopherol is a powerful antioxidant, but in high concentrations, it can become a pro-oxidant. When vitamin E reacts with a free radical, it becomes a radical itself, and if there is not enough ascorbic acid for its regeneration, it will remain in this highly reactive state and support the autoxidation of linoleic acid [53].

Although not much evidence is found, it is proposed that carotenoids can also display pro-oxidant effects especially through autoxidation in the presence of high concentrations of oxygen-forming hydroxyl radicals [54]. Flavonoids may also serve as pro-oxidants. The occurrence of $O_2$, iron, and copper damages biological molecules [55].

| **Enzymatic antioxidants** | **Location** | **Properties** |
|---|---|---|
| Superoxide dismutase (SOD) | Mitochondria and cytosol | Dismutation of superoxide radicals |
| Catalase (CAT) | Mitochondria and cytosol | Removes hydrogen peroxide |
| Glutathione peroxidase (GSH)) | Mitochondria and cytosol | Removes hydrogen peroxide and organic hydroperoxide |
| **Non-enzymatic antioxidants** | **Location** | **Properties** |
| Vitamin C | Aqueous phase of cell | Acts as a free radical scavenger and recycles vitamin E |
| Vitamin E | Cell membrane | Major chain-breaking antioxidant in cell membrane |
| Uric acid | Product of purine metabolism | Scavenger of OH radicals |
| Carotenoids | Membrane tissue | Scavengers of ROS and singlet oxygen quencher |
| Glutathione | Non-protein thiol in cell | Serves multiple roles in the cellular antioxidant defense |
| Lipoic acid | Endogenous thiol | Effectual in recycling vitamin C, and also a functional glutathione substitute |
| Metals ions sequestration: transferrin, ferritin, lactoferrin | Mitochondria and cytosol | Scavenger of free radical and inhibitor of lipid peroxidation |
| Nitric oxide | Mitochondria and cytosol | Chelating of metal ions, and responsible for Fenton reactions |
| Ubiquinones | Mitochondria | Reduced form serve as functional antioxidants |
| Bilirubin | Product of heme metabolism in blood | Extracellular antioxidant |

**Table 1.** Major enzymatic and non-enzymatic antioxidants (Adapted and modified from [56]).

## 3. Antioxidant defense mechanism

Free radicals are constantly being generated in the body through various mechanisms and are also being removed by endogenous antioxidant defensive mechanisms that act either by scavenging free radicals, by decomposing peroxides, or by binding with pro-oxidant metal ions (Tables 2 and 3; Figure 4).

Antioxidants are classified into three categories [56–58] as follows:

http://dx.doi.org/10.5772/61884

1. Primary antioxidants: It is involved in the prevention of oxidant formation. They act by suppressing the formation of free radicals (examples: glutathione peroxidase, catalase, selenoprotein, transferrin, ferritin, lactoferrin, carotenoids, etc.).
2. Secondary antioxidants: These exhibit scavengers of ROS. They act by suppressing chain initiation and breaking chain propagation reactions (radical scavenging antioxidants).
3. Tertiary antioxidants: They act by repairing the oxidized molecules (some proteolytic enzymes, enzymes of DNA, etc.) through sources like dietary or consecutive antioxidants.

The human body employs three general categories of antioxidants to safeguard against free radicals. They are endogenous antioxidants, dietary antioxidants, and metal-binding proteins [16].

### 3.1. Endogenous antioxidants

These are categorized into primary antioxidants and secondary antioxidants. SOD, catalase, and glutathione peroxidase are the primary antioxidant enzymes that inactivate the ROS into intermediates [13]. Secondary antioxidant enzymes (glutathione reductase, glucose-6-phosphate dehydrogenase, glutathione-s-transferase, and ubiquinone) detoxify ROS and supply the NADPH and glutathione for primary antioxidant enzymes for proper functioning. Metals such as copper, iron, manganese, zinc, and selenium up-regulate the antioxidant enzyme activities [14, 15].

### 3.2. Exogenous antioxidants

Many polyphenolic compounds such as flavonoids, isoflavones, flavones, anthocyanins, coumarins, lignans, catechins, isocatechins, epicatechins, and phenolic acids have gained importance as antioxidant drugs [16]. Dietary antioxidants act through scavenging free radicals to break the chain reaction responsible for lipid peroxidation. Vitamins C and E, carotenoids, and flavonoids are the dietary antioxidants. These vitamins are also known as chain-breaking antioxidants [16]. The metal-binding proteins (albumin, ferritin, and myoglobin) inactivate the transition metal ions that catalyze the production of free radicals [17, 18].

Antioxidant enzymes – catalase, SOD, glutathione peroxidase, glutathione reductase, and thioredoxin – act against the ROS. The non-enzymatic antioxidants are the scavengers of ROS and RNS [59].

### 3.3. Cellular antioxidant system

Lipid peroxidation is slowed down by the activity of chemical compounds that contain monohydroxy/polyhydroxy phenol acting as antioxidants. These compounds have low activation energy to donate the hydrogen atom and, therefore, cannot initiate the secondary free radicals. The free radical electrons are stable and, thus, slow down the oxidation. Prevention of excessive ROS and repair of cellular damage are essential for the life of cells, and cells, in turn, contain many antioxidant systems to prevent the oxidative injury [60, 61].

### 3.4. Mechanism of action of antioxidants

$$R + AH \rightarrow RH + A$$
$$RO + AH \rightarrow ROH + A$$
$$ROO + AH \rightarrow ROOH + A$$
$$R + A \rightarrow RA$$
$$RO + A \rightarrow ROA$$
$$ROO + A \rightarrow ROOH$$
$$\text{Antioxidant} + O_2 \rightarrow \text{Oxidized antioxidant.}$$

### 3.5. Mode of action of antioxidants

1. Primary or chain-breaking antioxidants: break chain reaction and the resulting radical are less reactive

$$ROO + AH \rightarrow ROOH + A$$
$$ROOH + A \rightarrow ROOA.$$

2. Secondary or preventive antioxidants:

They may act either by

- Chelating/deactivating metals,
- Scavenging singlet oxygen (highly toxic), or
- Removing ROS.

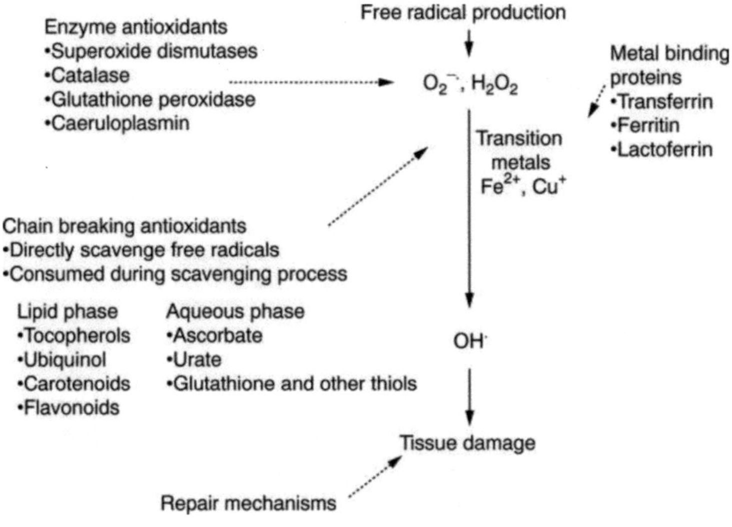


**Figure 4.** A schematic diagram of the antioxidant defense mechanism (Adapted from [5]).

| ROS scavengers | ROS protective enzymes | Sequestration of transition metal ions which form ROS |
|---|---|---|
| Glutathione | Superoxide dismutase | Transferrin |
| Uric acid | Catalase | Ferritin |
| Ascorbic acid | Glutathione peroxidase | Metallothionein |
| Albumin | Glutathione reductase | Ceruloplasmin |

**Table 2.** Antioxidant defensive agents (Adapted from [66]).

| Enzymatic antioxidants | Reactions catalyzed |
|---|---|
| Superoxide dismutase (SOD) | $O_2$-+ $O_2$ + 2H → $2H_2O_2$ + $O_2$ |
| Catalase (CAT) | $H_2O_2$ → $H_2O$ +1 /$2O_2$ |
| Glutathione peroxidase (GPX) | $H_2O_2$+GSH → $H_2O$ +GSSG |
| Glutathione reductase (GR) | GSSG+NAD(P)H → 2GSH+ $NAD(P)^+$ |

**Table 3.** Major ROS scavenging antioxidant enzymes (Adapted from [67]).

Antioxidants are present with protective efficiency. If there is an electron-donating group, especially a hydroxyl group loaded on *o*- or *p*-positions of the phenolic compounds, it makes the compound polar, and, therefore, antioxidant activities and metal chelating ability are increased. These groups make the phenols more easily donate hydrogen atoms to activate free radicals to interrupt the chain reaction of autoxidation. Antioxidants of natural origin such as polyphenols (tannins, flavonoids, and chalcones) act by donating an electron to the intermediate radicals formed in oxidative stress or tissue damage, which helps in the inhibition of lipid peroxidation. A computational study also supports that the compounds having more electron donating potentials are better inhibitors of hydroperoxides that suggest many of the antioxidant agents [62–67].

## 4. Antioxidants: Health and diseases

Several human pathologies such as neurodegenerative diseases, cancer, stroke, and many other ailments are believed to be caused by ROS. Antioxidants are assumed to prevent the harmful effects of ROS and therefore treat oxidative stress-related diseases (Figure 5).

Antioxidant approach to disease management holds potential as most of the diseases are mediated through ROS; also with the rapid advancement of civilization, industrialization, and overpopulation, there has been a significant rise in oxidative stressors. Epidemiological researches strongly suggested that foods containing antioxidants and scavengers have a potential protective effect against disorders caused by ROS [66]. Many chronic diseases can be prevented, and disease progression can be slowed by increasing the body natural antioxidant

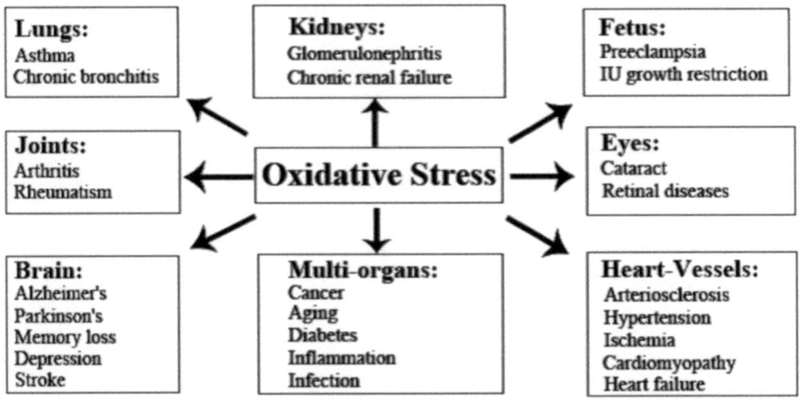


**Figure 5.** Oxidative stress-induced diseases in humans (Adapted from [65]).

defenses or by supplementing with dietary antioxidants. Natural antioxidants such as flavonoids, tannins, and polyphenols act by donating electrons to intermediate radicals and help in inhibition of lipid peroxidation. Antioxidants are essential to prevent the formation and oppose the actions of reactive oxygen and nitrogen species, which are generated in vivo and cause damage to DNA, lipids, proteins, and other biomolecules. The antioxidant system contains exogenous antioxidants (dietary sources) and endogenous antioxidants.

### 4.1. Exogenous antioxidants as drugs

Many polyphenolic compounds such as flavonoids, isoflavones, flavones, anthocyanins, coumarins, lignans, catechins, isocatechins, epicatechins, and phenolic acids have gained importance as antioxidant drugs.

### 4.2. Role of dietary nutrients in defensive mechanism

Protein and amino acids play an important role in the synthesis of antioxidant enzymes. Small peptides like GSH and carnosine and nitrogenous metabolites like creatine and uric acid directly scavenge the reactive metabolites [67]. iNOS expression and synthesis in various cells are controlled by taurine and taurine chloramines. Deficiency of dietary protein can have a harmful effect on the antioxidant system of the cell. Arginine and tetrahydrobiopterin deficiency directly affect the superoxide enzyme production. Decreased protein intake affects the availability of zinc, which is a cofactor of SOD. Similarly, a high-protein diet exhibits oxidative stress. Homocysteine increases inducible and constitutive NOS synthesis and stimulates ROS generation in polymorphonuclear leukocytes and monocytic cells [68–70].

#### *4.2.1. Lipids*

There is a generation of ROS due to the intake of polyunsaturated fatty acids which are neutralized by vitamins C and E and carotenoids. There is an increase in the risk of cardiovascular diseases due to high intake of polyunsaturated fatty acids. On the other hand, a high-saturated-fat diet increases the risk of iNOS activity in the liver and colon. Fish oil decreases the cardiovascular risk by reducing triacylglycerol production in plasma as it contains ω-3 PUFA that is the inhibitor of ROS, iNOS expression, and NOS synthesis [71].

#### 4.2.2. *Vitamins*

Vitamins exhibit anti-atherogenic and anti-inflammatory properties. Vitamin A inhibits iNOS in vascular muscle cells, endothelial cells, cardiac myocytes, and mesangial cells. Vitamins D3, K2, and niacin inhibit iNOS activity in the neuronal cells (macrophage, microglia, and astrocytes) Lipid peroxidation of the membrane is prevented by vitamin E as it inhibits the ROS generation. Irradiation decreases the concentration of vitamin C and folate, thus leading to ROS generation. It has been reported that vitamin B12 and folic acid reduce radical-induced radiation damage and improve leukocyte counts. DNA damage and hepatocellular carcinoma are prevented by vitamin C and choline. Vitamins B12, B6, and folate are essential for the synthesis of cystathionine synthase and cystathionase (B6) and methionine synthase (B12). These vitamins prevent cardiovascular diseases in humans and rodents. NADP, NADH, FAD, nicotinamide, and riboflavin protect the cells from ROS generation. NADPH and FAD are essential for glutathione reductase. NADPH is required for catalase activity [70–75].

#### 4.2.3. *Micronutrients and minerals*

Copper, zinc, and manganese, the important trace elements in our body, serve as cofactors of SOD enzyme. Deficiency of either copper or zinc increases the cytochrome $P_{450}$ activity in microsomes of the liver and lungs, and thus increases the generation of ROS and iNOS expression [76]. Selenium possesses potential antioxidant activity as it is a cofactor of glutathione transferase enzyme and other selenoproteins.

### 4.3. Phytochemicals

Many medicinal plants contain phytochemicals like phenolic and polyphenolic compounds such as flavonoids, isoflavones, flavones, anthocyanins, coumarins, lignans, catechin, isocatechin, gallic acid, and esculatin that possess antioxidant activities [77]. These phytochemicals are present in many plants and herbs like grapes, berry crops, tea, herbs, nutmeg, and tea. Many medicinal plants contain phenolics like gallic acids and other active constituents. *Terminalia chebula, T. bellerica, T. muelleri, Phyllanthus emblica, Hemidesmus indicus, Cichorium Intybus, Withania somnifera, Ocimum sanctum, Mangifera indica,* and *Punica granatum* are known to have potential antioxidant activities [78].

## 5. Antioxidant therapy

Recent human studies exploring the efficiency of antioxidants in prevention and treatment of various diseases are reviewed (Table 4).

| Disease studied | Antioxidant used | Reference | Reference no. |
|---|---|---|---|
| Mortality: Primary/Secondary Prevention | Beta-carotene, vitamin A, vitamin C, vitamin E, and selenium | Bjelakovic et al. | [79] |
| Fatty liver disease | Vitamin A, carotenoids, vitamin C, vitamin E and selenium | Lirussi et al. | [80] |
| Amyotrophic Lateral Sclerosis (SLA) | Vitamin E 500 mg twice daily | Orrell et al. | [81] |
| Multiple Sclerosis | Omega-6 fatty acids (11-23 g/day linoleic acid) | Farinotti et al. | [82] |
| Alcoholic Liver Disease | S-adenosyl-L-methionine | Rambaldi et al. | [83] |
| Oncology Treatments | Selenium | Dennert et al. | [84] |
| Eye Related Macular Disease | Beta-carotene and alpha-tocopherol. | Evans et al. | [85] |
| Pregnancy and Pre-eclampsia | Vitamin C and vitamin E supplements | Poston et al. | [86] |
| Cardiovascular Risk Profile | Dietary antioxidants | EJ Brunner et al. | [87] |
| Neonatal Growth Under Parenteral Nutrition (PN) | Cysteine, cystine or its precursor N-acetylcysteine | Soghier et al. | [88] |
| Melatonin and Cognitive Impairment or dementia | Melatonin | Jansen et al. | [89] |
| Alzheimer Disease | Vitamins C or E | Gray et al. | [90] |
| Parkinson Disease | Tocopherol, CoQ10, and glutathione. | Weber et al. | [91] |
| . Cancer | Lipid-soluble antioxidant vitamins, | Kirsh et al. | [92] |
| Asthma | Vitamin C, manganese etc. | Patel et al. | [93] |
| Cardiovascular Diseases | Vitamins C and E | Berhendt et al. | [94] |
| Ischemia-Reperfusion Injury | Vitamin C. | Pleiner et al | [95] |
| Chronic Obstructive Pulmonary Disease (COPD) | Polyphenol-rich pomegranate juice (PJ) | Cerda et al. | [96] |
| Pancreatitis | Selenium, L-methionine, and vitamins C and E, | Kirk et al. | [97] |
| Rheumatoid Arthritis | Vitamins A, C, E or selenium or their combination | Canter et al. | [98] |
| Kidney Diseases | Vitamin E | Ong-ajyooth et al. | [99] |
| Liver Diseases | Antioxidant therapy | Gabbay et al. | [100] |
| Diabetes Type I and II | Probucol and statins | Endo et al. | [101] |

**Table 4.** The efficiency of antioxidants in prevention and treatment of various diseases.

Many of these studies, either due to the small patient sample size, with uncontrolled admissions and treatment criteria, or due to relevant bias of the clinical studies failed to give precise information on effectiveness and practical advantage in taking antioxidants.

Antioxidants therapies have been in progress these days. Edaravone (for ischemic stroke), N-acetylcysteine (for acetaminophen toxicity), alfa-lipoic acid (for diabetic neuropathy), and some flavonoids (for chronic venous insufficiency) as well as baicalein and catechins (for osteoarthritis) have clinical importance. The evidence from human epidemiological studies about the beneficial effects of dietary antioxidants and preclinical in vitro and animal data are compelling. Attention needs to be drawn on focusing more on disease-specific, target-directed, highly bioavailable antioxidants [102]. In the recent years, due to the increase in the consumption of food and medicinal products, we are exposed to the adverse effects of various compounds noticed in the above products. For example, in our animal experimental studies, we have determined induction of oxidative stress induced by the compound cinnamaldehyde, a food flavor and also an anticancer drug [103–105]. As a therapeutic measure, addition of vegetables and fruits, the great sources of vitamins or antioxidants, in our routine diet might protect our health from toxic effects of food chemicals or drugs to a certain extent [106].

### 5.1. Oxidative stress test

In this advanced materialistic life, monitoring the levels of free radicals and oxidative stress is important in case of clinical practice. FORD (Free Oxygen Radicals Defense) is an easy, cheap, and reliable diagnostic device to monitor oxidative stress [19, 107]. It discriminates the high risk of oxidative damage on sick or healthy individuals, monitoring with precise laboratory parameters in the clinical situation at the baseline and in the follow-up of a medical prescription.

FORD (Free Oxygen Radicals Defense) is a colorimetric test based on the influence of antioxidants present in plasma to reduce the activity of free radicals. The principle of the assay is that at an acidic pH (5.2) and in the presence of a suitable oxidant solution (FeCl3), 4-amino-n, n- diethylaniline, the FORD chromogen, can form a stable and colored radical cation. Antioxidant molecules (AOH) present in the sample which are able to transfer a hydrogen atom to the FORD chromogen radical cation, reduce it, quenching the color and producing a discoloration of the solution which is proportional to their concentration in the sample. This instrument will be helpful in understanding the problem of the individual bioavailability of each antioxidant molecule which can be monitored during the administration, with a pre-post measure of the oxidative balance. In order to achieve the evidence of the oxidative background related to the outcome of specific symptoms and diseases, epidemiological studies can be encouraged, and the role of nutrition and targeted antioxidant therapy can be better defined.

## Author details

Shalini Kapoor Mehta[1] and Sivakumar Joghi Thatha Gowder[2*]

*Address all correspondence to: sivakumargowder@yahoo.com

1 MS Ramaiah Medical College, Bengaluru, Karnataka, India

2 College of Applied Medical Sciences, Qassim University, Buraidah, Saudi Arabia

## References

[1] Halliwell B, Gutteridge JM. The definition and measurement of antioxidants in biological systems. Free Radic Biol Med 1995; 18:125–126.

[2] Halliwell B. Biochemistry of oxidative stress. Biochem Soc Trans 2007; 35:1147–1150

[3] Khlebnikov AI, Schepetkin IA, Domina NG. Kirpotina LN, Quinn MT. Improved quantitative structure-activity relationship models to predict antioxidant activity of flavonoids in chemical, enzymatic, and cellular systems. Bioorg Med Chem 2007; 15:1749–1770.

[4] Pham-Huy LA, He H, Pham-Huy C, Free radicals, antioxidants in disease and health. Int J Biomed Sci 2008; 4:89–96.

[5] Young IS, Woodside JV. Antioxidants in health and disease. Clin Pathol 2001; 54:176–186.

[6] Halliwell B. Establishing the significance and optimal intake of dietary antioxidants: the biomarker concept. Nut Rev 1999; 57:104–113.

[7] Sies H. Oxidative stress: oxidants and antioxidants. Expl Physiol 1997; 82:291–295.

[8] Branien AL. Toxicology and biochemistry of butylated hydroxyanisole and butylated hydroxytoluene. J Am Oil Chemists' Soc 1975; 52:59–63.

[9] Darmanyan AP, Gregory DD, Guo Y, Jenks WS, Burel L, Eloy D, Jardon P. Quenching of singlet oxygen by oxygen- and sulfur-centered radicals: Evidence for energy transfer to peroxyl radicals in solution. J Am Chem Soc 1998; 120:396–403.

[10] Block G, Patterson B, Subar A. Fruit, vegetables, and cancer prevention: a review of the epidemiological evidence. Nut Cancer 1992; 18:1–29.

[11] Singh RP, Khanna R, Kaw JL, Khanna SK, Das M. Comparative effect of benzanthrone and 3-bromobenzanthrone on hepatic xenobiotic metabolism and anti-oxidative defense system in guinea pigs. Arch Toxicol 2003; 77:94–99.

[12] Halliwell B. The antioxidant paradox. Lancet 2000; 355:1179–1180.

[13] Shu YZ. Recent natural products based drug development: a pharmaceutical industry perspective. J Nat Prod 1998; 61:1053–1071.

[14] Vertuani S, Angusti A, Manfredini S. The antioxidants and pro antioxidants network: an overview. Curr Pharm Des 2004; 10:1677–1694.

[15] Gale CR Dietary antioxidants and dementia. Int Psycho Geriatr 2001; 13:259–262.

[16] Dempster WS, Sive AA, Rosseau S, Malan H, Heese HV. Misplaced iron in kwashiorkor. Eur J Clin Nutr 1995; 49:208–210.

[17] Wu G, Flynn NE, Flynn SP, Jolly CA, Davis PK. Dietary protein or arginine deficiency impairs constitutive and inducible nitric oxide synthesis by young rats. J Nutr 1999; 129:1347–1354.

[18] Rassaf T, Preik M, Kleinbongard P, Lauer T, Heiss C, Strauer BE, Feelisch M, Kelm M. Evidence for in vivo transport of bioactive nitric oxide in human plasma. J Clin Invest 2002; 109:1241–1248.

[19] Iannitti T, Palmier B. Antioxidant therapy effectiveness: an up to date. Eur Rev Med Pharmacol Sci 2009; 13:245–278.

[20] Rahman K. Studies on free radicals, antioxidants and co-factors. Clin Inverv Aging 2007; 2:219–236.

[21] Gamble PE, Burke J. Effect of water stress on the chloroplast antioxidant system. Plant Physiol 1984; 76:615–621.

[22] Ratnam DV, Ankola DD, Bhardwaj V, Sahana DK, Kumar MN. Role of antioxidants in prophylaxis and therapy: A pharmaceutical perspective. J Control Release 2006; 113:189–207.

[23] Liou W, Chang LY, Geuze HJ, Strous GJ, Crapo JD, Slot JW. Distribution of Cu Zn superoxide dismutase in rat liver. Free Rad Biol Med 1993; 14:201–207.

[24] Marklund S. Human copper-containing superoxide dismutase of high molecular weight. Proc Natl Acad Sci U S A.1982; 79:7634–7638.

[25] Karlsson K, Sandstrom J, Edlund A, Edlund T, Marklund SL. Pharmacokinetics of extracellular superoxide dismutase in the vascular system. Free Radic Biol Med 1993; 14:185–190.

[26] McIntyre M, Bohr DF, Dominiczak AF. Endothelial function hypertension—the role of superoxide anion. Hypertension 1999; 34:539–545.

[27] Kirkman HN, Galiano S, Gaetani GF. The function of catalase-bound NADPH. J Biol Chem 1987; 262:660–665.

[28] Takahashi K, Cohen HJ. Selenium-dependent glutathione peroxidase protein and activity: immunological investigations on cellular and plasma enzymes. Blood 1986; 68:640–646.

[29] Nakane T, Asayama K, Kodera K, Hayashibe H, Uchida N, Nakazawa S. Effect of selenium deficiency on cellular and extracellular glutathione peroxidases: immunochemical detection and mRNA analysis in rat kidney and serum. Free Radic Biol Med 1998; 25:504–511.

[30] Holben DH, Smith AM. The diverse role of selenium within selenoproteins: a review. J Am Diet Assoc 1999; 99:836–843.

[31] Gibson DG, Hawrylko J, McCay PB. GSH-dependent inhibition of lipid peroxidation: properties of a potent cytosolic system which protects cell membranes. Lipids 1985; 20:704–710.

[32] Palace VP, Khaper N, Qin Q, Singal PK. Antioxidant potentials of vitamin A and carotenoids and their relevance to heart disease. Free Radic Biol Med 1999; 26:746–761.

[33] Jee J, Lim S, Park J, Kim C. Stabilization of all-*trans* retinol by loading lipophilic antioxidants in solid lipid nanoparticles. Eur J Pharm Biopharm 2006; 63:134–139.

[34] Turunen M, Olsson J, Dallner G. Metabolism and function of coenzyme. Q Biochim Biophys Acta 2004; 1660:171–199.

[35] Kand'ár, R, Žáková P, Mužáková V. Monitoring of antioxidant properties of uric acid in humans for a consideration measuring of levels of allantoin in plasma by liquid chromatography. Clin Chim Acta 2006; 365:249–256.

[36] Steenvoorden DPT, Henegouwen, GMJB. The use of endogenous antioxidants to improve photoprotection. J Photochem Photobiol 1997; 41:1–10

[37] Pietta P. Flavonoids as antioxidants. J Nat Prod 2000; 63:1035–1042.

[38] Barros AIRNA, Nunes FM, Gonçalves B, Bennett RN, Silva AP. Effect of cooking on total vitamin C contents and antioxidant activity of sweet chestnuts (*Castanea sativa* Mill.). Food Chem 2011; 128:165–172.

[39] Burton GW, Traber, MG. Vitamin E: Antioxidant activity, biokinetics, and bioavailability. Annu Rev Nutr 1990; 10:357–382.

[40] Halpner AD, Handelman GJ, Belmont CA, Harris JM, Blumberg JB. Protection by vitamin C of oxidant-induced loss of vitamin E in rat hepatocytes. J Nutr Biochem 1998; 9:355–359.

[41] Vervoort LM, Ronden JE, Thijssen HH. The potent antioxidant activity of the vitamin K cycle in microsomal lipid peroxidation. Biochem Pharmacol 1997; 54:871–876.

[42] Rice-EvansCA, Miller NJ, Paganga G. Structure-antioxidant activity relationships of flavonoids and phenolic acids. Free Radic Biol Med 1996; 20:933–956.

[43] Procházková D, Boušová I, Wilhelmová N. Antioxidant and prooxidant properties of flavonoids. Fitoterapia 2011; 82:513–523

[44] Krimmel B, Swoboda F, Solar S, Reznicek G. OH-radical induced degradation of hydroxybenzoic- and hydroxycinnamic acids and formation of aromatic products: A gamma radiolysis study. Radiat Phys Chem 2010; 79:1247–1254.

[45] Terpinc P, Polak T, Šegatin N, Hanzlowsky A, Ulrih NP, Abramovič H. Antioxidant properties of 4-vinyl derivatives of hydroxyl cinnamic acids. Food Chem 2011; 128:62–68.

[46] Paiva SA, Russell RM. β-Carotene and other carotenoids as antioxidants. J Am Coll Nutr 1999; 18:426–433.

[47] Tabassum A, Bristow RG, Venkateswaran V. Ingestion of selenium and other antioxidants during prostate cancer radiotherapy: A good thing? Cancer Treat Rev 2010; 36:230–234.

[48] Prasad AS, Bao B, Beck FW, Kuck O, Sarkar FH. Antioxidant effect of zinc in humans. Free Radic Biol Med 2004; 37:1182–1190.

[49] Antioxidant Enzyme Systems. Available from: http://www.newsmedical.net/health/Antioxidant-Enzyme-Systems.aspx [Last retrieved on 2015 March 02].

[50] Joung T, Nihei K, Kubo I. Lipoxygenase inhibitory activity of octyl gallate. J Agric Food Chem 2004; 52:3177–3181.

[51] Evan AP, Gardner KD. Nephron obstruction in nordihydroguaiaretic acid induced renal cystic disease. Kidney Int 1979; 15:7–19.

[52] Duarte TL, Lunec J. Review: When is an antioxidant not an antioxidant? A review of novel actions and reactions of vitamin C. Free Rad Res 2005; 39:671–686.

[53] Cillard J, Cillard P, Cormier M, Girre L. α-Tocopherol prooxidants effect in aqueous media: Increased autoxidation rate of linoleic acid. J Am Oil Chem Soc 1980; 57:252–255.

[54] Carocho M, Ferreira ICFR. A review on antioxidants, prooxidants and related controversy: Natural and synthetic compounds, screening and analysis methodologies and future perspectives. Food Chem Toxicol 2013; 5:115–125.

[55] Galati GO, Brien PJ. Potential toxicity of flavonoids and other dietaryphenolics: Significance for their chemopreventive and anticancer properties. Free Radic Biol Med 2004; 37:287–303.

[56] Halliwell B. Role of free radicals in the neurodegenerative diseases: therapeutic implications for antioxidant treatment. Drugs Aging 2001; 18:685–716.

[57] Gutteridge JM, Halliwell B. Antioxidants in Nutrition, Health and Disease. Oxford University Press, Oxford, UK 1994.

[58] Noguchi N, Watanabe A, Shi HL. Diverse functions of antioxidants. Free Radic Res 2000; 33:809–817.

[59] Buettner GR. The pecking order of free radicals and antioxidants: lipid peroxidation, alpha-tocopherol, and ascorbate. Arch Biochem Biophys. 1993; 300:535–543.

[60] German JB. Food processing and lipid oxidation. Adv Exp Med Biol 1999; 459:23–50.

[61] Bose KS, Vyas P, Singh M. Plasma non-enzymatic antioxidants-vitamin C, E, betacarotenes, reduced glutathione levels and total antioxidant activity in oral sub mucous fibrosis. Eur Rev Med Pharmacol Sci 2012; 16:530–532.

[62] Singh RP, Khanna R, Kaw JL, Khanna SK, Das M. Comparative effect of benzanthrone and 3-bromobenzanthrone on hepatic xenobiotic metabolism and anti-oxidative defense system in guinea pigs. Arch Toxicol 2003; 77:94–99.

[63] Buettner GR. The pecking order of free radicals and antioxidants: lipid peroxidation, alpha-tocopherol, and ascorbate. Arch Biochem Biophys 1993; 300:535–543.

[64] Vertuani S, Angusti A, Manfredini S. The antioxidants and pro-antioxidants network: an overview. Curr Pharm Des 2004; 10:1677–1694.

[65] Lien AP, Hua H, Chuong PH. Free radicals, antioxidants in disease and health. Int J Biomed 2008; 4:89–96.

[66] Block G, Patterson B, Subar A. Fruit, vegetables, and cancer prevention: a review of the epidemiological evidence. Nut Cancer 1992; 18:1–29.

[67] Gill SS, Tuteja N. Reactive oxygen species and antioxidant machinery in abiotic stress tolerance in crop plants. Plant Physiol Biochem 2010; 48:909–930.

[68] Mohanty P, Ghanim H, Hamouda W, Aljada A, Garg R, Dandona P. Both lipid and protein intakes stimulate increased generation of reactive oxygen species by poly morphonuclear leukocytes and mononuclear cells. Am J Clin Nutr 2002; 75:767–772.

[69] Garcion E, Nataf S, Berod A, Darcy F, Brachet P. 1,25-Dihydroxyvitamin D3 inhibits the expression of inducible nitric oxide synthase in rat central nervous system during experimental allergic encephalomyelitis. Brain Res Mol Brain Res 1997; 45:255–267.

[70] Gurujeyalakshmi G, Wang Y, Giri SN. Suppression of bleomycin-induced nitric oxide production in mice by taurine and niacin. Nit Oxide 2000; 4:399–411.

[71] Sano M, Fujita H, Morita I, Uematsu H, Murota S. Vitamin K2 (menatetrenone) induces iNOS in bovine vascular smooth muscle cells: no relationship between nitric oxide production and gamma-carboxylation. J Nutr Sci Vitaminol 1999; 45:711–723.

[72] Chow CK, Reddy K, Tappel Al. Effect of dietary vitamin E on the activity of glutathione peroxidase in vitro and in vivo studies. J Clin Invest 1969; 48:1957–1966.

[73] Yunzhong F, Yefu L, Weiqun C. The effect of vitamin B_(12) and folic acid on radiation damage Nitrogen metabolism. Acta Nutrimenta Sinica 1983:3.

[74] Yunzhong F, Ycfu L, Bin H, Shafei H, Weiqun C, Lijun W, Rong W. Studies on the effect of vitamin B_ (12) and folic acid on radiation damage4. Body weight, leucocyte counts and mortality of rats. Acta Nutrmenta Sinica 1984:4

[75] Hammermueller JD, Bray TM, Bettger WJ. Effect of zinc and copper deficiency on microsomal NADPH-dependent active oxygen generation in rat lung and liver. J Nutr 1987; 117:894–901.

[76] Youdim KA, Shukitt-Hale B, MacKinnon S, Kalt W, Joseph JA. Polyphenolics enhance red blood cell resistance to oxidative stress: in vitro and in vivo. Biochim Biophys Acta 2000; 1523:117–122.

[77] Aqil F, Ahmad I, Mehmood Z. Antioxidant and Free Radical Scavenging Properties of Twelve Traditionally Used Indian Medicinal Plants. Turk J Biol 2006; 30:177–183.

[78] Kirlin WG, Cai J, Thompson SA, Diaz D, Kavanagh TJ, Jones DP. Glutathione redox potential in response to differentiation and enzyme inducers. Free Radic Biol Med 1999; 27:1208–1218.

[79] Bjelakovic G, Nikolova D, Gluud I, Simonetti RG, Gluud C. Antioxidant supplements for prevention of mortality in healthy participants and patients with various diseases. Cochrane Database Syst Rev 2008; (2): CD007176.

[80] Lirussi F, Azzalini I, Orando S, Orlando R, Angelico F. Antioxidant supplements for non-alcoholic fatty liver disease and/or steatohepatitis. Cochrane Database Syst Rev 2007; 1: CD004996.

[81] Orrell RW, Lane RJ, Ross M. Antioxidant treatment for amyotrophic lateral sclerosis/ motor neuron disease. Cochrane Database Syst Rev 2007; 1:CD002829.

[82] Farinotti M, Simi S, Di Pietranton J C, Mc Dowell N, Brait L, Lupo D, Filippini G. Dietary interventions for multiple sclerosis. Cochrane Database Syst Rev 2007; 1: CD004192.

[83] Rambaldi A, Gluud C. S-adenosyl-L-methionine for alcoholic liver diseases. Cochrane Database Syst Rev 2006; 2: CD002235.

[84] Denner TG, Horneber M. Selenium for alleviating the side effects of chemotherapy, radiotherapy and surgery in cancer patients. Cochrane Database Syst Rev 2006; 3: CD005037.

[85] Evans JR, Henshaw K. Antioxidant vitamin and mineral supplements for preventing age-related macular degeneration. Cochrane Database Syst Rev2008; 1: CD000253.

[86] Poston L, Briley AL, Seed PT, Kelly FJ, Shennan AH. Vitamins in pre-eclampsia (vip) trial consortium. Vitamin C and vitamin E in pregnant women at risk for pre-eclampsia (VIP trial): randomized placebo-controlled trial. Lancet 2006; 367:1145–1154.

[87] Brunner EJ, Rees K, Ward K, Burke M, Thorogood M. Dietary advice for reducing cardiovascular risk. Cochrane Database Syst Rev 2007; (4):CD002128

[88] Soghier LM, Brion LP. Cysteine, cystine or Nacetylcysteine supplementation in parenterally fed neonates. Cochrane Database Syst Rev2006; (4): CD004869.

[89] Jansen SL, Forbes DA, Duncan V, Morgan DG. Melatonin for cognitive impairment. Cochrane Database Syst Rev 2006; (1): CD003802.

[90] Gray Sl, Anderson ML, Crane PK, Breitner JC, McCormic KW, Bowen JD, TerI L, Larson E. Antioxidant vitamin supplement use and risk of dementia or Alzheimer's disease in older adults. J Am Geriatr Soc 2008; 56:291–295.

[91] Weber CA, Ernst ME. Antioxidants, supplements, and Parkinson's disease. Ann Pharmacother 2006; 40:935–938.

[92] Kirsh VA, Hayes RB, Mayne ST, Chatterjee N, Subara F, Dixon LB, Albanes D, Andriole GL, Urban DA, Peters U; PLCO Trial. Supplemental and dietary vitamin E, beta-carotene, and vitamin C intakes and prostate cancer risk. J Natl Cancer Inst 2006; 98:245–254.

[93] Patel BD, Welch AA, Ingham SA, Luben RN, Dayne, Khaw KT, Lomas DA, Wareham NJ. Dietary antioxidants and asthma in adults. Thorax 2006; 61:388–393.

[94] Behrend TD, Beltrame J, Hikit IH, Wainstein M, KInlay S, Selwyn AP, Ganz P, Fang JC. Impact of coronary endothelial function on the progression of cardiac transplant-associated arteriosclerosis: effect of anti-oxidant vitamins C and E. J Heart Lung Transplant 2006; 25:426–433.

[95] Pleiner J, Schaller G, Mittermayer F, Marsik C, Macallister RJ, Kapiotis S, Ziegler S, Ferlitsch A, Wolzt M. Intra-arterial vitamin C prevents endothelial dysfunction caused by ischemia-reperfusion. Atherosclerosis 2008; 197:383–391.

[96] Cerdá B, Soto C, Albaladejo MD, Martínez P, Sánchez-Gascón F, Tomás-Barberán F, ESPÍN JC. Pomegranate juice supplementation in chronic obstructive pulmonary disease: a 5-week randomized, double-blind, placebo-controlled trial. Eur J Clin Nutr 2006; 60:245–253.

[97] Kirk GR, White JS, MCkie L, Stevenson M, Young I, Clements WD, Rowlands BJ. Combined antioxidant therapy reduces pain and improves quality of life in chronic pancreatitis. J Gastrointest Surg 2006; 10:499–503.

[98] Canter PH, Wider B, Ernst E. The antioxidant vitamins A, C, E and selenium in the treatment of arthritis: a systematic review of randomized clinical trials. Rheumatology 2007; 46:1223–1233.

[99] Ong-Ajyooth L, Ong-Ajyooth S, Parichatikanond P. The effect of alpha-tocopherol on the oxidative stress and antioxidants in idiopathic IgAnephropathy. J Med Assoc Thai 2006; 89:164–170.

[100] Gabbay E, Zigmond E, Pappo O, Hemed N, Rowe M, Zabrecky G, Cohen R, Ilan Y. Antioxidant therapy for chronic hepatitis C after failure of interferon: results of phase II randomized, double-blind placebo controlled clinical trial. World J Gastroenterol 2007; 13:5317–5323.

[101] Endo K, Miyashita Y, Sasaki H, Ebisuno M, Ohira M, Saiki A, Koide N, OYama T, Takeyoshi M, Shirai K. Probucol and atorvastatin decrease urinary 8-hydroxy-2'-de-

http://dx.doi.org/10.5772/61884

oxyguanosine in patients with diabetes and hypercholesterolemia. J Atheroscler Thromb 2006; 13:68–75.

[102] Firuzi O, Miri R, Tavakkoli M, Saso L. Antioxidant therapy: current status and future prospects. Curr Med Chem 2011; 18:3871–3888.

[103] Gowder SJT, Devaraj H. Effect of food flavor cinnamaldehyde on the antioxidant status of rat kidney. Basic Clin Pharmacol Toxicol 2006; 99:379–382.

[104] Gowder SJT, Devaraj H. Food flavor cinnamaldehyde - induced biochemical and histological changes in the kidney of male albino wistar rat. Environ Toxicol Pharmacol 2008; 26:68–74.

[105] Gowder S, Devaraj H. A review on the nephrotoxicity of food flavor cinnamaldehyde. Curr Bioact Compd 2010; 6:106–117.

[106] Gowder SJT. An updated review of toxicity of bisphenol A (BPA) with special reference to the kidney. Curr Mol Pharmacol 2013; 6:163–172.

[107] Palmieri B, Sblendoria V. Oxidative stress tests: overview on reliability and use. Eur Rev Med Pharmacol Sci 2007; 11:383–399.

Section 2

# Natural Compounds as Antioxidants

# Chapter 5

# Antioxidants and Natural Compounds

Davood Maleki, Aziz Homayouni Rad,
Leila Khalili and Baitullah Alipour

Additional information is available at the end of the chapter

http://dx.doi.org/10.5772/60809

**Abstract**

Oxidative stress happens in body when the production of oxidants exceeds the antioxidant capacity of body system and plays a role in pathogenesis of several chronic diseases such as diabetes, cardiovascular diseases, stock, and renal failure. Tea, sesame seed, and burdock root (*Arctium lappa* L.) may improve oxidative stress and suppress the side effects of diabetes, rheumatoid arthritis and osteoarthritis. Total extract of black tea and its fractions can increase antioxidants such as Super oxide Dismotase (SOD), Glutathione Peroxides (GPX) and total antioxidants and can decrease oxidants like malondialdehyde (MDA). So, these herbal compounds can improve oxidative stress in diabetes rats. The injection of total extract and 20% fraction of black tea had positive effect on blood lipid profile in diabetic rats. Sesame seeds improved the antioxidants capacity in arthritis; therefore, decreased pain. Burdock root, in arthritis patients improved the antioxidants capacity and decreased the intensity of the pain. It can be concluded that the positive effects of these herbal components are due to the presence of antioxidants. The aim of this chapter is to review the antioxidant capacity of tea, sesame seed, and burdock root as well as to discuss their biological effects in human being.

**Keywords:** antioxidants, food, natural compounds, oxidative stress

## 1. Introduction

Oxidative stress occurs in the body when the production of oxidants exceeds the antioxidant capacity of the body system and plays a role in the pathogenesis of several chronic diseases such as diabetes, cardiovascular diseases, stroke, and renal failure. Using antioxidants, for instance, vitamin E, vitamin C, and flavanones, can improve antioxidant capacity in the body. Degenerative diseases such as cardiovascular diseases, diabetes, stroke, cancers, and renal disorders are global health problems and it can be predicted that lack of control and attention will involve a lot of costs. Lack of physical activity, nonhealthy diet patterns, and eating salty fast foods are the main causes of these chronic diseases; while change in life style, eating healthy foods like fruits and vegetables can prevent chronic diseases. Herbal extracts and drinks were used to prevent or control diseases from ancient times. Several investigations' results show that there are several components such as flavanones, isoflavanones, and other antioxidants in herbal components that can improve antioxidants in blood and, so, can prevent chronic diseases. Tea, sesame seed, and burdock root (*Arctium lappa* L.) are the important and effective herbal components in this regard. Our results in Tabriz University of Medical Sciences showed the effectiveness of herbal components in improving oxidative stress and suppressing the side effects of diabetes, rheumatoid arthritis, and osteoarthritis. The results indicate that total extract of black tea and its fractions can increase antioxidants such as Super Oxide Dismutase (SOD), Glutathione Peroxides (GPX) and total antioxidants and can decrease oxidants like malondialdehyde (MDA). So, these herbal compounds can improve oxidative stress in diabetes rats. In a study by Alipour *et al.* [1], it has been concluded that injection of total extract and 20% fraction of black tea had positive effect on blood lipid profile in diabetic rats. In one study, using sesame seeds in arthritis improved the antioxidant capacity; therefore, it decreased pain in the mentioned patients. Burdock root, in arthritis patients, improved the antioxidant capacity and decreased the intensity of the pain. It can be concluded that the positive effects of these herbal components are due to the presence of antioxidants. The aim of this chapter is to review the antioxidant capacity of tea, sesame seed, and burdock root, as well as to discuss their biological effects in human beings.

## 2. Oxidative stress

When the production of free radical moieties exceeds the antioxidant capacity of a cellular system, oxidative stress takes place. Radicals will attack and damage proteins, lipids, and nucleic acids if cellular antioxidants do not remove them. The oxidized or nitrosylated products of free radical attack may decrease biological activity, leading to loss of energy metabolism, cell signaling, transport, and other major functions. These altered products also lead to proteosome degradation and further decrease cellular function, resulting in cell death through necrotic or apoptotic mechanisms [2].

### 2.1. Background

Free radicals were defined as a Pandora's Box of evils that may account for cellular damage, mutagenesis, cancer, and, last but not least, the degenerative process of biological aging [3]. Mittal and Murad [4] suggested that the superoxide anion, through its derivative, the hydroxyl radical, stimulates the activation of guanylate-cyclase and formation of the "second messenger" cGMP. Related effects were reported for the superoxide derivative hydrogen peroxide as it was discovered that nitric oxide (NO) has independent role as a regulatory molecule in the control of smooth muscle relaxation and in the inhibition of platelet adhesion [5]. It is recognized that the production of T-cell growth factor and interleukin-2, an immunologically important T-cell protein, is increased in activated T-cells by superoxide anion or low micromolar concentrations of hydrogen peroxide. Studies have shown that the expression of the hemeoxygenase (HO-1) gene is induced by hydrogen peroxide and it has also induction effects on several genes in bacteria, as well as stimulation of the transcription factor nuclear factor κB (NF-κB) in mammalian cells. There is a large amount of proof showing that living organisms have not only adapted to an unfavorable coexistence with free radicals but also have developed mechanisms for the advantageous use of free radicals. The following includes vital physiological functions that involve free radicals or their derivatives: regulation of vascular tone, sensing of oxygen tension and regulation of functions that are controlled by oxygen concentration, improvement of signal transduction from various membrane receptors including the antigen receptor of lymphocytes, and oxidative stress responses that ensure the maintenance of redox homeostasis [5]. In consideration of the role that oxidative stress has been found to play in numerous disease conditions, the field of redox regulation is also receiving growing attention from clinical colleagues. These conditions show the biological significance of redox regulation. The balance between the advantageous and hurtful effects of free radicals is clearly an important aspect of life. The science of biological "redox regulation" is a rapidly growing area of research that has effect on various disciplines including physiology, cell biology, and clinical medicine [5].

### 2.2. Oxidative stress biomarkers

Oxidative stress biomarkers' measurement is an important step in order to understand the pathogenesis of and developing treatments for some diseases like diabetes. Measurements of the reduction of antioxidant reserves, changes in the activities of antioxidant enzymes, free radical production, and presence of protein, lipid, and DNA free radical adducts are several methods that may be accepted. As enzyme activities and cellular antioxidants are likely to display transient changes, for the purposes of clinical assessment, measurements of end products of free radical attack may be the most reliable determination of the occurrence of oxidative stress. An indication of the stress level experienced in a cell or tissue can be provided by the enzymes responsible for detoxifying free radicals or regenerating antioxidant molecules and they can be measured by in vitro activity assays; however, changes in transcription can also provide evidence of cell stress. For example in long-term diabetes, catalase, GSH reductase, GSH peroxidase, and SOD decrease in complication-prone tissue [2].

### 2.3. Plasma markers of oxidative stress

The body's total antioxidant capacity functions in order to protect cells from excess production of reactive oxygen/nitrogen species (RONS) [6]. Antioxidant capacity includes endogenous (e.g. uric acid, superoxide dismutase, catalase, glutathione peroxidase) and exogenous (e.g., carotenoids, tocopherols, ascorbate, bioflavonoids) compounds. The exogenous antioxidants are consumed in the diet mainly from fruits and vegetables [7]. Dietary habits can alter an individual's susceptibility to oxidative damage since these exogenous compounds contribute to antioxidant capacity.

Oxidative stress can be defined by formation of RONS that exceed the body's antioxidant capacity [8]. Cellular metabolism as well as environmental (e.g., cigarette smoke, ozone, certain nutrients) and physiological (e.g., physical and mental stress) challenges can form RONS in the body. Cellular harms and disease generation that may accompany oxidative stress are related to those macromolecules (nucleic acid, protein, and lipid) that are targeted by RONS, the frequency and duration of attack, and the tissue-specific antioxidant defenses present [9]. For instance, RONS reacting with DNA can create wide strand breakage and degradation of deoxyribose, an effect caused by formation of hydroxyl radicals [10]. Over time such changes may lead to disease because of the alterations in nucleotide roots. The presence of 8-hydroxydeoxyguanosine (8-OHdG), an irregular intermediate in nucleotide metabolism, in urine and blood shows oxidative DNA damage. Biomarkers of protein oxidation represent amino acid modifications such as phenylalanine residues alteration to o-tyrosine or tyrosine to dityrosine, as well as overall modifications such as the alteration to carbonyl derivatives. Aromatic and sulfllydryl-containing residues are chiefly vulnerable to oxidation, often causing loss of catalytic or structural function in the affected proteins, makeing them susceptible to proteolytic degradation [11]. Proteins oxidation has been associated with disease like cancer, diabetes, and cardiovascular disease [12]. Lipid peroxidation has been recognized as a main mechanism of cellular damage in humans along with DNA and protein oxidation [13]. Polyunsaturated lipids are mainly vulnerable to damage in an oxidizing environment and may make lipid peroxides that can react to form malondialdehyde (MDA). MDA that can be measured in plasma is regularly used to estimate lipid-specific oxidative stress [14]. While MDA has been a practical indicator of oxidative stress for clinical studies [15], other lipid biomarkers, such as F2-isoprostanes, have gained agreement in recent years [16].

### 2.4. Oxidative stress and disease

Oxidative stress plays a role in several pathophysiologic conditions, for example, malignant diseases, diabetes, atherosclerosis, chronic inflammation, human immunodeficiency virus (HIV) infection, ischemia reperfusion injury, and sleep apnea. These diseases are divided into two types. In the first type, diabetes mellitus and cancer display a pro-oxidative shift in the systemic thiol/disulfide redox state and impaired glucose clearance, proposing that skeletal muscle mitochondria may be the major site of elevated reactive oxygen species (ROS) creation. These conditions may be referred to as "mitochondrial oxidative stress," which, without healing intervention may lead to massive skeletal muscle wasting, indicative of aging-related wasting. The second type refers to "inflammatory oxidative conditions," because it is usually associated with stimulation of NAD(P)H,$H^+$ oxidase activity by cytokines or other agents. In this case, changes in intracellular glutathione levels or increased ROS levels are often associated with pathological changes revealing a disregulation of signal cascades and/or gene expression, demonstrated by altered expression of cell adhesion molecules [5].

### 2.5. Oxidative stress and antioxidants

Several free radical species are produced in the body to carry out specific functions. $O_2^-$, $H_2O_2$, and NO are three free radical reactive oxygen species (ROS) that are crucial for body physiology. However, they can accelerate the process of aging, mediate cellular degeneration in disease states, and produce highly active singlet oxygen, hydroxyl radicals, and peroxy-nitrite that can attack proteins, lipids, and DNA. Antioxidants by donating at least one hydrogen atom to a free radical can result in the termination of radical chain reactions. Extra production of free radicals can cause oxidative stress. Acute hyperglycemic episodes such as an oral glucose tolerance test or a meal can increase oxidative stress in diabetic patients and decrease the antioxidant ability of plasma in both normal and diabetic subjects. Elevated basal levels of free radical production and decreased antioxidants are strengthened by elevated plasma glucose. Analysis of individual vitamin and enzyme components of the antioxidant system in man shows that the levels of vitamins A and E and catalase activity are decreased in both type 1 and 2 patients compared with controls. Whereas GSH-metabolizing enzymes are decreased in type 1 but not type 2 patients, SOD activity is lower in type 2 but not type 1 [2].

## 3. Antioxidants

### 3.1. Definition

Antioxidants are defined as any compound that can donate at least one hydrogen atom to a free radical so that it can result in the termination of radical chain reactions. A substitute type of antioxidant is defined by its ability to prevent the initiation of a free radical chain reaction rather than to terminate them. This latter type of antioxidant includes ceruloplasmin, transferrin, and albumin that are usually dependent upon the ability to bind metal ions. Cells must preserve the levels of antioxidants, often defined as antioxidant potential, through dietary uptake or de novo synthesis. Excess production of free radicals can reduce the intracellular antioxidants, resulting in oxidative stress [2].

### 3.2. Classification

Ingold [17] classified all antioxidants into two groups, namely primary or chain-breaking antioxidants, and secondary or preventive antioxidants.

#### *3.2.1. Primary or chain-breaking antioxidants*

Primary antioxidants can react with lipid radicals to convert them to more stable products. The major lipid radical at normal oxygen pressures is the alkylperoxy radical ROO, which is an oxidizing agent and is readily reduced to the related anion and then converted to a hydroperoxide by an electron donor, or which may be directly converted to a hydroperoxide by a hydrogen donor, AH. Alkyl radicals are in general reducing agents and are scavenged by electron acceptors. Inhibition by electron acceptors is not significant in most food systems, but it can become important in biological tissues since the oxygen pressure is much lower in healthy tissues.

#### *3.2.2. Secondary or preventive antioxidants*

Secondary antioxidants are the compounds which decelerate the degree of autoxidation of lipids by processes other than that of interrupting the autoxidation chain by changing free radicals to more stable species. These may function by several mechanisms including compounds that bind metal ions, scavenge oxygen, decompose hydroperoxide to non-radical species, absorb UV radiation, or deactivate singlet oxygen. Secondary antioxidants show antioxidant activity if a second minor component exists in the sample.

## 4. Natural antioxidants in foods

Reports in recent years both in the popular and scientific press have stressed the value and advantages of natural ingredients as food preservatives. There is an implied assumption of safety for compounds that occur naturally in foods and that have been consumed for many centuries. Natural antioxidants in foods may be from (a) endogenous compounds in one or more components of the food; (b) substances formed from reactions during processing; and (c) food additives isolated from natural sources.

Natural antioxidants may function in one or more of the following ways: (a) as reducing agents, (b) as free radical scavengers, (c) as complex of pro-oxidant metals, and (d) as quenchers of the formation of singlet oxygen. The compounds are most commonly phenolic or polyphenolic from plants sources. The most common natural antioxidants are flavonoids (flavanols, isoflavones, flavones, catechins, and flavanones), cinnamic acid derivatives, coumarins, tocopherols, and polyfunctional organic acids [18].

### 4.1. Tea antioxidants and oxidative stress

Tea (from the plant *Camellia sinensis*), consumed by over two-thirds of the world's population, is the most desirable beverage next to water. About three billion kilograms of tea are produced and consumed annually [19-21]. Thease are the enzymes in tea which catalyze oxidation. During fermentation in which tea pectins are demethylated, polyphenolic compounds are decomposed which as a result of the quinone appearance, turn into colorful agents including theaflavin and thearubigin, both of which are plentiful in black tea [22-24]. More than 600 volatile agents have been documented in tea, most of which are yellow in color and have a characteristic scent. Linalool is the chief essence in tea, others of lesser importance being dihydroactinide iolido paravinile phenol, hexenol, hexenal, aldehydes, phenyl ethyl alcohols, phenols, and geraniols [22]. According to the preparation method, the degree to which it is fermented and the steps it goes through during production, different types of tea consumed all over the world are classified into at least six categories. The less processed the tea, the greater the polyphenols content will be, which the extent of oxidation accounts for [25].

1. White tea: White tea is manufactured only from the buds or first leaves of *C. sinensis*. It is the least processed type of tea and is simply steamed and dried without a prior withering stage; therefore, the concentrations of EGCG and also methylxanthines (like caffeine) are enriched in white tea compared with green and black tea.

2. Yellow tea: It usually refers to a special tea processed in a similar way to green tea; but the drying process takes place at a slower rate. The damp tea leaves are allowed to sit and yellow. Its taste resembles that of green and white teas.

3. Green tea: To manufacture green tea, first the fresh leaves are steamed, then primary drying-rolling, rolling, and secondary drying-rolling, final drying-rolling, and at last drying are performed. No fermentation takes place in this type of tea.

4. Oolong tea: Fresh leaves undergo solar withering at the first step, then indoor withering and rolling, pan firing, rolling, mass breaking and drying are the steps taken, to produce oolong tea. In this kind of tea, partial fermentation occurs after the rolling.

5. Black tea: The manufacturing process for black tea includes withering of fresh leaves, rolling, fermenting, and drying. Thorough fermentation is done in black tea.

6. Pu-Erh: Pu-Erh reffers to old tea with extreme fermentation [22, 25-27].

Regular intake of tea is associated with low risk of certain types of cancer, coronary heart disease, atherosclerosis, stroke, reduced mutagenicity and inflammation, protection against neurodegenerative diseases, and increasing insulin sensitivity since it can improve antioxidant status in vivo conditions [28,29]. Tea is a great source of antioxidants especially flavonoids. Animal studies have strongly supported the idea of tea being an efficient suppressor of oxidative stress [30]. Several studies have shown that different types of tea are potentially effective in reducing oxidative stress and related diseases [31]. Attempts have been made to manufacture products containing tea bioactive compounds for prevention and treatment of the aforementioned diseases. To design such products, the effective compounds of tea and their safe doses must be first identified. For instance, EGCG (Epicatechingallate) has been revealed to act as a pro-oxidant when administered in high doses, leading to apoptosis. Moreover, compounds other than catechins may exert the desired effects too [32-34]. To determine the compounds acting as antioxidants in black tea, Alipour et al., in 2009 [35] performed a study in which diabetic rats were supplemented total extract of black tea and its fractions. Total extract and fractions were attained by hydromethanol method and solid phase extraction using Sep-Pak, respectively. Results of this study indicated that injection of total extract and 20% fraction of black tea decreased malondialdehyde (MDA) and increased total antioxidant, Super oxide Dismutase (SOD), Glutathione Peroxides (GPX), and Glutathione in diabetic rats. To determine the major substances in the 20% fraction, Analytical HPLC, Preparative HPLC (High Performance Liquid Chromatography), and NMR (Nuclear Magnetic Resonance) (CNMR and HNMR) were employed. Caffeine, EpicatechinGallate, Quercetin, and Kampferol were the main compounds capable of fighting oxidative stress, determined in 20% fraction of tea [31]. Caffeine is a strong antioxidant; its activity is equal to that of glutathione and exceeding that of vitamin C [36,37]. The free radical scavenging capacity of flavonoids is due to the 3', 4' dihydroxyl and 3' hydroxy in the β ring of its structure [38]. The 20% fraction of black tea has been shown to be more effective than the other fractions, which may be explained by the high concentration of the aforementioned compounds in it and absence of polyphenol antagonists in the very extract prepared [35]. Tea polyphenols have been found to induce expression of phase II enzymes and endogenous antioxidants that protect cells from oxidative stress. Phase II enzymes have vital antioxidant properties in fighting reactive oxygen species and xenobiotics (foreign substances), including potential carcinogens. Induction of phase II detoxifying and antioxidant enzymes is mediated through cis-regulatory DNA sequences known as antioxidant-response elements (AREs) that are found in the promoter or enhancer region of the gene. The major ARE transcription factor nuclear factor E2-related factor 2 (Nrf2) is a key agent in the initiation of antioxidant and detoxifying enzymes, such as heme oxygenase-1 (HO-1), glutathione S-transferases (GSTs), and reduced nicotinamide adenine dinucleotide phosphate:quinone oxidoreductase [39]. Nrf2 binds to Kelch-like ECH-associated protein 1 (Keap1) under nonstressed conditions. Keap1 in complex with cullin3, Rocl, and E2 proteins provides ubiquitination followed by proteasomal degradation. When oxidative stress occurs, oxidation of Keap1 leads to inability to bind Nrf2 protein by forming intramolecular disulfide bonds. Then, Nrf2 migrates into the nucleus and binds a protein of Maf family (like sMaf) and CBP/p. This complex is formed on ARE promoter region of certain genes leading to transcription activation. Phosphorylation of by protein kinases which may be activated by oxidants is one way to provide Nrf2 migration in nucleus [40].

### 4.2. Sesame and oxidative stress

In recent years, there has been a growing attention paid to natural antioxidants of plants and their use is gaining importance as nutraceuticals and phytoceuticals as they have significant effect on the status of human health and disease prevention [41]. For thousands of years sesame (*Sesamum indicum L.*) is an important traditional health food that has been used to improve nutritional status and prevent various diseases in Asian countries. The results of several studies support the hypothesis that sesame seed and its lignans may have antioxidant and hypocholesterolemic effects [42-45]. Alipour et al. in 2012 [46] showed that sesame seed supplementation decreased serum TC, LDL-C, and lipid peroxidation, and increased antioxidant status in hyperlipidemic patients. A positive effect of sesame seed was seen in improving lipid profile and oxidative stress in patients with knee OA, indicating that sesame seed might be of help to reduce oxidative stress in OA patients [47]. Sesame seeds are not only rich in oil and protein, but also in lignans (e.g., sesamin and sesamolin) [48] especially phenolic lignans such as sesamol. Antioxidant effects of lignans were shown in former studies [49,50]. Furthermore, it is obvious that diets containing polyphenols and flavonoids increase catalase and SOD activity, decrease MDA and improve lipid profile [51,52]. Nakai et al. [52] revealed that sesame metabolism by CYP450 in liver results in inversion of methylene dioxyphenyl to dihydrophenyl (strong radical scavenger). Previously, scientific evidence has showed the fact that protective effects of sesame seed are because of the suppression of oxygen species production [53]. Moreover, sesame lignans have an ability to increase vitamin E levels in various tissues [54,55] and increase gamma-tocopherol levels that could lead to the suppression of different free radicals (those that usually increase in age-related diseases) [56]. In a study on fifty patients with knee OA, supplementation with sesame seed improved clinical signs and symptoms and indicated that sesame might be a viable adjunctive therapy in treating OA [57]. Epidemiological studies have shown that high HDL-cholesterol concentration could potentially contribute to its anti-atherogenic properties, including its ability to inhibit LDL-oxidation and protect endothelial cells from the cytotoxic effects of oxidized LDL [58]. Presently, sesame seed powder administration has shown to elevate HDL-C levels in hypercholesteremic animals [59]. Several studies reported that high antioxidant properties of sesame seed appear to be related to its lignans- sesamol, sesamolinol, pinoresinol, and sesaminol [60-64], as well as vitamin E [65]. The tocopherol of sesame seed has been shown to be mainly y-tocopherol, with only less amounts of a-tocopherol [66]. In addition, Kahae et al. [67] indicated that $\gamma$ -tocopherol in sesame seed exerts vitamin E activity equal to that of $\alpha$-tocopherol through a synergistic interaction with sesame seed lignans. These compounds also have inhibitory effects on membrane lipid peroxidation, the microsomal peroxidation induced by ADP-$Fe^{3+}$/NADPH [61], and the oxidation of LDL induced by copper ions [64].

### 4.3. Burdock and oxidative stress

The examination of burdock (*Arctium lappa* L.) protective effects on oxidation of low-density lipoprotein (LDL) and nitric oxide production showed that methanolic extracts of burdock (MEB) and their major components, chlorogenic acid (CHA) and caffeic acid (CA), have antioxidant effects against oxidative damages. For many decades in Taiwan and Japan, burdock (*Arctium lappa* L.) has been consumed as a vegetable and beverage. Furthermore, burdock is also used as a folk medicine, such as a diuretic and antipyretic [68]. Analyses of its components [69], investigation of desmutagenic effect, and hepatoprotective efficacy [70] have been described. Chen (2004) [68] finds that burdock has significant free radical scavenging activity, which was mainly attributed to chlorogenicacid (CHA) and caffeic acid (CA). Duh in 1998 [71] presented that burdock, with or without heat treatment, acts as a primary and secondary antioxidant, as well as active oxygen scavenger. In addition, burdock displayed potential inhibitory action on microbial growth [72]. Although burdock showed the biological activity mentioned above, whether it has any protective effects on biomolecules and nitric oxide production remains uncertain. The protective activities of CA and CHA on LDL oxidation and nitric oxide production have been established [73,74]. Several reports showed that oxidatively damaged LDL as an atherogenic agent is clearly a main risk factor for cardiovascular disease. Thus, the bioactive activity of MEB on LDL oxidation and nitric oxide production needed to be examined. MEB inhibited oxidation of phospholipid, protein and deoxyribose, which are components of the cells. So, MEB can protect cells and tissues against oxidative damage. CA and CHA, which were major compounds present in MEB [69], showed notable inhibitory effect on oxidative damage of liposome, deoxyribose, and protein. Maghsoumi et al. in 2014 [75] suggested that *Arctium lappa L.* root tea improves inflammatory status and oxidative stress in patients with knee osteoarthritis.

### 4.4. Lentil and oxidative stress

Several studies have shown that legumes consumption has many effects in health improvement, control, and protection against metabolic diseases such as type 2 diabetes and CVDs [76]. Lentils (Lens culinaris), the most consumed legume grains, are good sources of dietary fiber, resistant starch, tannins, β-glucan, functional antioxidant ingredients, a wide range of phenolic acids including gallic acid, proanthcyanidins, prodelphinidin, procyanidins, catechins, epicatechin, kampferol, quercetin, cinapic acid, and apigenin [77]. Lentil contains about 28% protein on a dry weight basis [78]. The effects of lentil sprouts (LS) on glycemic parameters are associated with its fiber content. The fiber content of lentil seed is 3.7 g per 100 g; lentil seed also has low glycemic index (21/2). After germination of the lentil seeds, the amount of fiber and protein are increased [79]. The enzymatic hydrolysis of lentil proteins have resulted in bile salts binding activity and the production of hydrolysates with ACE-inhibitory activity [80, 81]. Furthermore, it has been recognized that specific fragments from legumin, vicilin, and convicilin with amino acid sequences contribute to the antioxidant and ACE-inhibitory activity of lentil hydrolysates [82]. Epidemiological studies recommend that lentils through biological activities including antioxidant, anticancer, angiotensin I-converting enzyme inhibition, reducing blood lipid, and reducing the risk of cardiovascular diseases, confer protection against chronic diseases [83,84]. Studies show that bioactive proteins of lentil decrease plasma levels of LDL-C, triglyceride content of the liver, and adipose tissue lipoprotein lipase activity; moreover, polyphenols of lentil could prevent angiotensin II-induced hypertension, and pathological changes including vascular remodeling and vascular fibrosis [85,86]. Lentils have a higher oxygen radical absorbing capacity (ORCA) value than most of the common fruits and vegetables [87]. In a study by Aslani *et al.* in 2014 [88], effects of lentil sprout (LS) consumption on glycemic parameters and anthropometric measurements in overweight and obese patients with type 2 diabetes were investigated. They found that LS consumption could have favorable effects on glycemic control in overweight and obese patients with type 2 diabetes. They showed that consumption of LS as supplementary treatment in type 2 diabetes could have favorable effects on HbA1c (Glycated hemoglobin), FBS (fasting blood glucose), QUICKI (quantitative insulin sensitivity check index), and HOMA-IR (homeostasis model assessment- Insulin resistance).

## 5. Biological effects of food antioxidants

The human organism, like that of animals, is oxygen-dependent. This suggests that oxygen, essential for life, works through a succession of mechanisms which indeed have their limits and their side effects. The survival of the species means that biochemical protection systems have developed in parallel with that promoting oxygen utilization. More generally, it may be considered that oxidation involves loss of one or more electrons that is of negative electric charges [89]. However, more complex oxidation reactions occur, with the production of toxic radicals, which, in the absence of antioxidant mechanisms, would soon destroy the vital elements of the cell [90]. The accumulation of hydroperoxides, for instance, requires the intervention of catalase, tocopherol, selenium, or reduced glutathione and its conversion

http://dx.doi.org/10.5772/60809

enzymes; their deficiency would promptly lead to an attack on membranes [91,92]. The multiplicity of in- vivo cellular reaction allowing aerobic life to develop implies extremely active binding as well as exchanges of oxygen molecules with other radicals. Whereas some of these metabolic sequences require radicals interacting with reactive oxygen the synthesis of prostaglandins, the metabolism of molecules with quinone structures, or the activity of macrophages, for instance these radicals may also be at the origin of chain reactions eliciting deleterious effects at the level of cell particles. The survival of tissues requires in turn that other molecules either terminate oxidative chain reactions or scavenge the excess of the generated free radicals [89].

## 6. Conclusion and future trends

There is now an agreement among scientists regarding the effect of uncontrolled oxygen radicals in the deterioration of health. Oxidative stress has been showed to play an important role in initiation and progression of disease. Thus, to prevent the very consequences of oxidative stress, it seems logical to take the necessary steps to reduce it. Antioxidants have been reported to be effective in achieving this goal. Some of these antioxidants are flavanols, isoflavones, flavones, catechins, flavanones and probiotics which are assessed for treatment and/or prevention of diseases such as diabetes [18, 93-97]. Routine methods for the determination of oxidation and peroxidation levels further need to be developed. It is recommended that well-designed, controlled clinical trials be done taking into account all the factors affecting the oxidative status of the patients and using sensitive and specific indicators of oxidative stress. By analyzing the results of studies, several nutritional factors that are effective in reducing oxidative stress markers should be recognized in order to be used as functional foods and supplements for controlling oxidative stress.

## Author details

Davood Maleki[1], Aziz Homayouni Rad[2], Leila Khalili[3] and Baitullah Alipour[4*]

*Address all correspondence to: alipourb@tbzmed.ac.ir

1 Hematology and Oncology ward, Urmia University of Medical Science, Urmia, Iran

2 Department of Food Science and Technology, Faculty of Nutrition, Tabriz University of Medical Sciences, Tabriz, I.R, Iran

3 Department of Nutrition, Faculty of Nutrition, Tabriz University of Medical Sciences, Tabriz, I.R, Iran

4 Department of Nutrition, Faculty of Nutrition, Tabriz University of Medical Sciences, Tabriz, I.R, Iran

## References


[1] Alipour B.A., Ostad Rahimi A.R., Delazar A., Mesgari M., Asnaashari S., Vatankhah A.M., Alipour Azhiri S., Safaeian A.A.R. The effect of total extract and its different hydromethanol fractions of Iranian black orthodox tea on blood lipid profile in type 1 diabetic rats. *Med J Tabriz Univ Med Sci*; 2008. p83-87.

[2] Vincent A.M., Russell J.W., Low P., EVA L. Feldman E.L. Oxidative stress in the pathogenesis of Diabetic Neuropathy. *Endocr Rev*; 2003. p612-628.

[3] Harman D. Aging: A theory based on free radical and radiation chemistry. *J Gerontol*; 1956. p298-300.

[4] Mittal C.K., Murad F. Activation of guanylate cyclase by superoxide dismutase and hydroxyl radical: A physiological regulator of guanosine 3',5'-monophosphate formation. *Proc Natl Acad Sci USA*; 1977. p4360-4364.

[5] Droge W. Free radicals in the physiological control of cell function. *Physiol Rev*; 2002. p47-95.

[6] Urso M.L., Clarkson P.M. Oxidative stress, exercise, and antioxidant supplementation. *Toxicology*; 2003. p41-54.

[7] Watson T.A., Callister R., Taylor R.D., et al. Antioxidant restriction and oxidative stress in short-duration exhaustive exercise. *Med Sci Sports Exerc*; 2005. p63-71.

[8] Finaud J., Lac G., Filaire E. Oxidative stress: Relationship with exercise and training. *Sports Med*; 2006. p327-358.

[9] Halliwell B., Cross C.E. Oxygen-derived species: Their relation to human disease and environmental stress. *Environ Health Perspect*; 1994. p5-12.

[10] Halliwell B. Oxygen radicals: A common sense look at their nature and medical importance. *Med Biol*; 1984. p71-77.

[11] Levine R.L., Stadtman E.R. Oxidative modification of proteins during aging. *Exp Gerontol*; 2001. p1495-1502.

[12] Dalle-Donne I., Giustarini D., Colombo R., et al. Protein carbonylation in human diseases. *Trends Mol Med*; 2003. p169-176.

[13] Halliwell B. Reactive oxygen species in living systems: Source, biochemistry, and role in human disease. *Am J Med*; 1991. p14-22.

[14] Nielsen F., Mikkelsen B.B., Nielsen J.B., et al. Plasma malondialdehyde as biomarker for oxidative stress: Reference interval and effects of life-style factors. *Clin Chem*; 1997. p1209-1214.

[15] Vassalle C., Lubrano V., L'Abbate A., Clerico A. Determination of nitrite plus nitrate and malondialdehyde in human plasma: Analytical performance and the effect of smoking and exercise. *ClinChem Lab Med*; 2002. p802-809.

[16] Milne G.L., Yin H., Brooks J.D., et al. Quantification of F2-isoprostanes in biological fluids and tissues as a measure of oxidant stress. *Methods Enzymol*; 2007. p113-126.

[17] Ingold, K.U. Inhibition of autoxidation. *Adv Chem Ser*; 1968. p296-305.

[18] Dan E. Pratt and Bertram J.F. Hudson. *Natural Antioxidants Not Exploited Commercially in Food Antioxidants* (ed. B. J.F. Hudson). Elsevier science publishers LTD, London and New York; 1990. p171-172.

[19] Yang C.S, Landau J.M. Effects of tea consumption on nutrition and health. *J Nutr*; 2000. p 2409-2412.

[20] Gupta S., Saha B., Giri A.K. Comparative antimutagenic and anticlastogenic effects of green tea and black tea: A review. *Mutation Res*; 2002. p37-65.

[21] Crespy V., Williamson G. A review of the health effects of green tea catechins in in vivo animal models. *J Nutr*; 2004. p3431-3440.

[22] Hara Y. *Green Tea, Health Benefits and Applications*. Marcell Dekker Inc, New York; 2001. p1-11, 27-43, 53-83, 97-119.

[23] Leung L.K, Su Y., Chen R., Zhang Z., Huang Y., Chen Z.Y. Theaflavins in black tea and catechins in green tea are equally effective antioxidants. *J Nutri*; 2001. p 2248-2251.

[24] Cadenas E., Packer L. *Handbook of Antioxidants*. 2nd edn., Marcel Dekker, Inc, 2002. p371-399.

[25] Santana-Rios, G., Orner, G.A., Amantana, A., Provost, C., Wu, S.-Y., Dashwood, R.H. Potent antimutagenic activity of white tea in comparison with green tea in the salmonella assay. *Mutation Res/Gen Toxicol Environ Mutagen*; 2001. p61-74.

[26] Kuo K.L., Weng M.S, Chiang C.T, Tsai Y.J, Lin-Shiau S.Y, Lin J.K. Comparative studies on the hypolipidemic and growth suppressive effects of Oolong, Black, Puerh, and Green tea leaves in rats. *J Agric Food Chem*; 2005. p480-489.

[27] Lin J.K, Sy L.S. Mechanisms of hypolipidemic and anti-obesity effects of tea and tea polyphenols. *Mol Nutri Food Res*; 2006. p211-217.

[28] Luximon-Ramma A., Bahorun T., Crozier A., Zbarsky V., Datla K.P., Dexter D.T., Aruoma O.I. Characterization of the antioxidant functions of flavonoids and proanthocyanidins in Mauritian black teas. *Food Res Int*; 2005. p357-367.

[29] Alipour B., Alipoor Aghiri S., Ostad Rahimi A., Delazar A., Mesghary M. Effects of total extract and its different hydromethanol fractions of Iranian black tea on inflam-

matory factors and glycosilatedhemoglubin (HbA1c) in STZ-induced diabetic rats. *Urmia Med*; 2011. p2710-2737.

[30] Alipour B., Homayouni A., Vaghef Mehrabany E. Oxidative stress, tea antioxidants and diabetes. In: *Oxidative Stress Environmental Induction and Dietary Antioxidants* (ed. Volodymyr I. Lushchak); 2012. p223-248.

[31] Alipour B., Rad A.H. A review on the therapeutical effects of tea. *Asian J Clin Nutri*; 2012. p1-15.

[32] Liao S, Kao Y.H, Hiipakka R.A. Green tea: Biochemical and biological basis for health benefits. *Vitam Horm*; 2001. p1-94.

[33] Mandel S., Youdim M.B. Catechin polyphenols: Neurodegeneration and neuroprotection in neurodegenerative diseases. *Free Radic Biol Med*; 2004. p304-317.

[34] Kao Y.H, Chang H.H, Lee M.J, Chen C.L. Tea, obesity, and diabetes. *Mol Nutr Food Res*; 2006. p188-210.

[35] Alipour B., Delazar A., Ostad Rahimi A., Alipoor Aghiri S., Mesghary M. The effect of total extract and hydromethanol fractions of Iranian black tea on oxidative stress of type1 diabetic rats. *Iranian J Diabe Lipid*; 2009. p130-136.

[36] Devasagayam T.P.A, Kamat J.P, Mohan H, Kesavan P.C. Caffeine as an antioxidant: inhibition of lipid peroxidation induced by reactive oxygen species. *BiochimicaetBiophysicaActa (BBA)-Biomembranes*; 1996. p63-70.

[37] Nikolic J., Bjelakovic G., Stojanovic I. Effect of caffeine on metabolism of L-arginine in the brain. *Mol Cell Biochem*; 2003. p125-128.

[38] Amic D., Davidovic-Amic D., Beslo D., Trinajstic N. Structure-radical scavenging activity relationships of flavonoids, *Croatic Chemica Acta CCACAA*; 2003. p55-61.

[39] Xiang-Dong Wang. *Carotenoids*. (ed. A. Catharine Ross, Benjamin Caballero, Robert J. Cousins, Katherine L. Tucker, Thomas R. Ziegler), Lippincott Williams & Wilkins, a Wolters Kluwer Business; 2014. p436.

[40] Lushchak V.I. Adaptive response to oxidative stress: Bacteria, fungi, plants and animals. *Comp Biochem Physiol*; 2011. p175-190.

[41] Noguchi C., Nikki E. Phenolic antioxidants: a rationale for design and evaluation of novel antioxidant drugs for atherosclerosis. *Free Radic Biol and Med*; 2000. p1538-1546.

[42] Kang M.H., Naito M., Tsujihara N., Osawa T. Sesamolin inhibits lipid peroxidation in rat liver and kidney. *J Nutr*; 1998. p1018-1022.

[43] Hirata F., Fujita K., Ishikura Y., Hosoda K., Ishikawa T., Nakamura H. Hypocholestero-lemic effect of sesame lignan in humans. *Atherosclerosis*; 1996. p135-136.

[44] Nakano D., Itoh C., Ishii F., Kawanishi H., Takaoka M., Kiso Y., et al. Effects of sesamin on aortic oxidative stress and endothelial dys-function in deoxycorticosterone acetate-salt hypertensive rats. *Biol Pharm Bull*; 2003. p1701-1705.

[45] Ikeda S., Tohyama T., Yamashita K. Dietary sesame seed and its lignans inhibit 2,7,8-trimethyl-2(2'-carboxyethyl)-6-hydroxychro-man excretion into urine of rats fed gamma-tocopherol. *J Nutr*; 2002. p961-966.

[46] Alipour B., Haghighian M.K., Sadat B.E., Asghari M. Effect of sesame seed on lipid profile and redox status in hyperlipidemic patients. *Int J Food Sci Nutr*; 2012. p674-678.

[47] Mahdieh Khadem Haghighian, Beitollah Alipour, Bina Eftekhar Sadat, Aida Malek Mahdavi, Abdolvahab Moghaddam, Amir-Mansour Vatankhah. Effects of sesame seed supplementation on lipid profile and oxidative stress biomarkers in patients with knee osteoarthritis. *Health Promot Perspect*; 2014. p90-97.

[48] Chen P.R., Chien K.L., Su T.C., Chang C.J., Liu T.L., Cheng H., Tsai C. Dietary sesame re-duces serum cholesterol and enhances anti-oxidant capacity in hypercholesterolemia. *Nutr Res*; 2005. p559-567.

[49] Kapadia G.J., Azuine M.A., Tokuda H., Takasaki M., Mukainaka T., Konoshima T., et al. Chemo-preventive effect of resveratrol, sesamol, se-same oil and sunflower oil in the Epstein–Barr virus early antigen activation assay and the mouse skin two-stage carcinogenesis. *Pharmacol Res*; 2002. p499-505.

[50] Toyokuni S., Tanaka T., Kawaguchi W., Fang N.R., Ozeki M., Akatsuka S., et al. Effects of the phenolic contents of Mauritian endemic plant extracts on promoter activities of anti-oxidant enzymes. *Free Radic Res*; 2003. p1215-1224.

[51] RalayRanaivo H., Rakotoarison O., Tesse A., Schott C., Randriantsoa A., Lobstein A., An driantsitohaina R. Cedrelopsisgrevei induced hypotension and improved endothelial vaso-dilatation through an increase of Cu/Zn SOD protein expression. *Am J Physiol Heart CircPhysiol*; 2004. p775-781.

[52] Nakai M., Harada M., Nakahara K., Akimoto K., Shibata H., Miki W., et al. Novel antioxida-tive metabolites in rat liver with ingested se-samin. *J Agric Food Chem*; 2003. p1666-1670.

[53] Jeng K.C.G., Hou R.C.W. Sesamin and Sesa-molin: Nature's therapeutic lignans. *Curr En-zymInhib*; 2005. p11-20.

[54] Kamal-Eldin A., Frank J., Razdan A., Tengblad S., Basu S., Vessby B. Effects of dietary phenolic compounds on tocopherol, cholesterol, and fatty acids in rats. *Lipids*; 2000. p427-435.

[55] Kamal-Eldin A., Pettersson D., Appelqvist LA. Sesamin (a compound from sesame oil) increases tocopherol levels in ratsfedadlibi-tum. *Lipids*; 1995. p499-505.

[56] Kontush A., Spranger T., Reich A., Baum K., Beisiegel U. Lipophilic antioxidants in blood plasma as markers of atherosclerosis: the role of alpha-carotene and gamma-tocopherol. *Atherosclerosis;* 1990. p117-122.

[57] Eftekhar Sadat B., Khadem Haghighian M., Alipour B., Malek Mahdavi A., Asghari Jafarabadi M., Moghaddam A. Effects of sesame seed supplementation on clinical signs and symptoms in patients with knee osteoarthritis. *Int J Rheum Dis;* 2013. p578-582.

[58] Assmann, G., Nofer, J. Atheroprotective effects of high-density lipoproteins. *Ann.Rev Med;* 2003. p321-341.

[59] Amal H. Hamza, Rasha H. Mahmoud. Soy milk and sesame seeds (Phytoestrogens) ameliorate cardiotoxcity induced by adriamycin in experimental animals. *Am J Res Commun;* 2003.

[60] Fukuda, Y., Nagata, T., Osawa, T., Namiki, M. Contribution of lignan analogues to antioxidative activity of refined unroasted sesame seed oil. *J Am Oil Chem Soc;* 1986. p1027-1031.

[61] Kang M.-H., Katsuzaki H., Osawa T. Inhibition of 2,29-azobis (2,4 dimethylvaleronitrile)-induced lipid peroxidation by sesaminols. *Lipids;* 1998. p1031-1036.

[62] Kang M.-H., Naito M., Tsujihara N., Osaw, T. Sesamolin inhibits lipid peroxidation in rat liver and kidney. *J Nutri;* 1998. p1018-1022.

[63] Kang M.-H., Kawai Y., Naito Y., Osawa1 T. Dietary defatted sesame flour decreases susceptibility to oxidative stress in hypercholesterolemic rabbits. *J Nutri;* 1999. p1885-1890.

[64] Kang M.-H., Naito M., Sakai K., Uchita K., Osawa T. Mode of action of sesame ligands in protecting low-density lipoprotein against oxidative damage in vitro. *Life Science;* 2000. p161-171.

[65] Lemcke-Norojarvi M., Kamal-Eldin A., Appelqvist L.-A., Dimberg L.H., Ohrvall M., Vessby B. Corn and sesame oils increase serum g-tocopherol concentrations in healthy Swedish women. *J Nutri;* 2001. p1195-1201.

[66] Fukuda Y., Osawa T., Namiki M., Ozaki T. Studies on antioxidative substances in sesame seed. *Agric Biol Chem;* 1985. p301-306.

[67] Kahae Yamashita, YÃoeÃoeKoohhara, KÃoeMikokatayama,MitsÃoeOnamiki. Sesame seed lignans and y-tocopherol act synergistically to produce vitamin E activity in rats. *Nutrition,* 1992. P2440-2446.

[68] Chen F.A., Wu A.B., Chen C.Y. The influence of different treatment on the free radical scavenging activity of burdock and variations of its active components. *Food Chem;* 2004. p479-484.

[69] Fuchigami M., Kishigami Y., & Sasaki A. Pectic polysaccharides in edible burdock root. *J Home Econ Japan*; 1990. p947-962.

[70] Lin S.C., Lin C.H., Lin C.C., Lin Y.H., Chem C.F., Chem I.C., et al. Hepatoprotective effects of ArctiumlappaLinne on liver injuries induced by chronic ethanol consumption and potentiated by carbon tetrachloride tetrachloride. *J Biomed Sci*; 2002. p401-409.

[71] Duh P.D. Antioxidant activity of burdock (ArctiumlappaLinne): Its scavenging effect on free-radical and active oxygen. *J Am Oil Chem Soc*; 1998. p455-461.

[72] Chow L.W., Wang S.J., Duh P.D. Antibacterial activity of burdock (Arctiumlappa-Linne). *Food Sci*; 1997. p195-202.

[73] Dinis T.C., Santosa C.L., Almeida L.M. The apoprotein is the preferential target for peroxynitrite-induced LDL damage protection by dietary phenolic acids. *Free Radic Res*; 2002. p531-543.

[74] Soliman K.F., Mazzio E.A. In vitro attenuation of nitric oxide production in C6 astrocyte cell culture by various dietary compounds. *Proc Soc Exper Biol Med*; 1998. p390-397.

[75] Maghsoumi-Norouzabad L, Alipour B, Abed R, Eftekhar Sadat B, Mesgari-Abbasi M, Asghari Jafarabadi M. Effects of Arctium lappa L. (Burdock) root tea on inflammatory status and oxidative stress in patients with knee osteoarthritis. *Int J Rheum Dis*; 2014.

[76] Englyst K.N., Vinoy S., Englyst H.N., Lang V. Glycaemic index of cereal products explained by their content of rapidly and slowly available glucose. *BJN*; 2003. p329-339.

[77] Zou Y., Chang S.K., Gu Y., Qian S.Y. Antioxidant activity and phenolic compositions of lentil (Lens culinaris var. Morton) extract and its fractions. *J Agric Food Chem*; 2011. p2268-2276.

[78] Roy F., Boye J.I., Simpson B.K. Bioactive proteins and peptides in pulse crops: Pea, chickpea and lentil. *Food Res Int*; 2010. p432-442.

[79] Asgari E., Rahmani k., Taslimi A. Physicochemical properties of conventional and-complementary foods made from wheat or lentil sprouts. *J Food Sci*; 2008. p33-44.

[80] Barbana C., Boucher A.C., Boye J.I. In vitro binding of bile salts by lentil flours, lentil protein concentrates and lentil protein hydrolysates. *Food Res Int*; 2011. p174-180.

[81] Barbana C., Boye J.I. Angiotensin I-converting enzyme inhibitory properties of lentil protein hydrolysates: Determination of the kinetics of inhibition. *Food Chem*; 2011. p94-101.

[82] Garcia-Mora P., Peñas E., Frias J., Martinez-Villaluenga C. Savinase the most suitable enzyme for releasing peptides from lentil (Lens culinaris var. Castellana) protein concentrates with multifunctional properties. *J Agric Food Chem*; 2014. p4166-4174.

[83] Shepherd J., Codde S.M., Ford I., Isles C.G., Lorimer A.R., McFarlane P.W., Mckillop J.H., Packhard C.J. Prevention of coronary heart disease with pravastatin in men with hypercholesterolemia. *N Engl J Med*; 1995. p1301-1307.

[84] Duane W.C. Effects of legume consumption on serum cholesterol, biliary lipids, and sterol metabolism in humans. *J Lipid Res*; 1997. p1120-1128.

[85] Boualga A., Prost J., Taleb-Senouci D., Krouf D., Kharoubi O., Lamri-Senhadji M., Belleville J., Bouchenak M. Purified chickpea or lentil proteins impair VLDL metabolism and lipoprotein lipase activity in epididymal fat, but not in muscle, compared to casein, in growing rats. *Eur J Nutr*; 2009. p162-169.

[86] Yao F., Sun C., Chang S.K. Lentil polyphenol extract prevents angiotensin II-induced hypertension, vascular remodeling and perivascular fibrosis. *Food Funct*; 2012. p127-133.

[87] Low P.A., Nickander K.K., Tritschler H.J. The roles of oxidative stress and antioxidant treatment in experimental diabetic neuropathy. *Diabetes*; 1997. p38-42.

[88] Zahra Aslani, Beitollahe Alipour, Zahra Bahadoran, Farzane Bagherzadeh, Parvin Mirmiran. Effect of lentil sprouts on glycemic control in overweight and obese patients with type 2 diabetes. *Int J Nutri Food Sci*; 2015. p10-14.

[89] Bermond P. Biological effects of food antioxidants. In: *Food Antioxidants* (ed. B.J.F. Hudson), Elsevier science publishers LTD, London and New York; 1990. p193-194.

[90] Polonovski M., Boulanger P., Macheboeuf M. Roche J. *Biochim Med*. Masson etCie, Paris; 1952.

[91] Chance B., Sies H., Broveris A. Hyperoxide metabolism in mammalian organs. Phys Rev; 1979. p527-605.

[92] Harvey J., Dworken M.D. Vitamin E reconsidered. Ann Intern Med; 1983. p253-254.

[93] Ejtahed H., Mohtadi-Nia J., Homayouni-Rad A., Niafar M., Asghari-Jafarabadi M., Mofid V. Probiotic yogurt improves antioxidant status in type 2 diabetic patients. *Nutrition*; 2012. p539-543.

[94] Vaghef-Mehrabany E., Alipour B., Homayouni-Rad A., Sharif S., Asghari-Jafarabadi M., Zavvari S. Probiotic supplementation improves inflammatory status in patients with rheumatoid arthritis. *Nutrition*; 2014. p430-435.

[95] Homayouni A., Payahoo L., Azizi A. Effects of probiotics on lipid profile: A review. *Am J Food Technol*; 2012. p251-265.

[96] Homayouni A., Amini A., Khodavirdivand Keshtiban A., Mortazavian A., Esazadeh K., Pourmoradian Samira. Resistant starch in food industry: A changing outlook for consumer and producer. *Starch/Stärke*; 2013. p1-13.

[97] Homayouni Rad A. Therapeutical effects of functional probiotic, prebiotic and synbiotic foods. *Tabriz Univ Med Sci*; 2008.

Chapter 6

# Antioxidants and Natural Compounds in Mexican Foods

José Luis Silencio Barrita, Sara Montaño Benavides and Santiago Sánchez

Additional information is available at the end of the chapter

http://dx.doi.org/10.5772/61626

**Abstract**

In Mexico there is a quality of climate and land suited to the cultivation and production of a wide variety of fruits and vegetables rich in natural antioxidants. Although these fruits and vegetables contain sufficient antioxidants, consumption is low, especially in at-risk populations such as children, pregnant women and the elderly. Fast food on the street and in restaurants is preferred to food at home, and more fruits are consumed than green vegetables. In virtually all social strata there is a marked preference for the consumption of fast food with a high content of saturated fat, salt, cholesterol, protein and simple sugars. The consumption of raw or cooked green vegetables has declined with at best the consumption of a serving a day when the World Health Organization (WHO) suggests at least 3–5 servings of vegetables a day. This decrease in the consumption of natural foods, and therefore in associated antioxidant components has been crucial in the development of chronic degenerative diseases such as obesity, cancer, diabetes and cardiovascular disease. Such diseases are the leading cause of death in the Mexican population according to data from the National Statistical Institute of Geography and Informatics (INEGI, 2013) [1]. On the other hand, there is an excessive consumption of food supplements containing these same natural antioxidants in a purer and more concentrated form than in natural food sources. Such supplements or additions were initially only consumed by athletes, but are now widely used by the general public without an understanding of the normal recommendations and possible toxic effects they may have on the user.

**Keywords:** antioxidants, Mexican foods, vegetables, fruits

## 1. Introduction

Although Mexico has a great variety of climates suited to the growth of a large amount of vegetables, their consumption is limited. National consumption is estimated at only a single serving of vegetable food per day, usually with the midday meal. The most widely eaten vegetable foodstuffs are: tomato, onion, lettuce, chili, green tomato, cucumber, avocado and potato. Some vegetables are used as a condiment in the preparation of food (seasoning) and others as a spice, as in the case of vegetables such as coriander, parsley, epazote, mint and others. The average diet of the Mexican people also includes corn tortillas (maize) and wheat in the form of bread, which become "tacos" when the tortilla is used as a base on the one hand, and "tortas" when bread is used on the other. Added to these are – in raw form – lemon, onion, cilantro and chili. The chili is added in the form of sauces based on tomatoes, or cooked or roasted green tomato.

The quantities of these sauces that are eaten vary, which also makes it difficult to calculate the total amount of antioxidants consumed daily. A person can eat between 10 and 12 tacos – with an average weight of 50g each – per serving, or a cake whose weight can vary between 200 and 500g and whose filling is variable, but may include tomato and onion slices, as well as chili and avocado.

Most of these "tacos" and "tortas" are sold on the street as fast food, and are readily available and cheap [2]. This type of food is eaten between four and five times per week on average, either at breakfast, lunch or dinner.

This type of diet is not particular to any socioeconomic level and can be seen in any social stratum. Other foods included in the daily diet are legumes such as beans, peas, lentils and beans, consumed in large amounts in broths or soups, pasta and rice.

Due to this great variability and irregularity of content, consumption of antioxidants in the population is very low and many that are present are often removed by the effects of heat or frying.

The low consumption of antioxidants in fresh food correlates well with the presence of slowly evolving illnesses and chronic degenerative diseases, which are major causes of morbidity and mortality, as described by the health sector. They are leading causes of death and disease from stroke, cancer, diabetes and obesity.

In the last three decades, the daily consumption of fast food has increased and its preparation eliminates or reduces the amount of antioxidants present. This type of food is represented by pizzas, hamburgers, roast chicken and chicken type "crispí" for example.

Nutrition in Mexico is currently polarized since some areas enjoy a sufficient supply of food while others have an insufficient supply that have been promoted for many years [2]. However, even in areas where there is a good supply of food, consumption is often restricted to foods that promote the development of chronic degenerative diseases. Such is the case with the huge consumption of sugars and flours in foods that have contributed to the rise in metabolic syndrome in the past decade and have therefore triggered diseases such as diabetes. In addition to the increased consumption of sugars, are large amounts of saturated fatty acids (saturated fat) and salt which are promoters of obesity and hypertension, which further contribute to the current state of public health in Mexico. These same food combinations produce an organoleptic effect, which has become, unfortunately, the basis of the national diet. The consumption of nutrients that have a major impact on the overall health of the individual, such as some vitamins, many minerals and long-chain polyunsaturated fatty acids is neglected. Of the latter, omega-3 fatty acids (n-3, Morris system) are the least consumed by Mexicans.

http://dx.doi.org/10.5772/61626

The most important sources, as will be discussed later, are marine fish and seafood (algae). However, in Mexico, little fish is eaten. It is estimated that the per capita consumption of fish is lower than 10 g per year, and does not meet international recommendations. The WHO has recommended the consumption of at least three servings of fish per week to cover more than 50% of the current recommendation. Vegetables are another important source, but the consumption of vegetables is affected by their low general popularity [4].

Due to these considerations and the properties of n-3 fats, it is likely that there is a positive correlation between the prevalence of chronic degenerative diseases and a deficiency in their consumption. In addition, symptoms observed in diabetic, hypertensive and obese patients are probably better explained by the failure to consume this type of fat, and especially by the physiological, vascular, blood and cellular functions this molecules that express.

Chronic noncommunicable diseases (CNCD) are one of the biggest challenges facing the Mexican health system. There are several factors involved: the large number of cases, their increasing contribution to overall mortality, forming the most frequent cause of premature disability, and the complexity and high cost of treatment. CNCDs are a heterogeneous group of conditions contributing to death through a small number of outcomes (diabetes, cardiovascular diseases and cerebral vascular disease). The deaths are the result of a process begun decades earlier. The natural evolution of diabetes and cardiovascular disease can be modified by actions that change the clinical course of the conditions that determine its incidence. These include being overweight and obese, abnormal concentrations of blood lipids, hypertension, smoking, a sedentary lifestyle, improper diet and metabolic syndrome. This introduces opportunities for prevention, the development of prognostic tools and the creation of macroeconomic models. And it is here where food, or the poor intake of foods, is linked to the high incidence and frequency of these chronic conditions. It is not only deficiencies in consumption that can lead to these clinical spectra, but also excessive consumption.

Variations in the prevalence of obesity, overweight, dyslipidemia, hypertension and metabolic syndrome in the period covered by National Health Surveys (1994–2006) are already documented. The percentage of the population with a higher weight than that desirable (>25 kg $m^{-2}$, body mass index [BMI]) rose 13 % in the period from 1994 to 2000 and the change was even greater (33.5%) between the years 2000 and 2006. The same increasing trend was observed in the prevalence of **metabolic syndrome**, a concept that identifies patients with increased risk of developing diabetes or cardiovascular disease in the medium term. The percentage of adults with metabolic syndrome (defined by the criteria of the National Cholesterol Education Program) increased by 27.8% between 1994 and 2000 and 39.7% cases corresponded to individuals under 40 years of age. The data suggest that the contribution to the mortality of CNCD will increase in the medium term [5, 6, 7, 8].

The data suggest that the contribution to mortality by CNCD will increase in the medium term. The percentage of adults with diabetes (diagnosis established by a doctor) grew by 25% in 2006. It has been projected that 11.7 million Mexicans will have diabetes by 2025. Each national survey shows a rise (4.0 to 5.8% in the period 1994–2000) and from 5.8 to 7% from 2000–2012. Type 2 diabetes is a major cause of premature disability, blindness, end-stage renal failure and non-traumatic amputations. It is one of the 10 most frequent causes of hospitalization in adults.

The last link in the chain is evaluated by the effect of CNCD on mortality. The percentage of mortality attributed to outcomes associated with CNCDs has shown continuous growth. In just four years (2000–2004), the proportion of overall mortality explained by diabetes and cardiovascular diseases rose from 24.9 to 28.7% in men and 33.7 to 37.8% in women. As a result, since 2000 ischemic heart disease and diabetes have been the two most frequent causes of death in Mexico [9, 10, 11].

In relation to cancer:

Cancer is a disease with a major psychological, social and economic impact. According to a preliminary analysis by Delgado Gutierrez et al., between 2000 and 2010, there was an average of 66, 000 deaths annually in Mexico from cancer, equivalent to 13% of annual national mortality. From the total number of deaths from cancer, women represent – on average – 51% annually. The economically active population (aged 15–64) comprises 43%, and the population aged 65 years or older, 54%. During this period, lung cancer was the cause of highest mortality (6, 701) followed by malignant neoplasms of the stomach (5, 298) and liver (4, 819). Among cancers of the reproductive organs, prostate cancer generated the highest number of deaths (4, 690), followed by breast (4, 321) and cervix (4, 236). These six tumors caused 46% of the total number of deaths from cancer [12, 13].

There have been efforts to elucidate the number of new cases of cancer at the national level, by recording histopathology of malignant neoplasms (RHNM). However, these have been isolated and institutional and have not been systematized. The RHNM represented an initial effort through a hospital registry, but it ceased operation in 2002. The idea was for it to evolve towards a population-based register, whereby it would be possible to measure the real extent of the disease. The commitment is still pending. Due to the size of Mexico, the high prevalence of risk factors and the rapid aging of the population, it is necessary to create several regional registries that are population-based, inter-agency and that have access to uniform and systematic information enabling the recording, year after year, of all the new patients who receive a diagnosis of a malignancy. We know that up to four out of 10 cases of cancer are preventable, and that 30 patients could be cured or achieve disease control if their cancer is detected early and they are referred to receive optimal treatment [13]. In Mexico, efforts in primary prevention and early detection are poorly-organized and the lack of an adequate budget results in a lack of timely performance and quality. This largely explains why more than 70 cases of cancer are diagnosed only in the advanced stages, reducing the chance of recovery and at the same time generating the broad needs of palliative care to improve the quality of life of the patients.

### 1.1. Morbidity

Globally, it is estimated that in developed countries, three in four children with cancer survive at least five years after their treatment, but in developing countries about 60 will die in the same period (National System of Epidemiological Surveillance SINAVE, 2011). In general, younger children have better survival rates, which might be because there is less delay in diagnosis, treatment and its success among teenagers and young adults. This is due in part to the fact that young children have greater monitoring through their parents and the health system, so any warning signs of developing disease, are noticed more quickly. In children, about five of the malignant tumors are related to hereditary mutations passed from parents to children, although not necessarily in all cases will the disease develop. An example of this is retinoblastoma, a cancer of the eye that occurs more frequently in children, and that parents identify through observing a whitish reflection in the eye of the child. In addition to the hereditary factor, there may be genetic mutations during fetal development that predispose individuals to cancer at early ages, as has been observed in leukemia, where one in every 100 children is born with a genetic anomaly and has greater risk, although only one in 8, 000 develops the disease (National Cancer Institute, NCI, 2014). Signs and symptoms that allow early detection of cancers in the population under 20 years, include the following (Secretary of Health of the State of Veracruz, 2014): prolonged fever without an obvious cause; an increase in volume of any part of the body; weight loss; unexplained red spots or bleeding anywhere on the body, and general malaise [14].

It is important to note that symptoms will depend on the type of cancer, but the above are indicative of those symptoms requiring the child to be presented to the health services. In Mexico, analysis of hospital morbidity from cancer (patients who died in a hospital from the disease) during 2012 showed that in both men and women under 20 years of age, cancer of the hematopoietic organs was the leading cause of death, (59.5% in women against 58.8% in men). For males, the second leading cause of hospital discharge for cancer is the lymphatic system and related tissues (8.8%), followed by the brain and other parts of the central nervous system (6.3%). However, in women, the order is reversed with brain cancer and other parts of central nervous system the second commonest cause (6.7%) and the lymphatic system and related tissues the third (5.9%) [14, 15].

It is not uncommon to think that the high incidences of chronic degenerative diseases are correlated with a deficiency in the consumption of some nutrients that could be the cornerstone of the protection of the patient, and within these it is very likely that an imbalance in the consumption of an antioxidant and pro-oxidant, or its excess, that are the factors that trigger these diseases. Such is the case in the poor consumption of omega 3 fatty acids, selenium, zinc, iron, and vitamin C.

## 2. Antioxidants

Human beings are exposed to a large number of "oxidizing agents" such as pollution, stress, cigarette smoking and some other chemical compounds used in improvement or preservation of food. In addition, our body produces so-called "free radicals" (FR), which can cause oxidation of membranes and DNA damage, triggering a series of undesirable reactions that lead to diseases such as cancer, cardiovascular problems and aging. Antioxidants are compounds which, by their chemical structure, can slow the formation of free radicals and prevent or treat the above-mentioned diseases caused by oxidative stress. This oxidative stress is a consequence of environmental pollution, stressful jobs, indiscriminate consumption of processed foods, excessive consumption of snacks and fatty foods and is encouraged by the presence of chronic diseases, self-medication and physical inactivity.

In order to quantify the antioxidant capacity of foods, the Oxygen Radical Absorbing (ORAC) test was developed. In this chapter, we show the concentrations of important antioxidants as measured in our laboratory, in raw and cooked foods, with a special emphasis on vegetables.

### 2.1. Micronutrients: Vitamins and minerals

In children under five years of age, dietary surveys indicate that the diet is energy deficient. The diet varies between 73 and 83% of needs, as well as having a low consumption of calcium, vitamin A, riboflavin and vitamin C (with averages of ingestion about 50% of recommended amounts). Although intakes of iron were greater than the dietary recommendations for other age groups, a large proportion came from corn and beans, foods that also contain high amounts of phytates and tannins, which inhibit the absorption of iron. This explains the presence of anemia in the population despite high intakes of iron. It should be noted that the methodology for collecting information of reminder 24 hours diet tends to underestimate the consumption of nutrients by between 15 and 20%, so it is possible that the deficiencies in energy and nutrients referred to were lower than reported. There were areas where the majority of the adult inhabitants were underweight, while in other areas, especially urban, there was a prevalence of being between 5 and 15% overweight. In particular, a group of workers in Mexico City was identify as 28% overweight. Also, the presence of anemia in women living in coastal areas was 20%. In rural regions and areas, the population had less variety in their diet, which was low in protein of animal origin and deficient in vitamins. The biggest nutritional problems were presented in the southern and southeastern regions of the country, followed by the downtown area, and to a lesser extent, on the coasts and the north of the country. Clinical signs of malnutrition and micronutrient deficiency are presented in severe conditions. Clinical signs of deficiency, such as hair pigmented and easy to boot (8%), scaly and pigmented skin (3%), alterations in the eyes (4.7%), smooth tongue (6.3%), cheilosis (6.6%) and edema (1.6%) have been documented in children. The latter is the most reliable sign for recognizing severe malnutrition. A presence of anemia was identified in 29% of preschool children, with 20% in rural and semi-rural areas, and 9% in urban areas.

## 3. Indicators of micronutrients

Anemia was determined using a portable photometer (HemoCue). In order to classify subjects with anemia, the following cut-off points of hemoglobin concentration in blood proposed by WHO were used: children aged 1–5 years, 110 g/L; children aged 6–11, 120 gL; (non-pregnant) women aged 12–49, 120 g/L; (pregnant) women aged 12–49, 110 g/L; men from 12–14 years, 16 mg/dL; zinc deficiency is defined as serum zinc concentration of 65 ug/dL; deficiency of vitamin A as retinol serum 20 ug/dL; folate deficiency as concentrations of folate in red blood cells of 140 ng/mL; and deficiency of vitamin C, such as serum concentrations of Ascorbic acid at 0.2 mg/dL [13].

http://dx.doi.org/10.5772/61626

Anemia and deficiencies of various micronutrients are the commonest nutrition-based problems in Mexico, primarily affecting children and women of fertile age. In Mexico, the most important deficiency in terms of magnitude is that of iron in children of all ages (from 12 months to 11 years), which ranges from 36 to 67%. In women, the prevalence of iron deficiency is over 40%, though it is not the largest of the deficiencies observed in this population. The prevalence of low concentrations of vitamin C, an indication of low dietary intake of this vitamin, is very high and fluctuates between 30 and 40% in the case of children, and reaches 50% in women. The prevalence of low zinc consumption is around 30% in children under five years and women, with little more than 20% of those in school [13, 15]. Marginal vitamin A deficiency is between 20 and 30% in children and less than 5% in women. Folic acid deficiency ranges from 10 to 20% in children and 8% in women.

Studies of micronutrient deficiencies in the 1999 ENN survey showed that the most common is a lack of iron, mainly affecting rural areas of the country. The highest prevalence rates were reported in children under six years of age, while in schoolchildren and women of fertile age no physiologic load had a lower prevalence (INSP, 1999) [15].

In some cases, the consumption of vitamins in tablet form is recommended, but only justified if there is clinical evidence of its deficiency. Otherwise, it is best to consume natural food sources that contain the relevant vitamins. However, in Mexico, an almost magical power is given to vitamins in such a way that some sectors consume them without medical advice and in ignorance of official recommendations and potential toxic effects. They are recommended by stars of television, and used in gyms by personal trainers and the general public, and since a prescription is not required for their purchase they are within the reach of virtually anyone who can visit a hypermarket.

Treatments with vitamins such as folic acid, vitamins C, A, $B_{12}$ and $B_1$ are widely used by the medical sector for the specific conditions that require them. They are necessary for normal metabolism and even for the metabolism of other drugs that are prescribed.

## 4. Vitamin A

Vitamin A is a generic description covering all compounds which show biological activity of retinol, such as some retinoids and carotenoids (also known as pro-vitamin A). Retinoids are isoprenoid compounds normally found in animal and plant tissues. Because of this, vitamin A is found in different forms, but you can tell that there esterifies fatty acids, primarily as retinyl palmitate. It is likely that several forms of the vitamin are present in the same food, but in differing concentrations [16].

### 4.1. Vitamin A structure

The vitamin A molecule contains a β-ionone ring with an unsaturated side chain containing an alcohol group at the end of the chain, characterizing the main vitamer. (Fig.1)

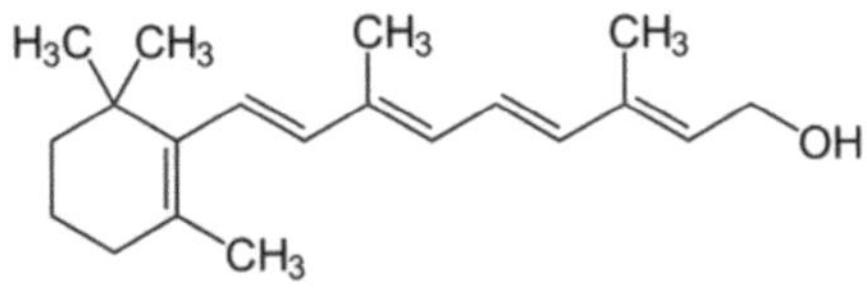


**Figure 1.** Structure of vitamin A (retinol)

### 4.2. Vitamer

Various forms of Vitamin A exist in food and are generally known as retinol alcohol ($A_1$). It is stored in tissues such as palmitic ester. The Isomer of retinol: 3-deshidroretinol (with 40% of the activity of $A_1$), is found in fish liver oils and has been shown to be converted to Vitamin $A_1$ [17] in rats.

The term vitamin A is broad as it includes all those compounds containing a β-ionone ring and designated as retinoids or derivatives of β-ionone, and that is why carotenoids are so called. The ring is very important as the activity of the vitamin depends on it. In fact, when the vitamin is exposed to sunlight, the ultraviolet light present in the sunlight breaks the ring and vitamin activity is lost. In vivo, retinol can be converted to its aldehyde, retinal, and this in turn can be converted to retinoic acid. This is an oxidation process wherein the alcohol (retinol) is oxidized to the aldehyde (retinal) and this in turn is oxidized to acid (retinoic acid). The reaction is thermodynamically irreversible and retinoic acid cannot be converted back to retinal, so the retinoic acid is rapidly oxidized and excreted. This action cannot be considered by itself as an example of an antioxidant reaction. However, the participation of vitamin E, beta carotene and vitamin C in the equilibrium of these reactions involving vitamin A is part of the antioxidant response. Vitamin A lies partly in cell membranes so that it is very likely be involved, like vitamin E, in free radical scavenging, itself becoming a free radical (possibly retinyl). This retinyl radical could purge, with vitamin C, free radicals seeking to initiate lipid peroxidation of the membrane. Retinol has been isolated from the retina of the eye and is the main form as vitamin works, in fact the only function biochemically proved, described by George Wald in 1967.

### 4.3. Sources

Vitamin A (retinol) can be ingested into the body or synthesized from plant carotenoids. Preformed vitamin A is present in animal tissues and the best sources are liver, milk, and the kidneys, where it is found in high concentrations in the form of fatty acid esters [18].

Vitamin activity occurs primarily in plant carotenes, which are precursors of vitamin A. Theoretically, a molecule of β-carotene should give rise to two molecules of retinol, however, under in vivo conditions this does not happen and only one molecule is obtained. Of the more than 100 carotenoids described so far, only 12 are converted to retinol and β-carotene is the most important of these. As a result, efficiency of the conversion of carotenoids in retinol is 1/8 (i.e. 100/12). Therefore, the amount of vitamin A from plant foods should be one-eighth of the concentration of the carotenoids present in the food. Of course this may change and the amount

http://dx.doi.org/10.5772/61626

could increase or decrease, depending on how many carotenoids or β-ionone derivatives are contained in foods. Thus, it is often recommended to ingest some retinol equivalents from a diet containing both animal and vegetable foods [19].

This situation is the result of inefficiency in the absorption of β-carotene from vegetables and fruits, its chemical structure, biological activity and the dynamics of oxidation of retinal to retinoic acid, which is quickly removed.

Table 1 shows the main sources (with highest concentrations) of vitamin A, although in Mexico there are other widely consumed sources with lower concentrations.

| | Food | Content (UI/100g) |
|---|---|---|
| 1 | Dehydrated carrot | 100, 000 |
| 2 | Cod liver oil | 85, 000 |
| 3 | Dried red pepper | 77, 000 |
| 4 | Roasted lamb liver | 74, 500 |
| 5 | Homemade paprika | 60, 604 |
| 6 | Fried beef liver | 53, 400 |
| 7 | Dehydrated sweet potato flakes | 47, 000 |
| 8 | Red pepper | 41, 610 |
| 9 | Chile powder | 34, 927 |
| 10 | Fried beef liver | 32, 700 |
| 11 | Canned pumpkin | 27, 383 |
| 12 | Parsley dried | 23, 340 |
| 13 | Dehydrated alfalfa | 22, 940 |
| 14 | Raw red pepper | 21, 600 |
| 15 | Liver cooked turkey simmered | 17, 500 |
| 16 | Liver cooked chicken simmered in grill or skillet | 16, 375 |
| 17 | Canned carrots, drained | 15, 000 |
| 18 | Fried pork liver in margarine | 14, 900 |
| 19 | Dried peaches, sulfur, crude | 14, 100 |
| 20 | Green dandelion, raw | 14, 000 |
| 21 | Carrots, raw | 11, 000 |
| 22 | Green, raw mustard | 9, 900 |
| 23 | Spinach, raw | 8, 100 |
| 24 | Sweet potato shell, cooked | 8, 100 |
| 25 | Beef cooked in water | 7, 800 |
| 26 | Green radish, raw | 7, 600 |
| 27 | Swiss, raw chard | 6, 500 |
| 28 | Peanut butter | 6, 400 |
| 29 | Dried peaches, raw | 5, 000 |
| 30 | Cantaloupe, raw | 3, 400 |

| | Food | Content (UI/100g) |
|---|---|---|
| 31 | Margarine | 3, 307 |
| 1 UI =0.300 ug retinol | | |

**Table 1.** Sources of Vitamin A

## 4.4. Functions

Preformed vitamin A is present in the diet as retinyl esters. Esters of vitamin A are hydrolyzed during digestion by pancreatic hydrolases, absorbed in their free form and re-sterified with fatty acids within the intestinal mucosa, entering the circulation by passive diffusion with chylomicrons formed. Beta-carotenes from plants are substrates for the synthesis of vitamin A, obtaining retinaldehyde, which is then reduced by a specific enzyme called aldehyde reductase (using NADH as a cofactor), and converted to retinol. Retinol from any source (animal or plant) is stored as retinyl esters in the parenchyma of the liver. The normal body retinol reserve varies from 300–900 mg (1 IU=0.300 ug trans-retinol). Prior to release, esters in the liver are hydrolyzed and the free alcohol binds to a specific binding protein retinol binding protein (RBP), which transports it to peripheral tissues. This protein is a single polypeptide chain of 182 amino acid residues with a molecular weight of 21 KD. It is synthesized initially as a 24 KD pre-RBP by parenchymal cells, and also by post-translational modification of a polypeptide removed (3.5 KD). Secretion is hormonally regulated by circulating estrogens and the same concentration of the vitamin. In healthy individuals, the concentration of RBP in plasma is equal to 40–50 ug/mL. This protein is catabolized in the kidney, however catabolism is reduced by the binding of the protein to retinal [20, 21].

The best known function of vitamin A is its role in vision, in which the prosthetic group is a series of carotenoid proteins providing the molecular basis for visual excitation [22, 23]. The photosensitive chromophore group responsible is 11-cis-retinal, present in a group of visual pigment cells of the eye known as "rods" and "cones". The latter have precision and accessory pigments at wavelengths with maximum absorption peaks in the red, green and blue parts of the spectrum. The "rods" have precision too and contain the pigment **rhodopsin** and cones contain the pigment **iodopsin**. In any case, 11-cis-retinal is bound to a lysine residue of a specific protein called opsin.

When light strikes these molecules, there is a light-induced isomerization transforming the 11-cis-retinal to the all-trans-retinal form. This reaction results in the dissociation of the protein-vitamin complex. This dissociation is coupled to nerve centers in the brain that respond to vision stimulation, so that each nerve impulse caused by each of the breaks of rhodopsin molecules produces an image point. The union of all these points in the brain leads to the display of an image. This complex and efficient phenomenon occurs millions of times per second each time we open our eyes. Thus we can say that the process of vision is a cyclical process as its constituents thereof are reconstituted. The all-trans retinol can convert enzymatically in the dark to the cis form, which is then esterified and stored, presumably among the lipids of epithelial cells. Rhodopsin regeneration includes the hydrolysis of retinyl esters to produce 11-cis-retinol which is transported (via RBP) to the outer part of the stick cell which is then oxidized to 11-cis-retinal [24].

The "extra-retinal" functions are less important than the visual function and are poorly understood. They are collectively referred to as vitamin systemic functions as they use retinoic acid more effectively than retinal, and secondly their biochemistry is very different and even more complex. Vitamin A is also involved in cell differentiation, morphogenesis, transmembrane transport and the immune response. A significant portion of absorbed vitamin A (40%) is oxidized or conjugated in the liver and secreted in bile through feces. Retinoids apparently affect cell differentiation in a manner analogous to steroid hormone action, by binding to nuclear chromatin for transcription processes signals. It has been proposed that with retinoic acid they synergistically stimulate the production of thyroid hormones and growth hormone in cultured pituitary cells [25, 26].

Its role in cell differentiation is based also because it shares regions of binding to ligands and DNA along with vitamin D3, steroid hormones and thyroid hormones (T3). These ligand-receptors for retinoic acid (RARa and RARβ) bind to promoter regions of specific genes activating transcription. It is also proposed that vitamin A plays a co-enzymatic role, as a carrier of sugars in the synthesis of glycoproteins, whose functions at the cell surface are adhesion, aggregation, recognition and other interactions.

Vitamin A is necessary for reproduction, but the biochemical basis for this is unknown. Experiments in rats indicate that retinal is more responsible for these actions in mammals than retinoic acid. Rats subjected to retinoic acid grow well and look healthy, but are unable to reproduce, and females are unable to get pregnant, abort and reabsorb their products while males show damage to the spermatogenesis process. Injections of retinol into the testes restores spermatogenesis, indicating their direct role [27].

Because vitamin deficiency results in damage to the differentiation of epithelial cells without damage to the proliferation, its role in the etiology of epithelial tumors, including carcinomas is questioned. In fact, squamous metaplastic changes seen in vitamin deficiency are similar to experimentally-induced precancerous lesions. Therefore it is proposed that retinol – as an anticarcinogen – acts against the high levels of carcinogens in tumors and prevents the expression of malignant phenotypes. Thus the hypothesis that the nutritional status of vitamin A can increase or decrease the risk of cancer is supported.

Although it also is not very clear, there is support for the idea that vitamin A is important in immunocompetence. Vitamin deficient animals are more susceptible to infection than those with a sufficient intake of the vitamin. Epidemiological studies indicate that a state of deficiency is closely related to a higher incidence of disease and increased mortality. Vitamin deficiency affects immunity through different routes. Retinoids act on the differentiation of immune cells, increasing lymphocyte mitogenesis and phagocytosis of monocytes and macrophages. Carotenoids affect the survival of activated NK cells and T helper cells, modifying the release of cytosine-type products from activated lymphocytes and monocytes [28].

### 4.5. Deficiencies

Liver and tissue levels of vitamin A compensate for low intakes of it in countries like Mexico. The speed with which the body's stores are mobilized, depends on the severity of the deficiency and how fast are released it between different tissues. Because the vitamin is present in different organs, a deficiency will include many nonspecific signs. One of these signs is related to vision, and there are ocular signs such as nyctalopia and xerophthalmia. The first of these responds quickly to therapy with the vitamin, while the second disorder involves permanent changes to the eye, which cannot be corrected. However, in cases of xerophthalmia, timely therapy can disrupt early-stage lesions before they become permanent injuries [29].

Vitamin A levels are decreased in night blindness, childhood blindness, hypothyroidism, infertility and teratogenesis, liver disease, xerosis, chronic infections, and pyrexia. A concentration lower than 10 mg of vitamin A/dL serum is indicative of severe hepatic impairment and absence of reserves. It also decreased in disseminated tuberculosis, carcinoid syndrome (very low) protein malnutrition, alpha-beta-lipoproteinemia and cystic fibrosis. Nyctalopia (night blindness) is a condition in which objects appear clear by day but are unclear at night. Poor dark adaptation is the best sign of vitamin A deficiency [30].

## 5. Vitamin E

### 5.1. Definition

Vitamin E is defined as all those forms exhibiting activity of α-tocopherol, which is the main representative of this group of compounds. The α-tocopherol is a derivative of phytyl and trimethyl hydroquinone alcohol.

The basic structure of this vitamin is chromanol ring and a phytyl tail in the case of tocopherols, and an unsaturated tail in the case of tocotrienols. See Figure 2.

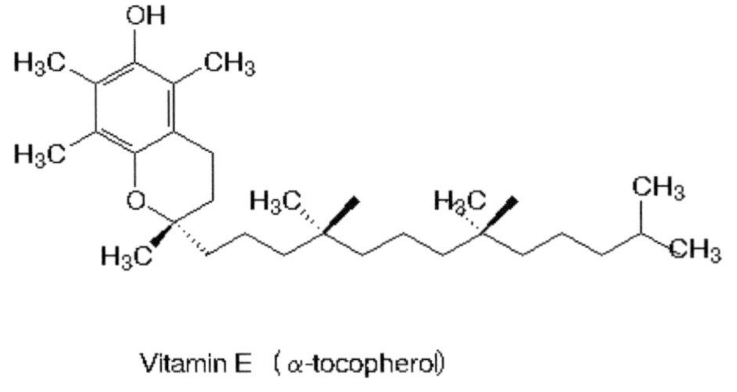


**Figure 2.** Vitamin E structure

Also known as tocopherol (α-tocopherol is the most powerful and abundant antioxidant in foods). These compounds are potent fat-soluble antioxidants that protect the integrity of cell

http://dx.doi.org/10.5772/61626

membranes (fig 3) and are associated with rich sources of vegetable oil. If no free vitamin E is present, free radicals oxidize fatty acids of the cell membrane and thus affect cellular structure [31]. This is the basis for the suggestion that Vitamin E important in preventing both aging and cancers. Vitamin E increases the absorption of vitamin A because it prevents oxidation in the intestine. In addition, its antioxidant action increases in the presence of zinc. In foods, vitamin E is found mainly in vegetable oils, wheat germ oil seeds (walnuts, peanuts, almonds, etc.), and green leafy vegetables, etc. Absorption is relatively poor (20–80%) and is linked to dietary lipids. Vitamin E is stored in liver and adipose tissue and is absorbed through micelles. Its use depends on the presence of adequate fat and pancreatic and biliary function. Once absorbed, it is captured by chylomicrons and very low-density lipoproteins (VLDL), which are found in the lymph. Vitamin E stored in the liver is subsequently reattached to VLDL and HDL, to be taken up by cells and is mainly directed to lipid membranes. It is excreted primarily via the bile and feces, and to a lesser extent in the urine [32].

### 5.2. Vitamin E sources

Vitamin E is found in large quantities in plants. It is present in high concentrations (0.1–0.3%) in wheat germ, maize, sunflower seed, rapeseed, soybean, alfalfa, and lettuce. Alpha-tocopherol is usually found naturally as beta and gamma tocopherols. The pure form of the vitamin is a pale yellow viscous oil [33].

| | **Food** | **Content (mg/100g)** |
|---|---|---|
| 1 | Wheat germ oil | 149.4 |
| 2 | Sunflower seed oil, 60% linoleic acid and higher | 44.9 |
| 3 | Almond oil | 39.2 |
| 4 | Cottonseed oil | 35.3 |
| 5 | Oil, linoleic acid above 70% | 34.1 |
| 6 | Alfalfa seeds | 33.0 |
| 7 | Rice bran oil | 32.3 |
| 8 | Margarine, salt, liquid oil | 28.4 |
| 9 | Dry almonds, with entire deck | 27.9 |
| 10 | Whole hazelnuts, shelled | 21.0 |
| 11 | Cod liver oil | 20.0 |
| 12 | Palm oil | 19.1 |
| 13 | Corn oil | 14.3 |
| 14 | Sunflower seed essence, dry shell | 13.0 |
| 15 | Olive oil | 11.9 |

| | Food | Content (mg/100g) |
|---|---|---|
| 16 | Peanut oil | 11.6 |
| 17 | Soy oil | 11.0 |
| 18 | Fat, vegetable | 9.9 |
| 19 | Roasted peanuts in shell, whole and savory | 9.7 |
| 20 | Wheat, soya blend and wheat flour | 7.0 |
| 21 | Commercial chips | 6.4 |
| 22 | Canned tuna in oil | 6.3 |
| 23 | Black, raw moras | 3.5 |
| 24 | Asparagus, boiled | 2.5 |
| 25 | Porridge | 2.3 |
| 26 | Green radish, raw | 2.2 |
| 27 | French fried shrimp, seasoned with breadcrumbs and batter | 1.9 |
| 28 | California avocado, raw shell | 1.7 |
| 29 | Rye, whole grain | 1.7 |
| 30 | Butter, regular, salt | 1.6 |

**Table 2.** Sources of Vitamin E

### 5.3. Functions

Vitamin E is absorbed from the gastrointestinal tract by a mechanism similar to other fat soluble vitamins, and it enters the blood initially via associated lymph chylomicrons, and then with lipoproteins, which correlates very well with the concentration of vitamin E and plasma lipid levels. It is stored in many tissues, residing mainly in the non-polar fraction of the lipid membranes in the center of the lipid bilayer. This feature protects against a body storage deficiency for long periods and is a basis for operation [34].

Vitamin E functions as an important natural antioxidant that prevents the spread of damage caused by free radicals to biological membranes. An important consequence of deficiency is that subjects suffer from anemia and peripheral nephropathy. The vitamin is an excellent "scavenger" of peroxyl radicals and preferentially protects polyunsaturated fatty acids (PUFAs) in the phospholipids of biological membranes [35].

When lipid hydroperoxides are oxidized to peroxyl radicals (ROO*) these react 1000 times faster with vitamin E (VIT E-OH) than PUFAs (RH). The hydroxyl group of vitamin E reacts with an organic peroxyl radical to form an organic hydroperoxide and the tocopheroxyl radical (VIT EO*). Hence vitamin E has become a free radical preventing autoxidation membrane lipids. See Figure 3.

http://dx.doi.org/10.5772/61626

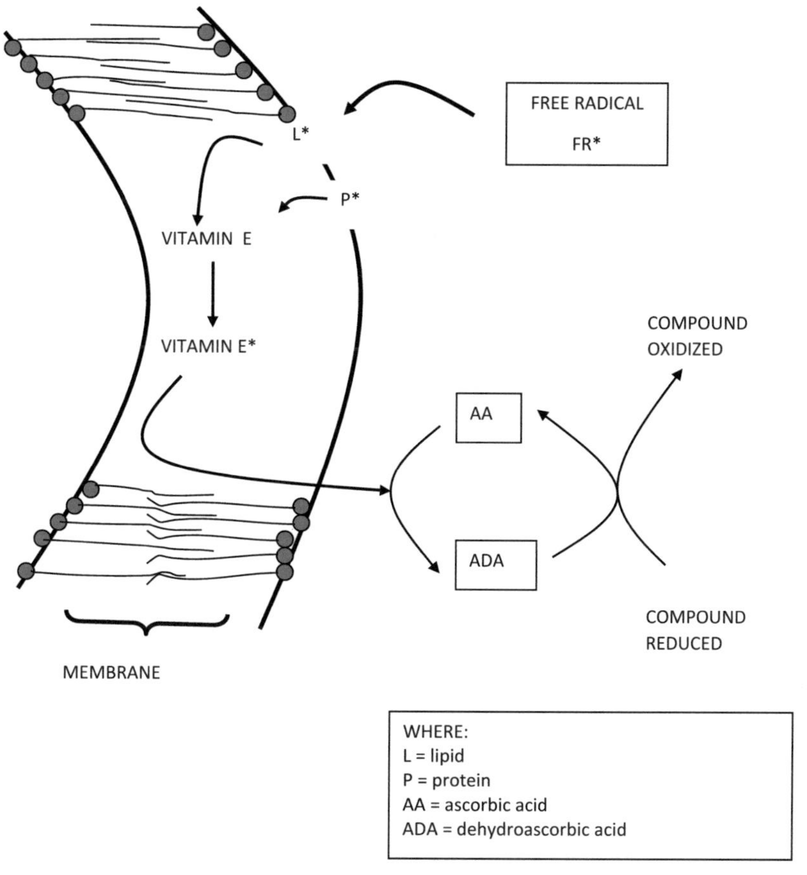


**Figure 3.** Functions of vitamin E

The formed tocopheroxyl radical expose its electron outside the lipid bilayer, into the aqueous part, to receive a part of any hydrogen donor reducting, such as vitamin C, which accepts and donates the electron hydrogen from vitamin E, being this in reduced form. This replenishes the vitamin so that its function becomes cyclical. Other compounds have been demonstrated to regenerate vitamin E and eliminate the symptoms of vitamin deficiency. These are selenium, glutathione, ubiquinone and other thiol groups (some amino acids with sulfhydryl groups such as cysteine). Thus the antioxidant activity of the vitamin depends not only on its concentration, but also on antioxidant compounds present in the intra- and extracellular water phase. The antioxidant status will also include the correct function related to maintaining an optimal state of oxide reduction for cell enzymes, which are also important activities of the enzymes catalase, glutathione peroxidase and superoxide dismutase.

Cells use superoxide dismutase, catalase and glutathione peroxidase to protect against reactive oxygen species. At the same time the body is careful to keep all iron and copper ions bound on protein transport or storage. Thus there is approximately three to four times more iron binding capacity in transferrin than in plasma, so there should be no free iron ions. Combining elements such as iron with transferrin protein cannot stimulate lipid peroxidation and the generation of hydroxyl radicals, and this is also true for copper bound to albumin or ceruloplasmin.

All of these are regarded as primary defense mechanisms against free radicals, but there are also secondary antioxidant defenses and alpha tocopherol is the most important of these because it acts as a chain-breaking antioxidant. Vitamin E is a lipid soluble molecule, found in biological membranes or LDL, and possesses a hydroxyl group whose hydrogen is easily removed. When alkoxy or peroxyl radicals are generated during lipid peroxidation, these combine preferentially with the antioxidant rather than an adjacent fatty acid.

The alpha tocopherol chain reaction ends and a new free radical is formed, which is less reactive and attacks the adjacent fatty acids less aggressively. There is evidence that this radical tocopherol migrates to the surface of the membrane and becomes alpha tocopherol through reaction with ascorbic acid. Therefore, vitamin C and alpha tocopherol help to minimize the effects of lipid peroxidation in LDL and membranes (see Figure 3).

### 5.4. Deficiencies

Vitamin E deficiency is very rare in humans and is unlikely to be of dietary origin due to the wide distribution of vitamin E in many foods. However, deficiency can occur to due to genetic abnormalities such as a deficiency in the synthesis of tocopherol transfer protein, malabsorption syndromes of fats as seen in children with chronic cholestasis, and pancreatic insufficiency in cystic fibrosis, short bowel syndrome, estatorrea chronic and total parenteral nutrition [36, 37, 38].

The primary manifestations of vitamin E deficiency include areflexia, ataxia, myopathy and pigmentary retinopathy. Hemolytic anemia develops in children if they are fed with a diet low in vitamin E during the first eight weeks of life. Vitamin E deficiency in chickens causes a generalized exudative diathesis, limited eye movement, and possibly paralysis.

### 5.5. Requirements and recommendations

The requirement decreases if the diet is high in polyunsaturated fatty acids and antioxidants. It has been estimated that the minimum daily requirement of tocopherol is 3–4 mg/day. However, it known that 7.3 mg eq tocopherol/day is required to reduce the peroxidation of membrane lipids [39]. The daily recommended intake is 10 mg for men and 8 mg for women. The recommended daily amount (RDA) is contained in 5 mL of polyunsaturated oil.

## 6. Vitamin C

A diet including a sufficient intake of fruit and vegetables is almost always associated with beneficial health effects and a decreased risk of disease. All fruit juices are beneficial, particularly citrus juices, with orange, lemon, tangerine and grape being the best sources of vitamin C. The whole of Mexico is suitable for the cultivation of these citrus fruits, and chili peppers, which also have a high Vitamin C content. Studies to date, indicate that vitamin C and fruit juices rich in vitamin C improve blood lipid profile, reduce oxidative stress and prevent atherogenic modification of LDL and platelet aggregation, in addition to improving HDL levels. Vitamin C has many important roles in health, ranging from its effects on the immune system to its intra- and extracellular molecular effects [39, 40].

Ascorbic acid (AA), commonly known as vitamin C, plays an important role in the human body, although its function at the cellular level is not yet clear. It is necessary for the synthesis of collagen, a protein that serves many functions in the body's connective tissues. Among the substances and structures containing collagen are bone, cartilage and surrounding material, as well as excipients and bonding materials between muscle, skin and other tissues. AA is also required for the synthesis of hormones, neurotransmitters and the metabolism of certain amino acids and vitamins. It is also required in the liver for the detoxification of toxic substances and in the blood for immunity. As an antioxidant, histamines and AA react with peroxides to reduce inflammatory symptoms [41].

Its antioxidant capacity is associated with reduced incidence of cancer. The requirements of vitamin C for adults are well defined, but these have not been uniform across different cultures, so its need has been defined in a culture-specific manner.

Some epidemiological data have stated its usefulness in reducing colds through increased consumption of foods rich in the vitamin, and people sometimes consume an "overdose" of it. Most reports mention that slight increases in levels of vitamin C in the blood reduce the risk of death in all conditions. Although vitamin C has many functions, its role in health is almost always discussed in relation to its function as an antioxidant and its effects on cancer, blood pressure, immunity, drug metabolism, and the urinary excretion of hydroxyproline [42].

Antioxidants have important roles in cell function and have been implicated in processes associated with aging, vascular disease, inflammatory damage and cancer. In its antioxidant role, AA is useful because it contributes to the maintenance of the vascular system and the reduction of atherogenesis through regulation of collagen synthesis and the production of nitric oxide and prostacyclin. In addition to an antioxidant role, AA acts at the molecular level as a cofactor for enzymes such as dopamine hydroxylase (EC 1.14.17.1), influencing the concentration of neurotransmitters, improving lysosomal protein degradation and mediating the consumption of monosodium glutamate (MSG) [43].

### 6.1. Sources

Vitamin C is one of the main constituents of fruits and vegetables, which also contain citric acid, oxalates and substances such as anthocyanins, carotenes and dyes that make it difficult to quantify amounts present when colorimetric methods are used.

There is currently a great interest in the consumption of natural foods, particularly with respect to the nutrient content of fruits, vegetables and vitamin C. This interest is due in part to the widespread use of vitamin C in the food and pharmaceutical industries. It is used as a supplement, additive, conservative, and antioxidant in processed foods. Table 4 shows the main foods that are good sources of vitamin C [44].

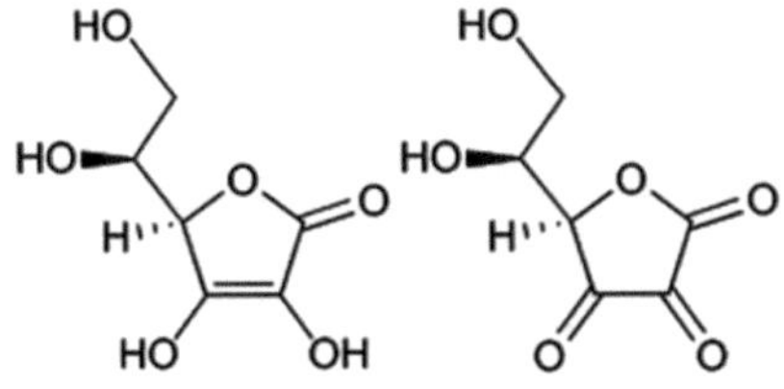


**Figure 4.** Structure. (1) Ascorbic acid and (2) dehydroascorbic acid

### 6.2. Requirements and recommendations

The recommended daily intake needed to avoid the appearance of symptoms of diseases caused by deficiency of vitamin C is 60–100 mg. Infants require a little over 100 mg / day. There is controversy over the required minimum amount of this vitamin. We must take into account that this vitamin is very heat labile in the presence of oxygen. Ascorbic acid is used in the treatment of scurvy, with dosage requirements best measured by determining the urinary excretion after a loading dose. Depending on the rate at which saturation is required, the recommended daily dose ranges between 0.2–2.0 g/day. In cases of vitamin C deficiency, tissue saturation is achieved with three daily doses of 700 mg c/u, over three days. Harris defined saturation of Vitamin C as sufficient storage in tissues so that excretion of 50 mg or more of ascorbic acid occurs in a period of 4–5 hours after a dose of 700 mg/day [45].

Decreased levels of vitamin C in smokers are explained by consumption of fewer sources of the vitamin. In smokers, 50% more than the recommended dose of vitamin is required. Because the RDA is defined as the average daily intake that is sufficient to meet the nutritional requirements of healthy individuals in a population group, it is necessary to continually reassess recommendations for vitamin C. All studies concerning recommended intake, suggest that 90–100 mg is sufficient for maximum reduced risk of chronic disease in non-smoking men and women, although some reports suggest amounts of up to 120 mg/day [46].

### 6.3. Antioxidant activity

Vitamin C is an important water-soluble antioxidant present in biological fluids. An antioxidant is defined as "any substance that when present at low concentrations compared with those of the oxidizable substrates (e.g., proteins, lipids and carbohydrates and even nucleic acids) significantly delays or prevents oxidation of that substrate." The definition given by the Panel of Dietary Antioxidants and Related Compounds in the Food Nutrition Board is that "a dietary antioxidant is a substance in foods that significantly decreases the adverse effects of reactive oxygen species (ROS), reactive nitrogen species (RNS) or both, on normal physiological function in humans."

Vitamin C rapidly clears reactive oxygen and nitrogen species such as superoxide, hydroperoxyl radicals, aqueous peroxyl radicals, oxygen "singlet", ozone, peroxy nitrite, nitrogen dioxide, nitroxide radicals, and hypochlorous acid, so it effectively protects other substrates from oxidative damage [46, 47].

Although AA reacts rapidly with hydroxyl radicals (constant speed > 109 L mol-1 s-1) it struggles to purge this radical preferentially over other substrates. This is because the hydroxyl radicals are very reactive and so combine with any substrate in their immediate environment, close to a speed limited by diffusion. Vitamin C may also act as a co-antioxidant to regenerate $\alpha$-tocopherol (vitamin E) from $\alpha$-tocopheroxyl radicals produced when purging liposoluble radicals. This is a potentially important role because in vitro experiments have shown that $\alpha$-tocopherol can act as a pro-oxidant in the absence of a co-oxidant such as Vitamin C. However, the in vivo interaction between the two vitamins is not yet clear. AA can regenerate urate, glutathione, and β-carotene in vitro from their oxidation products with an unpaired electron (urate radicals, glutathionyl radicals, and cations of β-carotene radicals) [48, 49].

Two important properties of vitamin C make it an ideal antioxidant. The first is its low ascorbate reduction potential (282 mV) and its oxidation product with an electron, the ascorbyl radical (2174 mV), which is derived from the functional group *en-diol*. These low reduction potentials of ascorbate and the ascorbyl radical, are potentially suitable for redox reactions, which is why vitamin C acts as a water soluble terminal antioxidant molecule. The second property that makes it an ideal antioxidant is its stability and the low reactivity of ascorbyl formed when the ascorbate radical purges reactive oxygen and nitrogen species.

| **Food** | **Concentration (mg / 100 g EP)** | **Moisture (%)** |
|---|---|---|
| Squash | 7.2±0.9 | 94.2 |
| Spinach | 8.5±0.2 | 91.6 |
| Potatoes | 74.6±6.7 | 79.7 |
| Cucumber | 93.0±7.1 | 96.0 |
| Green tomato | 222.8±10.7 | 90.9 |
| Poblano chile | 191.0±7.8 | 91.9 |
| Green pepper | 195.5±9.5 | 94.3 |
| Nopales | 268.9±30.1 | 95.2 |
| Cambray onion | 17.0±0.7 | 93.0 |
| Carrot | 50.4±5.6 | 87.2 |
| White cabbage | 184.7±17.2 | 93.2 |
| Grapefruit | 261.3±10.7 | 87.8 |
| Mango | 319.6±5.3 | 84.6 |
| Watermelon | 56.2±8.9 | 92.5 |
| Banana | 333.7±6.3 | 76.5 |

| Food | Concentration (mg / 100 g EP) | Moisture (%) |
|---|---|---|
| Orange | 279.8±39.4 | 84.0 |
| Mamey | 31.6±1.2 | 66.8 |
| Plum | 331.1±17.9 | 88.1 |
| Grape | 66.1±13.6 | 84.3 |
| Apple | 458.1±17.8 | 84.5 |
| Beet | 70.8±6.8 | 85.3 |
| Lemon | 39.4±2.5 | 88.3 |
| Avocado | 256.2±61.9 | 84.3 |
| Sweet lime | 306.8±23.4 | 89.8 |

**Table 3.** Content of ascorbic acid in Mexican fruits and vegetables [49]

EP= edible portion

# 7. Phenolic acids and polyphenolic compounds

Phenolic compounds are a large family of nutraceuticals that have beneficial health properties, ranging from inhibiting the spread of cancer, prevention of atherosclerosis, strokes and heart attacks, and inflammation. They are divided into three groups:

**a) Isoflavones**: a group of compounds present in some vegetables, especially soybeans. Within this family of isoflavones are daidzein, glycitein and genistein. Isoflavone intake plays an important role in reducing the risk of disease. Its antitumor, anticancer, antioxidant and immune response-enhancing actions have been shown. It also important in reducing cardiovascular risk and symptoms associated with effects of the menopause. Isoflavones exist in foods in their conjugated form. When ingested, these conjugated isoflavones undergo hydrolysis by β-glycosidases in the intestine to produce bioactive aglycones (daidzein and genistein). These can absorb or metabolize much more than other specific metabolites such as equol. The further metabolism of aglycones appears to be strongly influenced by diet. When in a high carbohydrate environment, increased intestinal fermentation occurs, resulting in more isoflavones being transformed into equol. This is relevant because the potency of equol is higher than its precursor daidzein. Intestinal bacteria also have an effect on the metabolism of isoflavones. When the intestinal flora is low (e.g. with the use of antibiotics in newborn babies) metabolism also falls. In germ-free animals it has been shown that isoflavones are not present in blood or bile. As endogenous estrogen (estradiol), isoflavones are metabolized in the gut and liver. Absorption occurs throughout the intestine and in secreted bile and urine. The excretion of isoflavones can vary greatly between individuals, being influenced by the fact that each person has their own specific intestinal microflora. Once absorbed, equol shows less affinity is limited to whey proteins and thus has a greater availability than estradiol [50, 51].

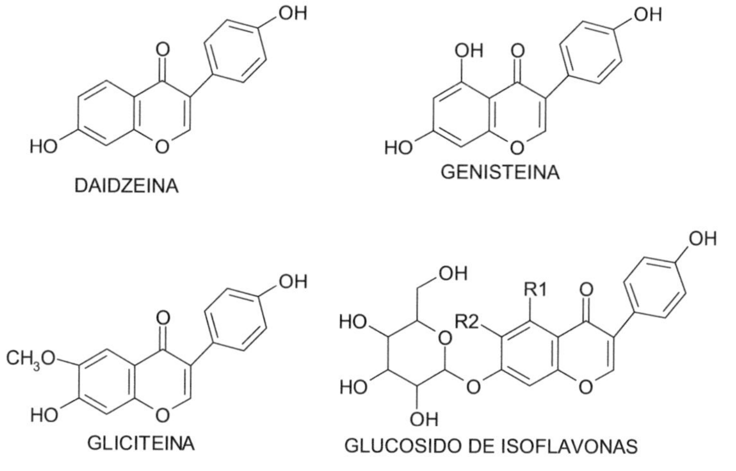


**Figure 5.** Isoflavones structure

**b) Flavonoids and anthocyanins:** Flavonoids and anthocyanins comprise a diverse class of pigments, which are normal found in cells or tissues that impart color to food. Flavonoids are similar to anthocyanins, which are water soluble vacuolar pigments present in the fluid in cells responsible for the majority of red, blue and violet flowers and leave the chemical structure colorations. Approximately 800 flavonoids are known and their number is increasing rapidly. One of the main groups is the flavonols: kaempferol, quercetin and mircetina. Another, less common group, is the flavones, comprising apigenin, luteolin and tricetina. In fruits, they are mostly found in the pericarp, so it is best to ingest the fruit unpeeled but washed properly beforehand. It is also important to note that many of these compounds are found in varying proportions in different types of wine, accounting for the preventive effect of moderate wine consumption on cardiovascular disease, cancer, and other degenerative diseases. Ingestion is recommended to maintain healthy tissues and promote a proper balance of hormones and antioxidants in the body, as many flavonoids appear to have an important role in human nutrition, and present very interesting medicinal properties, such as being antioxidant, anticancer, antithrombotic and reducing blood cholesterol levels [28].

**c) Phytochemicals such as phytosterols, Phytosterols**: plant sterols are widely distributed in nature and have a structure very similar to cholesterol. It has long been known that these sterols have a hypocholesterolemic effect when ingested in the range of 1–3 g/day, and are considered as important factors in preventing cardiovascular disease. Consumption is recommended for individuals with mild or moderate hypercholesterolemia.

## Phytosterols; structures, main sources and importance

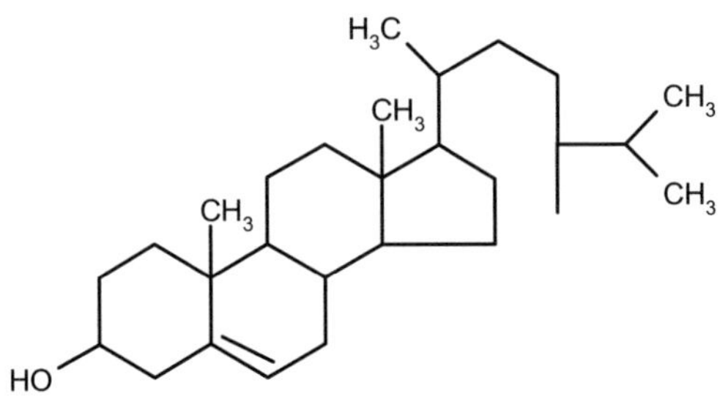


**Figure 6.** Campesterol or 24α-methyl-5-cholesten-3β-ol, 24α-metilcolesterol, 24(R)-metilcolesterol, (24R)-5-ergosten-3-β-ol. Chemical formula $C_{28}H_{48}O$, MW 400.68 and melting point of 157–158°C. [52, 53]

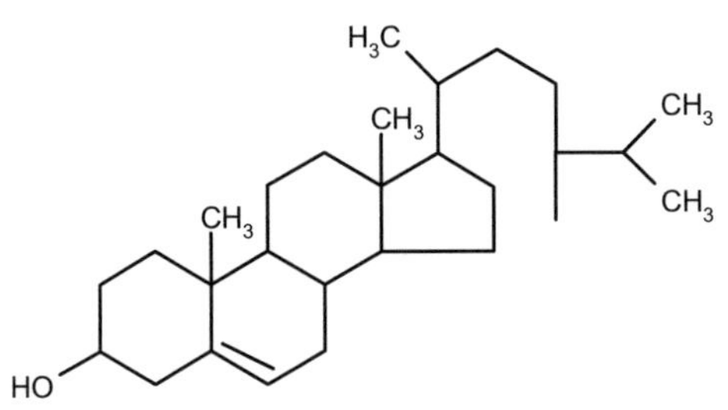


**Figure 7.** Sitosterol known as β-sitosterol, estigmast-5-en-3β-ol, 24-β-etilcoles-5-en-3β-ol. Chemical formula $C_{29}H_{50}O$, MW 414.71, mp 140 °C. [52, 53]

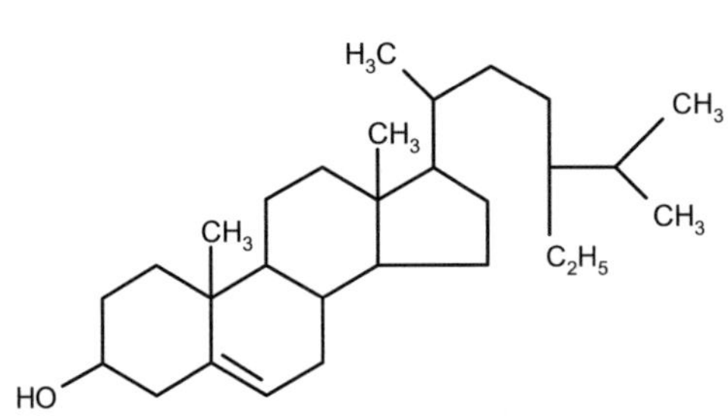


**Figure 8.** Stigmasterol, also called (22E)-estigmasta-5, 22-dien-3β–ol or 24α-etilcolesta-5, 22-E-dien-3β-ol. Chemical formula $C_{29}H_{48}O$, MW 412.69, mp 170°C. [52, 53]

http://dx.doi.org/10.5772/61626

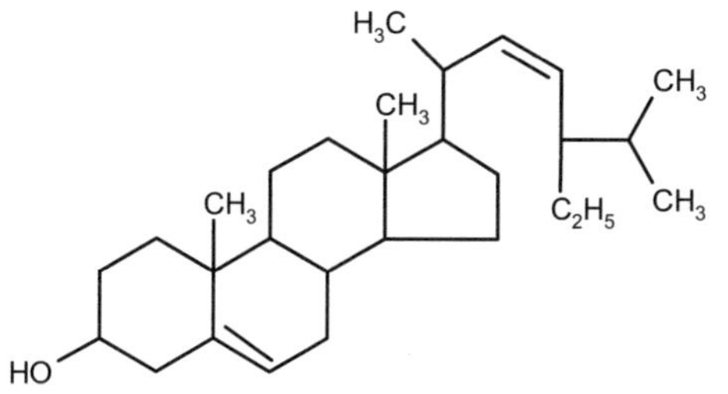


**Figure 9.** Fucosterol, also known as [24 (28) E]-estigmasta-5, 24 (28)-diene-3β-ol, [24 (24') E] -estigmasta-5, 24 (24') -dien-3β-ol or 24E-etildenecolesta-5, 24 (28) -diene-3β-ol. Chemical formula $C_{29}H_{48}O$, PM 412.69. [52, 53]

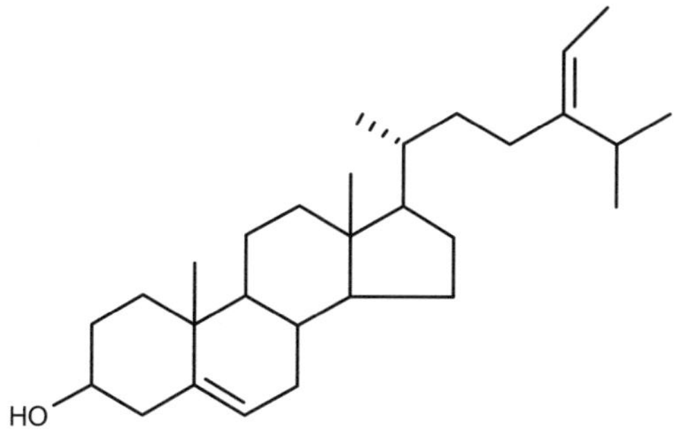


**Figure 10.** 5-avenasterol (5-avenasterol), called isofucosterol, 28-isofucosterol, 29-isofucosterol, 24Z-etilidenecolesta-5, 24 (28) -diene-3β–oλ, [24(28Z)] –estigmasta-5, 24 (28) -diene-3β-ol or [24(24')-Z]-estigmasta-5, 24(28')-diene-3β-ol, chemical formula $C_{29}H_{48}O$, PM 412.69. [52, 53]

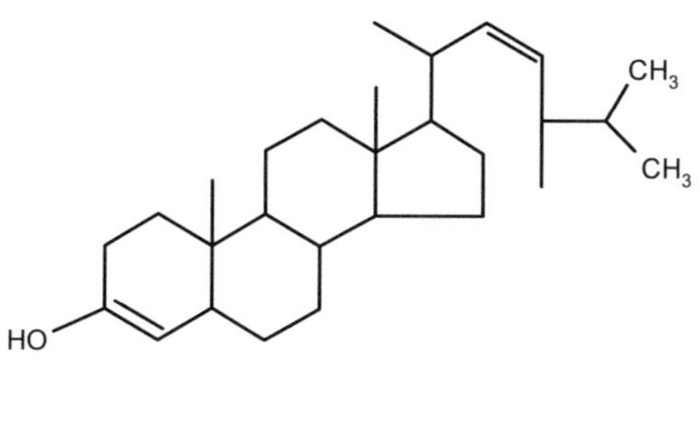


**Figure 11.** Spinasterol or bessisterol, hitodesterol, 7.22-estigmastidienol. Chemical formula $C_{29}H_{48}O$, MW 412.67 and melting point 168–169°C. [52, 53]

### 7.1. Sources

The main sources of phytosterols are varied and include fruits such as apple, apricot, banana, cherry, fig, grapefruit, navel oranges, Granada, and strawberry; vegetables such as tomatoes, potatoes, spinach, and carrot; vegetable oils such as avocado, cashew, beaver, brown, coffee, corn, cottonseed, linseed, mustard seed, palm, peanut, pine nut, pumpkin seed, rapeseed flower, rice bran, soybean, sunflower, and wheat germ; grains such as corn, rice bran, sorghum, and wheat; legumes such as beans, peanuts, and soybeans; spices like basil, dill, ginger, white mustard, oregano, paprika, red pepper, poppy seed, and turmeric among others [51].

The hypocholesterolemic effect of phytosterols is attributed to three metabolic actions: inhibition of intestinal cholesterol absorption by competition for the formation of cholesterol mixed micelles; reduction of cholesterol esterification in enterocytes by inhibiting the activity of acyl CoA cholesterol acyl-transferase enzyme, and stimulation of the flow of cholesterol from the intestinal lumen into enterocytes to increase the activity and expression of a conveyor. The joint action of phytosterols on these mechanisms produces a decrease in total plasma cholesterol and LDL cholesterol without changing the levels of HDL cholesterol. Since phytosterols are more lipophilic than cholesterol itself – a property derived from the characteristics of greater length and sidechain complexity – sterols and stanols competitively displace cholesterol from the mixed micelle formed by the action of phospholipids and bile salts in the intestinal lumen. Thus, upon contact with the mixed micelle formed by brush border microvilli on intestinal cells, phytosterols take the place of cholesterol. Cholesterol does not emulsify (moved from the micelle), cannot be absorbed and is eliminated by bowel movements. Phytosterols and phytostanols have low absorption in the intestine, so for the transfer of fatty acids and monoglycerides from the micelle to intestinal cells, which results in the disassembly of the mixed micelle, sterols and stanols would be released along with the unabsorbed cholesterol, finally being excreted in the stool. This would be the first level action of sterols. Intestinal absorption of phytosterols is extremely low (less than 0.5–1.0%) and that of phytostanols is even lower. However, when these sterols are absorbed, inhibition of ACAT (second level action) occurs, whereby cholesterol is not efficiently esterified and incorporated into chylomicrons and stimulates the flow into the intestinal lumen of non-esterified cholesterol. Sterols produce an over-expression of genes encoding proteins of the ABC transporter structure, thus accelerating the flow of cholesterol (third level action) [54].

### 7.2. Importance

The importance of these natural components is such that several authors state that the inclusion of phytosterols in the diet decreases the concentration of cholesterol in the serum. However, there are very few studies that prove these assertions. In 1999, the Food and Drug Administration (FDA) declared esters of phytosterols as to be Generally Recognized as Safe (GRAS) for use as ingredients in spreads based on vegetable oils in amounts not exceeding 20%, i.e. 20g per 100g of product. [55] In 2004, the Commission of the European Community allowed the use of phytosterols in food, restricted to consumption to 3 g/d maximum, and these components must be declared in the list of ingredients by percentage and g/100g or g/100ml of product.

# 8. Carotenoids

## 8.1. Carotene

Carotenes are chemical compounds whose absorption spectrum (β-carotene) shows two absorption peaks between 400 nm and 500 nm, corresponding to the colors blue and green, with the reflected red-orange-yellow light gives the compounds their characteristic color. They can reduce the chances of heart attacks, function as soluble antioxidants, and increase the efficiency of the immune system. They have been shown to reduce the likelihood of incidence of some cancers, but may increase the risk of lung cancer in smokers.

According to WHO, dietary factors are involved in about 30% of cancers in Western countries and in 20% of cancers in developing countries. There is ample evidence implicating the involvement of these compounds in such diseases. Lycopene (the pigment that gives tomatoes and watermelons their color) is one that has long been considered as a strong candidate for cancer prevention. The risk of prostate cancer decreases as the consumption of lycopene, α-carotene, β-carotene, β-cryptoxanthina, lutein and zeaxanthin (Figure 12) increases. The consumption of tomatoes, squash, spinach, watermelon and lemons is inversely related to the risk of prostate cancer [57].

Dark green and orange plants are good sources of β-carotene. Most β-carotene is administered as a synthetic supplement containing a single molecule called all-trans-β-carotene. The β-carotene-contained in foods consist of two molecules: all-trans-β-carotene and 9-cis β-carotene, although the proportion of both varies from one source to another. Initially, no significant difference between natural and synthetic beta carotene was observed, but it is possible that natural beta carotene has differences from the synthetic form. The absorption of carotenoids is affected by digestion of pigments, the presence of hydroxyl groups, and the degree of esterification. Astaxanthin absorption (AXT) occurs mainly in the pyloric caeca, while canthaxanthin (CTX) is absorbed throughout the gastrointestinal tract. AXT is transported mainly by high density lipoproteins (HDL). Vitellogenin is involved in the transport of AXT in liver, muscle and ovaries in mature females. The liver acts as body's main buffer of compounds transported by lipoproteins and consequently a high proportion of absorbed carotenoids are metabolized and excreted in the bile, making them unavailable for muscular pigmentation. When ingested, β-carotene is converted to vitamin A in the mucosa of the small intestine, and is stored mainly in the liver as retinol esters. β-carotene may also be absorbed and stored in fatty tissue without being modified, producing a slightly yellow or orange coloration on the palms of the hands and soles of the feet [57, 58].

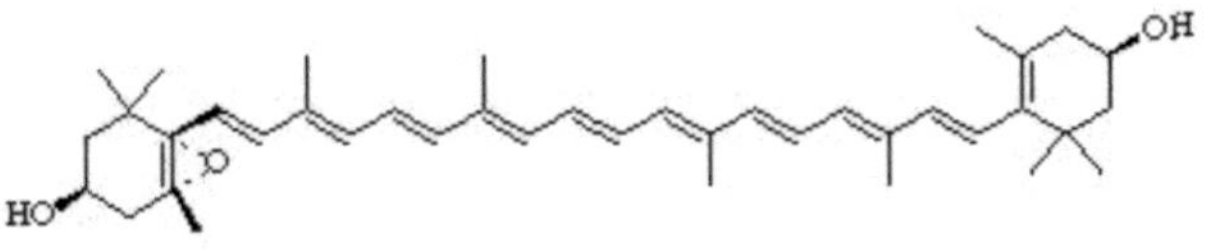


**Antheraxanthin:** Present in many plants, especially maize

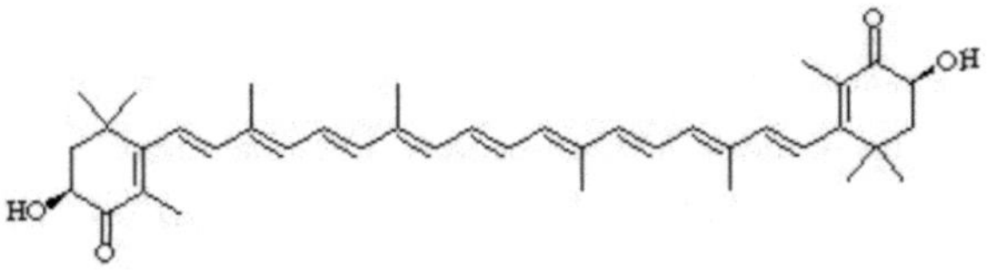


**Astaxanthin:** Present in salmon, shrimp, lobster, flamingo feathers, and algae

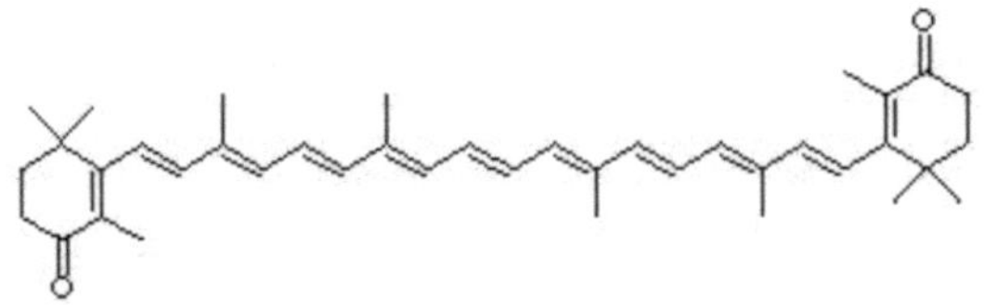

**Canthaxanthin:** Present in salmon, shrimp, chanterelle and other mushrooms, algae, flamingo feathers

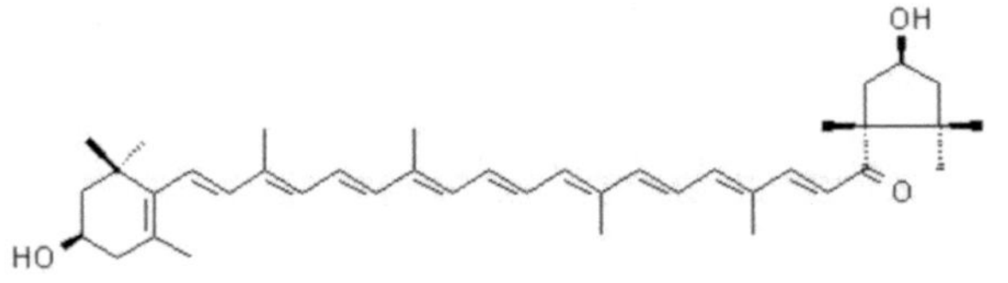


**Capsanthin:** Present in peppers, paprika

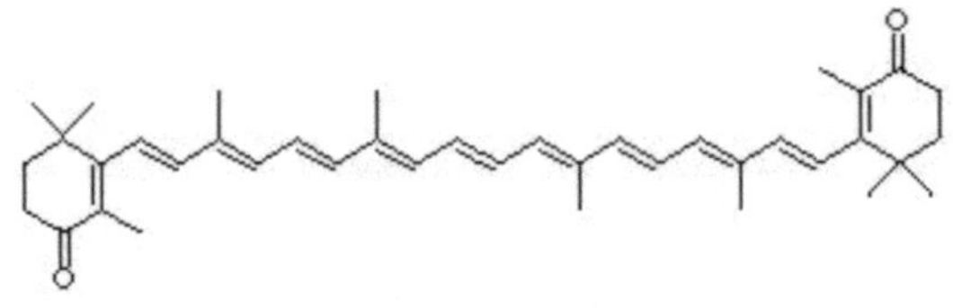

**α-carotene:** Present in: carrots, most green plants

http://dx.doi.org/10.5772/61626

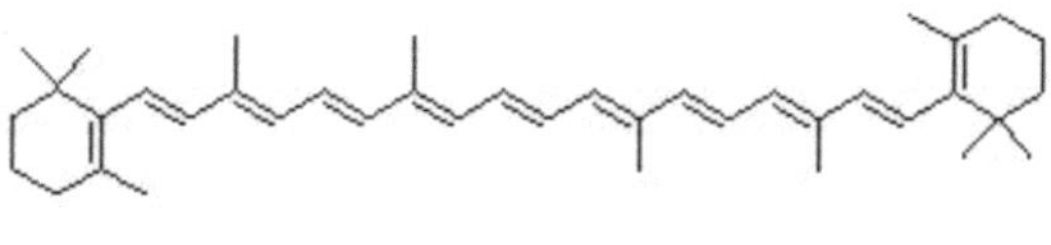

**ß-carotene:** Present in carrots and most other plants

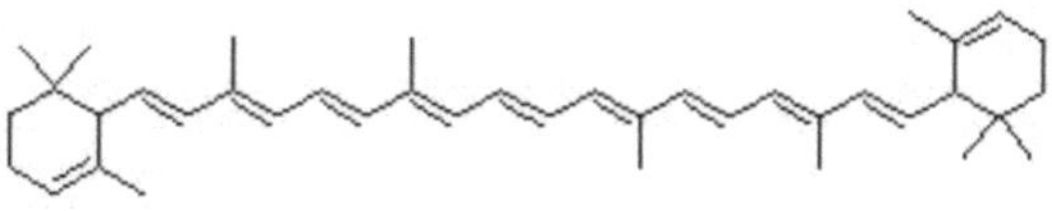

**ε-carotene:** Present in most green plants

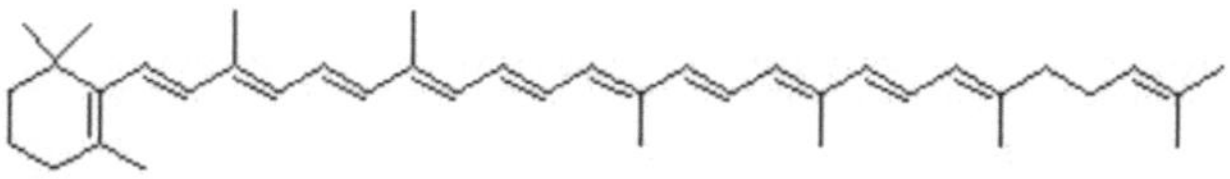

**γ-carotene:** Present in many plants, often with ß-carotene

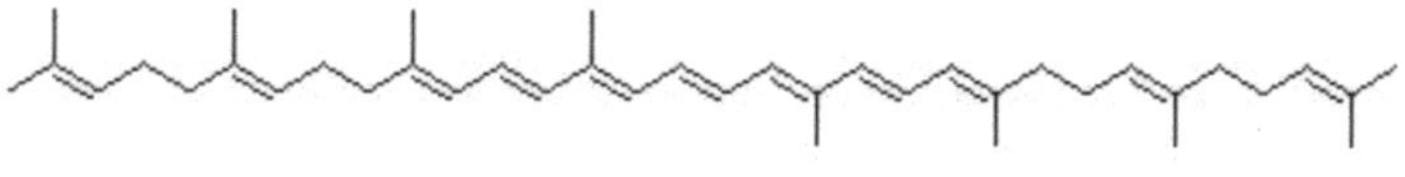

**ζ-carotene:** Present in many plants

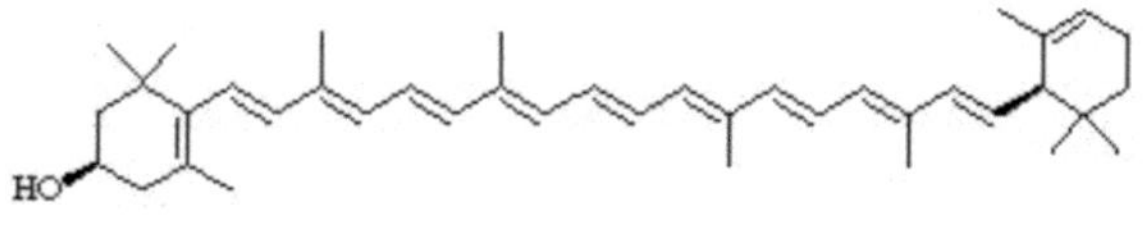


**α-cryptoxanthin:** Present in many colored plants, including maize and papaya

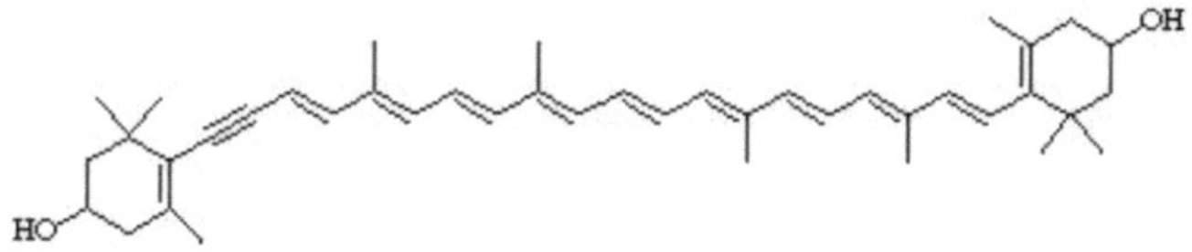


**Diatoxanthin:** Present in algae and corals

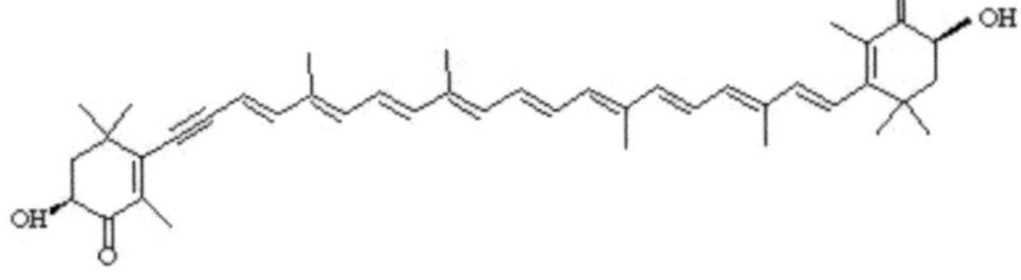


**7, 8-didehydroastaxanthin:** Present in salmon and crustaceans

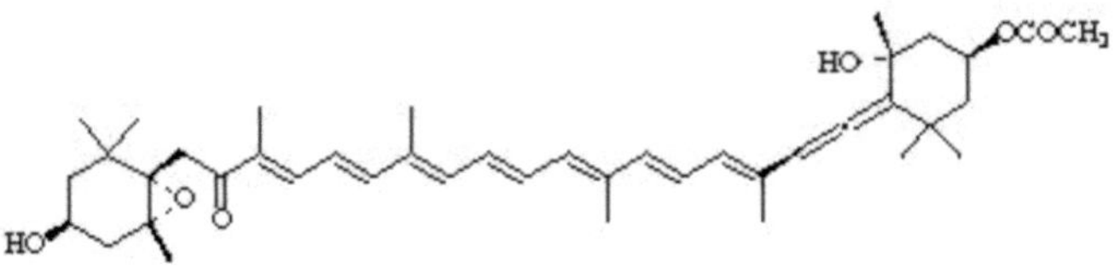


**Fucoxanthin**

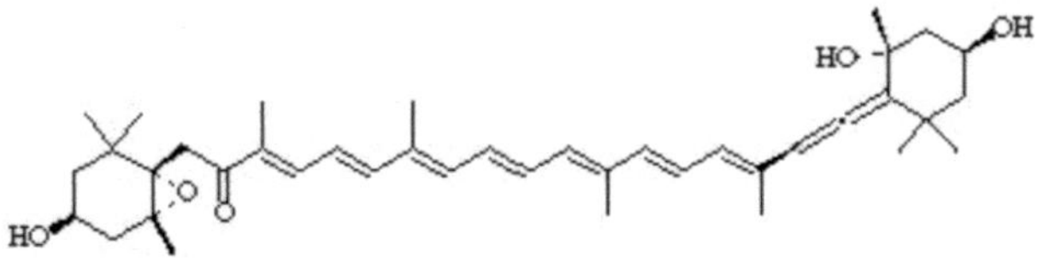


**Fucoxanthinol**: Present in many algae and seaweed

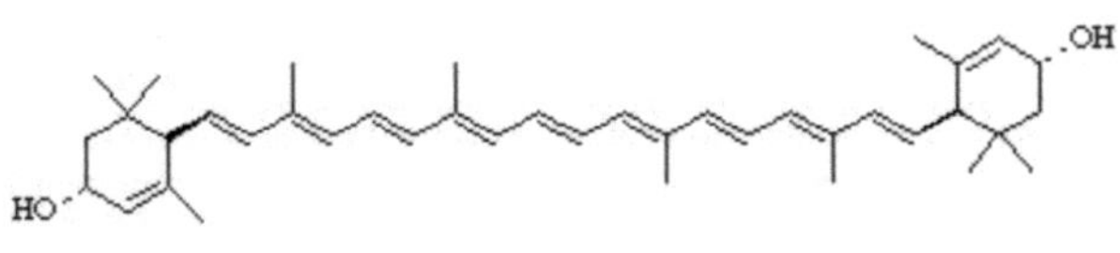


**Lactucaxanthin:** Present in algae

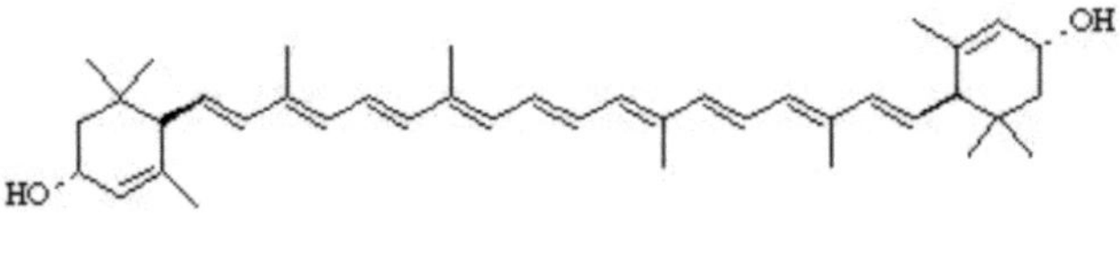


**Lutein:** Present in many green plants

http://dx.doi.org/10.5772/61626

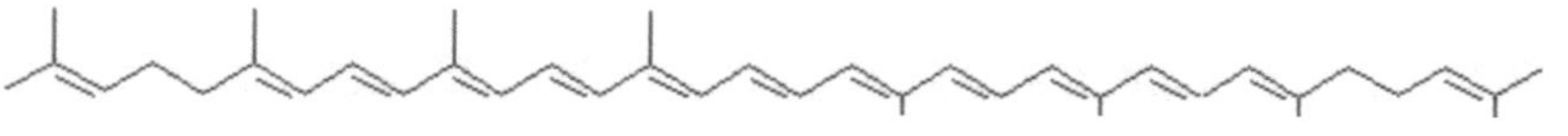

**Lycopene:** Present in many plants, especially in tomato

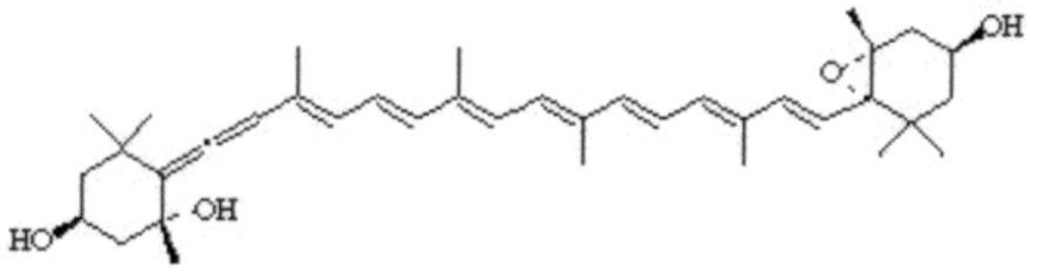

**Neoxanthin:** Present in the chloroplasts of most plants

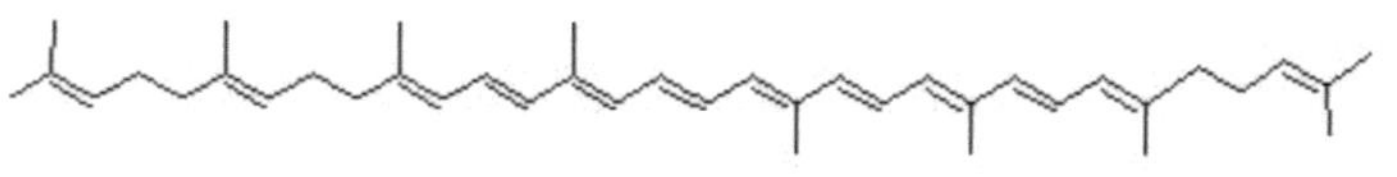

**Neurosporene:** Present in many plants, an intermediate compound between carotene and lycopene

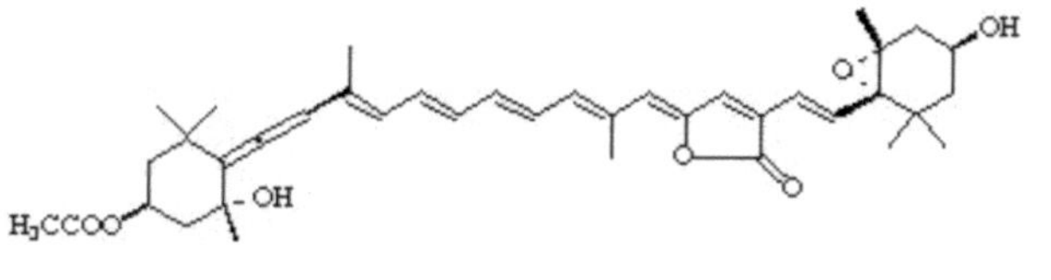

**Peridinin:** Present in the chloroplasts of most plants

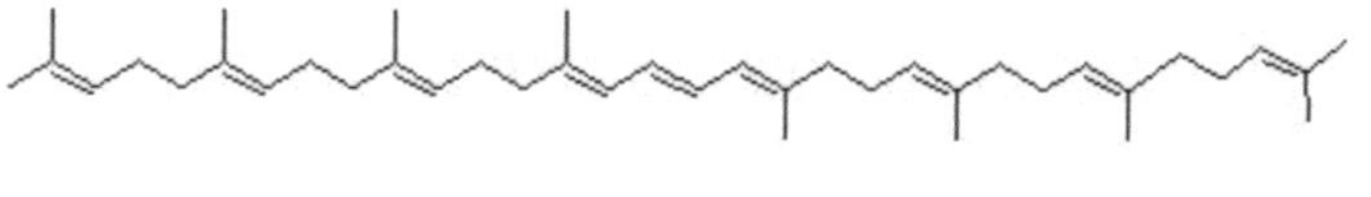

**Phytoene:** Present in many plants

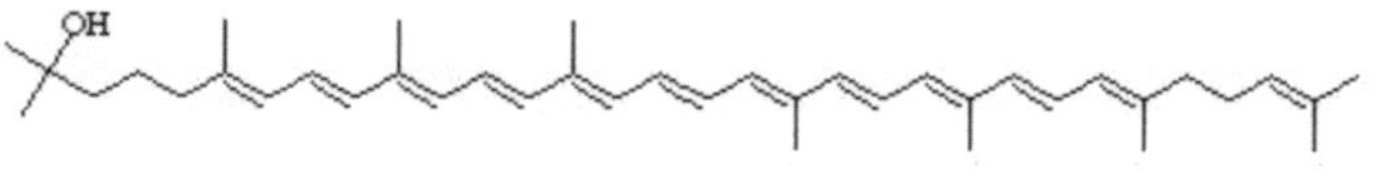

**Rhodopin:** Present in many red bacteria

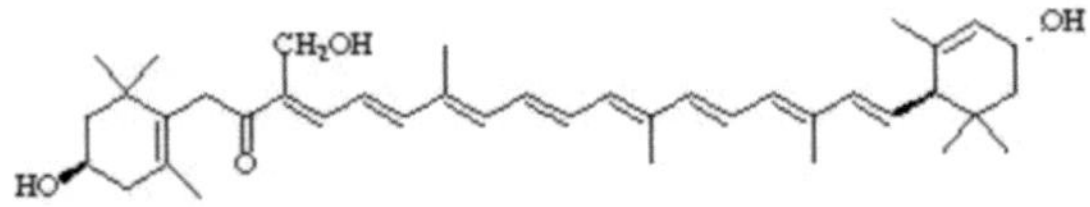


**Siphonaxanthin:** Present in many red algae

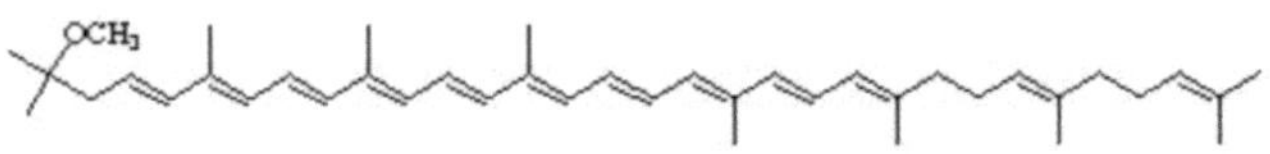


**Spheroidene:** Present in many red bacteria

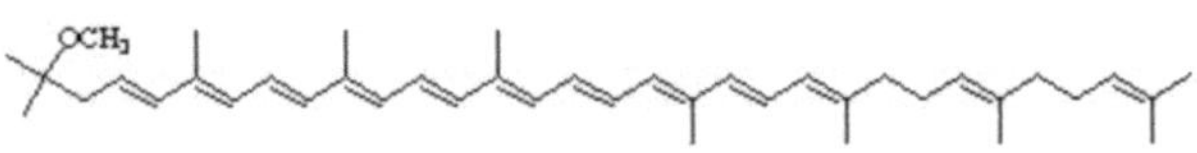


**Spheroidenone:** Present in many red bacteria

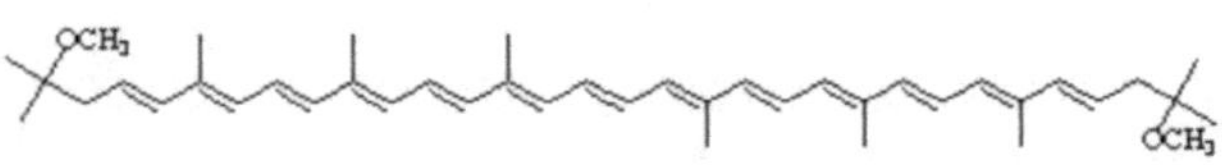


**Spirilloxanthin:** Present in many red bacteria

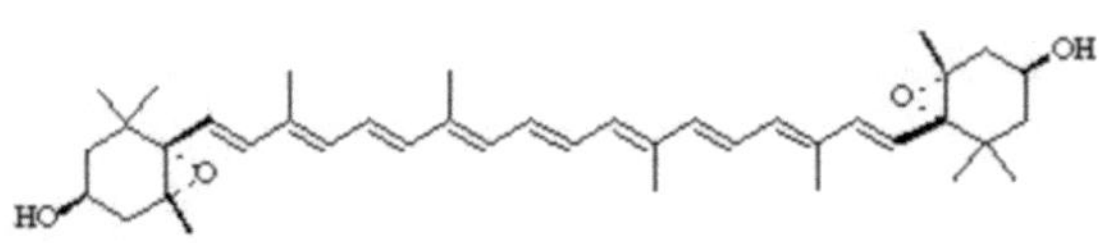


**Violaxanthin:** Present in many plants, especially from the *Viola* family (pansies)

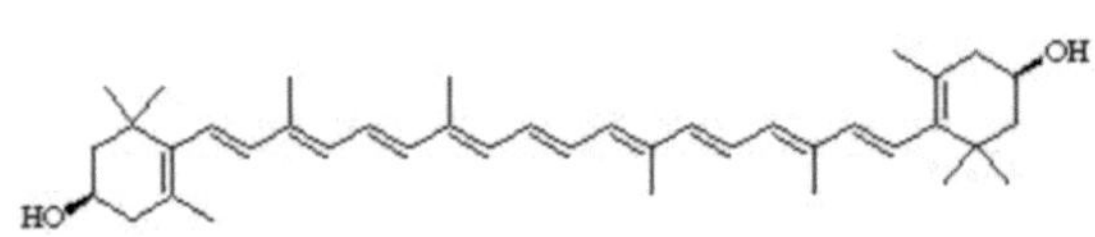


**Zeaxanthin:** Present in many plants, especially in maize

**Figure 12.** The most important structure of carotenes [56]

http://dx.doi.org/10.5772/61626

### 8.2. Lutein and related compounds

Lutein is the most abundant carotenoid in nature, and the main xanthophyll of the protein complexes responsible for uptake of light and pigmentation in plants. It is widely distributed in vegetables, flowers, leaves and fruits, and is involved in the processes of photosynthesis in plants.

### 8.3. Sources

Lutein is found in grains and dark green vegetables like kale, spinach, broccoli, cabbage, celery, spinach, and asparagus. Corn also contains good amounts as do wheat and fruits like oranges, melon, guava and pear. Table 5 contains the most important approximate values of this component.

| Food | Content (mg/service) |
|---|---|
| Cooked spinach. | 20.4 |
| Raw spinach | 3.8 |
| Cooked cabbage | 14.6 |
| Corn | 2.2 |
| Black beans | 2.2 |
| Broccoli | 1.6 |
| Romaine lettuce | 1.3 |
| Green beans | 0.8 |
| Egg | 0.3 |
| Orange | 0.2 |

**Table 4.** Lutein content in selected foods

### 8.4. Importance

Low concentrations of lutein increase the risk of macular degeneration, one of the principle causes of blindness in the elderly. As one of the main macula pigments, lutein is necessary for light absorption and for its antioxidant actions. Light entering the eye can cause photo-oxidation of cells of some eye components, but the pigments overlying the macula offer protection to it. Along with zeaxanthin it filters wavelengths between blue and green in the visible light spectrum. This filtration reduces both the chromatic aberrations produced by this wavelength and the oxidation induced by blue light (similar to damage from UV light). This photo-toxic damage contributes to the development of cataracts and macular degeneration [59].

In Mexico, people over 65 years of age who consuming acetylsalicylic acid daily, have an increased risk of macular degeneration. Usually these are patients are suffering from health problems such as diabetes, obesity and cardiovascular disease. The pattern is further complicated by a high incidence of renal failure and hypertension. Related macular degeneration (ARMD) in these patients is characterized by the progressive loss of central vision to eventual total blindness. Globally, there are currently about 80 million people with blindness caused by ARMD, with 80% of these patients having the wet (or exudative) form and 10–12% have the dry form (geographic atrophy). The main risk factors are age, smoking, race, pale iris, arterial hypertension, cardiovascular disease, hypercholesterolemia, low levels of antioxidants, a diet low in antioxidants, and high exposure to UV light of any kind [60, 61].

# 9. Minerals or inorganic nutrients

## 9.1. Role of specific antioxidant minerals

### *9.1.1. Selenium*

Selenium is a rare element in the crust of the earth and is almost always found in its native form in metal sulfides. As selenide it is found in association with various minerals such as copper pyrites. In addition to its biological significance to humans and animals, selenium is useful in coloring glass, and in ceramic glazes which give a reddish color. It is an excellent semiconductor and as a result is very useful in the electronics industry.

The chemical nature of selenium, which relates it to sulfur, is biologically important as selenium replaces sulfur in molecules with a biological function such as certain amino acids (seleno-amino acids such as selenomethionine and selenocysteine) and hence selenoproteins form part of the human body (Figure 13). In a similar way to sulfur forming hydrogen sulfide ($H_2S$), selenium forms Hydrogen selenide ($H_2Se$) [62].

#### *9.1.1.1. Metabolism*

It has been observed that there is relationship between the functions of selenium and vitamin E and methionine. When vitamin E interacts with selenium and methionine, a greater effect is achieved against the action of free radicals, both in mitochondria and microsomes. The main metabolic function of selenium is as a prosthetic group of glutathione peroxidase (GPx) and it forms part of its active site. This seleno-enzyme, discovered in 1973, destroys hydrogen peroxide and other oxidantsorganic hydroperoxides, which prevent oxidative damage to cell membranes. By regulating the concentration of glutathione peroxidase and selenium, lipid peroxidation and oxidation of the sulfhydryl groups of the erythrocyte cell membrane — a process that can cause hemolysis — is prevented. Glutathione (Figure 14) is a major component of the red cell membrane such that the balance between the formation of disulfide bridges (GS-SG) and the presence of reduced glutathione (GSH) maintains functionality of the cell membrane [63].

http://dx.doi.org/10.5772/61626

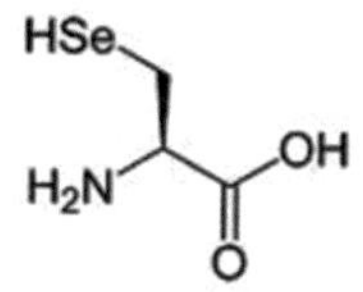

*Selenocysteine*

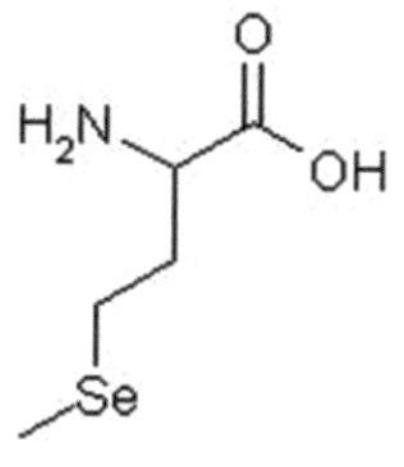

*Selenomethionine*

**Figure 13.** Chemical structure of seleno-amino acids

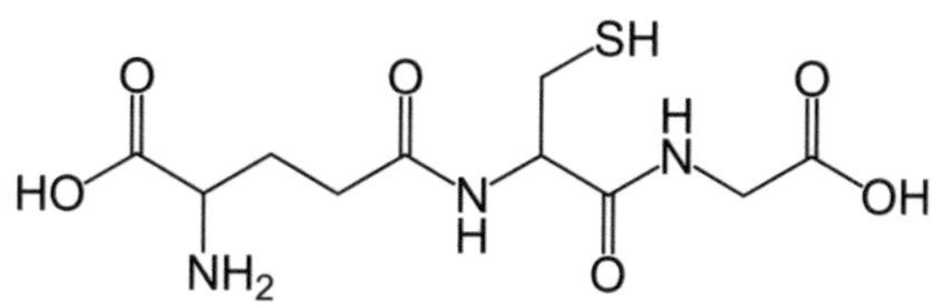

**Figure 14.** Structure of glutathione (GSH)

The metabolism of selenium in humans is different from that in other animals. The activity of the glutathione peroxidase (GPx) selenium unit is higher in the blood of some laboratory animals in higher primates. For example, blood samples from humans contain approximately 0.1 mcg Se/mL of blood. The GPx/selenium index is very similar in the blood of rats and in that of some species of monkey, but in the blood of rhesus monkeys and humans this index is low. The correlation between blood selenium and GPx activity in the blood of subjects from different areas of the world is significant and linear. This observation comes from studies in China, New Zealand and the United States (Oregon and South Dakota). Selenium accumulates more rapidly in subjects consuming selenomethionine by ingesting selenates, by enabling selenium to form the methionine complex of many proteins, including hemoglobin, more easily. There are two selenoproteins in erythrocytes, while plasma has three: GPx, selenoprotein P, and albumin. Any of these three selenoproteins can serve as a transport protein. In GPx and albumin, selenium is bound to methionine at the active site, and in the case of selenoprotein P, to cysteine [64].

Figure 15 shows the general form of selenium metabolism from its ingestion to excretion. It can be seen that the digestion of selenium in food and water, regardless of the chemical form, starts in the stomach, forming selenite, selenates and free inorganic selenium, and then selenomethionine and selenocysteine are formed through the action of digestive proteases, which are absorbed in the intestine. They are then transported by selenoprotein P to the various tissues and cells that require them.

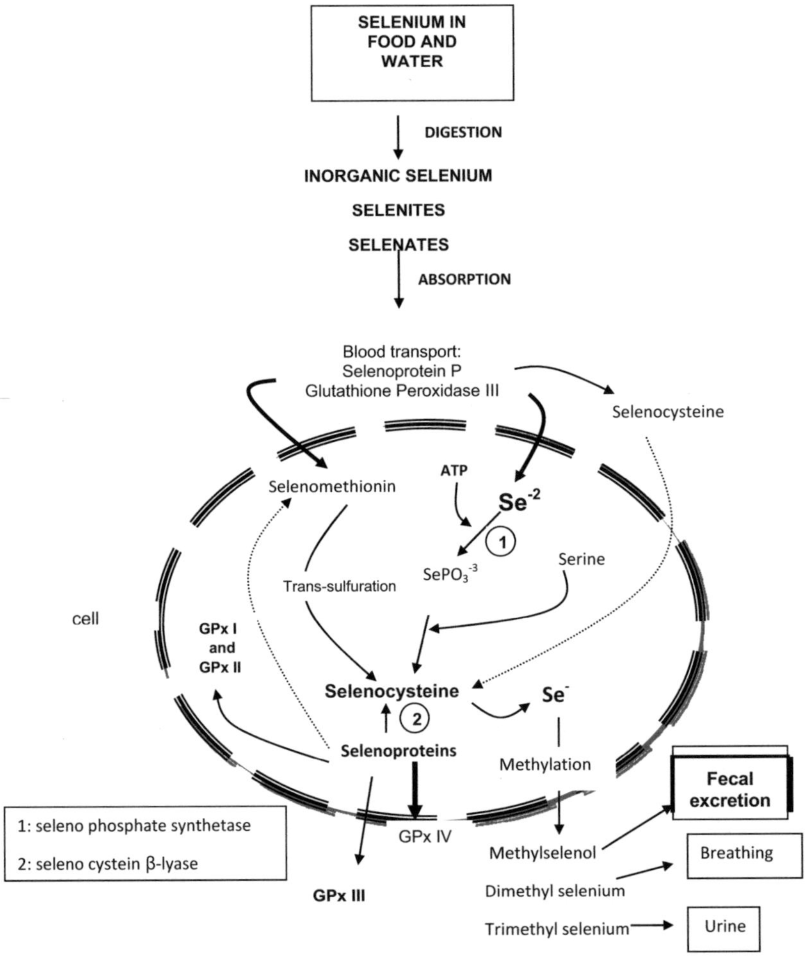


**Figure 15.** Origin and fate of selenium compounds [62]

Inorganic selenium ($Se^{-2}$) becomes intracellular phosphate and is incorporated into amino acids (such as selenocysteine). Once incorporated in amino acids, selenium is used in the synthesis of specific selenoproteins such as GPx I and II in the cytoplasm, extracellular GPx III, and membrane bound GPx IV. Unused selenium is methylated and excreted in the urine, feces and breath [63, 64].

http://dx.doi.org/10.5772/61626

*9.1.1.2. Requirements and recommendations*

The first recommended daily intakes of selenium were made in 1989 in the United States and were 55 and 70 mcg for adult men and women respectively; 40–50 mcg during puberty, 20–30 micrograms for children, and 10–15 mcg per day for infants. Pregnancy increases the recommendation by 10 mcg, and breastfeeding increases it even more. The requirement for selenium is increased with the amount of unsaturated fatty acids (monounsaturated and polyunsaturated) in the diet because of its role in the protection of these fatty acids from free radical attack.

To date, the total extent of selenium requirement has not been fully established. The reference value for the Mexican population is a suggested daily intake (SDI) of between15–20 mcg in the first year of life of, and 20–55 mcg/d for adult men and women. For pregnant women, an SDI of 60 mcg/d independent of age has been established and for lactating women the SDI is 59 mcg/d. The upper limit of daily intake in adults, without distinguishing between males and females, and pregnant or lactating women, is 400 mcg. For children in the first year of life, an upper limit of 60 mcg is recommended.

*9.1.1.3. Source*

Some foods that have been identified as important sources of selenium are shown in Tables 6 and 7, but it should be remembered that the concentration of selenium in food depends largely on its availability in the soil and in water used for irrigation. The concentration in the blood of humans depends on the average dietary intake, which varies according to region and tends to follow the pattern of the selenium content of livestock and soil in the region.

| Food | mcg/100 g |
|---|---|
| Brazil nut | Approx 2000 |
| Avocado | 11.73 |
| Cauliflower | 6.57 |
| Lettuce | 6.62 |
| Chili | 2.76 |
| Watermelon | 9.05 |
| Egg | 4.73 |
| Pork | 12.37 |
| Beef | 8.49 |
| Chicken | 8.44 |
| Tuna | 10.86 |
| Oyster | 5.65 |
| Sardine | 44.86 |
| Black beans | 3.47 |
| Walnuts | 13.53 |
| Cow milk | 5.15 |
| Bottled Water | 4.56 |

**Table 5.** Selenium content of some selected foods [62]

In the case of infants, the ingestion of selenium comes from the mother's milk of. In milk, the selenium content and glutathione peroxidase activity (GPX) is directly dependent on the nutritional status of the mother. After several years of selenium supplementation in Finnish mothers, a significant increase in the amount of selenium in milk and serum was observed. The average content of selenium in human milk is 70 mcg/L from a daily consumption of 85 mcg of selenium. Given a production of 750 mL milk/day, the average daily consumption in children is almost 55 mcg. The effects of supplemental selenium depend on the chemical form in which it is provided to mothers, but is estimated that consumption of 50–70 mcg/day is needed in pregnant women, and that this concentration allows saturation of plasma GPx. The importance of providing selenium in the recommended amounts through breast milk is explained by its antioxidant properties and function in the protection of cell membranes, as these are constantly under "attack" from highly oxidizing free radicals that can destroy them. In an infant, this effect could be devastating and can still affect overall health when pubescent or adult. Selenium deficiency in women can cause health complications from adolescence, along with problems arising from a serious lack of other inorganic nutrients such as iron and zinc that influence reproductive performance [62, 64].

In the case of children who are not breastfeeding, infant formulas with an appropriate selenium concentration are available on the market to meet requirements.

| vegetables | | fruits | | beef | | pork | | chicken | | marine | | others | |
|---|---|---|---|---|---|---|---|---|---|---|---|---|---|
| chard | 34,12 | blueberries | 13,61 | res needles | 1,1 | dried meat | 7,38 | wing | 4,06 | **cod** | **8,54** | Barbecue | 53,66 |
| celery | 21,45 | Capulín | 12,74 | Arrachera | 67,68 | Chicharrón | 11,99 | heart | 2,51 | scallops | 8,75 | snails | 28,75 |
| watercress | 33,88 | Chico zapote | 24,91 | head | 56,13 | Chicharrón pressing | 7,61 | liver | 6,18 | tent (carpa) | 12,42 | **Chinicuiles °°** | **0,9** |
| beetroot | 20,89 | plum | 11,56 | queue | 43,01 | chop | 9,06 | Molleja | 2,77 | Charales | 9,81 | quail | 18,64 |
| onion | 48,59 | prune | 24,96 | heart | 62 | heart | 14,77 | **drumsticks** | **11,03** | **Huachinango** | **54,42** | rabbit | 28,48 |
| squash | 17,86 | Coconut | 33,89 | Criadilla | 2,16 | ribs | 8,5 | breast | 9,51 | Mojarra | 39,33 | Escamoles | 48,7 |
| Chilaca | 36,64 | peach | 9,6 | **brisket** | **0,79** | fillet | 5,73 | neck | 7,96 | Salmón | 17,23 | egg quail | 12,37 |
| Chilacayote | 41,42 | strawberry | 11,2 | fillet | 0,91 | **liver** | **66,11** | **rump** | **2,46** | Surimi | 17,02 | lamb tongue | 6,24 |
| Chile cuaresmeño | 16,17 | soursop | 10,77 | tongue | 6,02 | sausage | 8,6 | remnant | 7,63 | tuna canned in water | 35,54 | turkey sausage | 72,05 |
| cabbage | 24,41 | **guava** | **9,53** | Milanese | 47,71 | **Milanese °°** | **5,64** | eggs | 4,78 | tuna with vegetables canned | 39,28 | sheep brains | 5,79 |
| squash blossom | 69,04 | fig | 28,95 | ground | 0,95 | ground | 8,66 | **oils** | | Sardine | 44,86 | turkey ham | 48,77 |
| **beans** | **82,04** | Kiwi | 38,08 | **belly** | **72,26** | Moronga | 7,43 | canola | 20,25 | **clam** | **6,61** | **turkey breast** | **99,99** |
| **Huauzontle** | **12,82** | lychee | 10,01 | res legs | 38,57 | pork legs | 14,67 | cartamo | 20,39 | squid | 23,05 | turkey thigh | 27,17 |
| Pore | 22,9 | Lime | 19,71 | neck | 1,04 | legs | 7,14 | canola linseed | 24,21 | **shrimp** | **74,41** | cofee | 4,37 |
| Quelite | 21,84 | Mamey | 18,34 | Pulp | 0,81 | kidneys | 8,52 | corn | 15,55 | octopus | 12,28 | sugar | 47,9 |
| radish | 18,68 | **blackberry** | **50,7** | Suadero | 0,87 | brains | 10,17 | sunflower | 11,54 | mussel | 65,2 | salt | 2,92 |
| purslane | 44,78 | pineapple | 10,71 | gut | 68,88 | beacon | 9,44 | walnut | 99,25 | ostion | 5,65 | corn tortilla | 3,49 |
| Xoconostle | 33,91 | dominican banana | 12,4 | liver | 7,84 | butter | 7,78 | peanuts | 72,25 | tuna canned oil | 10,86 | bread | 3,38 |
| mushrooms (setas) | 28,95 | grapefruit | 37 | kidney | 12,1 | ham | 4,5 | almond | 55,04 | catsup | 3,04 | cow milk | 5,15 |
| garlic | 28.33 | Tuna | 25,99 | | | loin | 12,37 | pistachio | 68,52 | water | 4,56 | cream | 3,89 |
| parsley | 56.44 | sweet potato | 18,19 | | | | | | | soda (cola) | 5,96 | chesse | 2,11 |

**Table 6.** Selenium content of Mexican foods (mcg Se/100g edible portion) [65, 66]

### *9.1.1.4. Selenium deficiency*

Selenium deficiency is rare in well-nourished populations and can be prevented by administering selenium. The first case of deficiency was reported in the Chinese province of Keshan. The disease occurs frequently in adolescence, with initial symptoms of symmetric stiffness,

http://dx.doi.org/10.5772/61626

swelling and pain in the interphalangeal joints of the fingers, followed by generalized osteoarthritis affecting the elbows, knees and ankles. Selenium deficiency was observed in malnourished patients when feeding over a prolonged parenteral time, and very low selenium concentrations were also observed in the serum of patients fed enterally. Endemic selenium deficiency causes myopathy, but it is unknown whether the deficiency is caused by a particular agent, such as a viral infection, or by some type of metabolic stress [65]. Other seleno-proteins are found in muscle, selenium carrier proteins, xanthine dehydrogenase and some bacterial enzymes. In bacterial systems, selenium-dependent enzymes have been identified such as glycine reductase. In microorganisms, the amino acidic selenium transfer RNA portion is incorporated. Selenium reduces the toxicity of mercury, cadmium and other toxic metals and functions as an important scavenger of toxic metals that are usually ingested in the diet or through the airways. Selenium deficiency prevents purging and thus the toxic action of these metals is unaffected [66].

*9.1.2. Zinc*

Because of the frequent and widespread consumption of zinc, deficiencies were not observed until 1955, when it was demonstrated in pigs. The appearance of para-keratosis was a sign that humans may be deficient in zinc, and it was noted in malnutritioned Chinese patients with very low zinc concentrations during World War II. In 1956, zinc deficiency in humans was demonstrated. Since 1961, it has been known that hypogonadism and dwarfism, endemic in rural populations of Iran, are caused by zinc deficiency and this was important in the study of zinc deficiencies in the clinical field and for the implications for public health. In 1963, the role of zinc in human health was unknown. In 1961, zinc deficiency was suspected when teens in Iran suffered from a syndrome resulting in stunted growth, sexual maturation, splenomegaly, and anemia and iron deficiency. Later, in 1963, studies in Egypt established that a zinc deficiency in humans caused a delay in growth and the presence of hypogonadism in men. Since its discovery just four decades ago, a dietary zinc deficiency has been described in children in many countries. Over this period, several researchers have shown that zinc deficiency causes the syndromes mentioned, mainly in children, sometimes coupled with a susceptibility to lower immune protection. Chinese researchers reported a growth failure due to lack of zinc in 30% of children, and 70% of children had low zinc concentrations in their plasma. Analysis of zinc supplementation in nine countries in Latin America and the Caribbean, eight in North America and Europe, five from Asia and the Middle East, and three from Africa was performed, showing a highly significant effect on improving size and body weight in children under 13 years of age. Over the past decade, the therapeutic uses of zinc are mentioned in Wilson's disease, in acute and chronic diarrhea in children, and in the treatment of colds and Alzheimer's disease. Zinc affects multiple aspects of the immune system. It is crucial for normal development and cell mediated immune function. Macrophages are also affected by zinc deficiency, phagocytosis, intracellular killing and cytokine production. Zinc deficiency also affects the functions of B and T cells, through disruption of basic biological functions at the cellular level. Zinc is needed for DNA synthesis and for RNA transcription in cell division and activation. Programmed cell death (apoptosis) is enhanced in the absence of zinc. The function and secretion of cytokines, the basic messengers of the immune system, are adversely affected by zinc deficiency. Zinc has the ability to function as an antioxidant, preventing cell membranes from being attacked by free radicals, and helping to improve the inflammatory process caused by injuries. Zinc deficiency causes a deterioration in lymphoid tissues and the thymus. It has been observed that immature T cells contain lower amounts of zinc. With a diet deficient in zinc, in a span of two weeks, a deterioration in the ability to prevent or counteract tumor development was observed. Zinc distributed in tissues, contains about 1.4–2.3 g, and 20% of this amount is in the skin, bones and teeth, although most is found in muscles. Only 2% of the total content is in the blood. Higher zinc concentrations are found in the sperm, prostate and epididymis. Based on the body containing around 0.3 mmol Zn/g (20 mg Zn/g), it is estimated that the total body concentration of Zinc in a newborn child is approximately 0.9 mmol (60 mg). During growth and maturation, the concentration of zinc in the human body increases to approximately 0.46 mmol / g (30 mg / g). The total content of zinc in the adult body is about 2.3 mmol (1.5 g) in women, and 3.8 mmol (2.5 g) in men. Zinc is present in the organs, tissues, fluids and secretions of the body. The zinc ion is mainly intracellular with almost 95% of zinc found within cells. Zinc is associated with all the organs but approximately 60–80% of cellular zinc is in the cytosol [67, 68].

*9.1.2.1. Function*

Zinc participates in the synthesis or degradation of major metabolites such as carbohydrates, proteins, lipids, and nucleic acid reactions. More than 200 enzymes are involved in the stabilization of the structure of proteins and nucleic acid, the integrity of subcellular organelles and transport processes, immune function, and regulation of gene expression, and proteins involved in such regulation are amino acid structures containing zinc. Zinc is also now identified as an antioxidant for the liner to combat the action of free radicals [69]. The metallothionein protein is a non-enzymatic low molecular weight cysteine-rich protein, with a higher zinc content and lower amounts of copper, iron, cadmium and mercury. It has no known biological action, but it acts in metal detoxification or inhibits thionine function, influencing the metabolism of sulfur amino acids. Zinc is abundant in the nucleus, where it stabilizes the structure of ribonucleic acids (RNA) and deoxyribonucleic (DNA) and it is required for the activity of RNA polymerases important in cell division. Zinc also acts on the chromatin proteins related to transcription and replication. Zinc is found in the crystal structure of bones, in bone enzymes and at the zone of demarcation. It is thought to be necessary for proper osteoblast activation, bone forming enzymes such as alkaline phosphatase, and calcification. When minimal bone resorption occurs, there is no zinc found in the bones [70].

At least 70 enzymes contain zinc, 15 are activated by zinc and, more importantly, almost all metabolic pathways require some zinc, e.g., carbonic anhydrase and carboxypeptidase A (pancreatic hexopeptidasa) require one molecule of zinc per mole of protein. The erythrocyte carbonic anhydrase, is 33% Zinc. One of the most important functions of zinc is its role in the synthesis of proteins and nucleic acids, as already mentioned. Zinc deficiency is reflected in the inability to incorporate thymidine into DNA. The ratio of zinc to insulin, and probably glucagon and ACTH, is of great importance even though their role in the function of the hormone is not well established. The correct ratio of zinc to insulin is necessary in order to achieve its function, along with the presence of other metals such as cadmium and cobalt. Zinc may act at the level of the beta cells of the pancreas in the storage and preparation of hormones. Zinc is involved in the secretion of the adrenal steroids, stimulated by adrenocorticotropic hormone (ACTH) [71]. Zinc variation in the body may influence the development of infection by affecting the growth of microorganisms or by altering the host response. It is known that the growth of *Aspergillus niger* is dependent on zinc and almost all living systems require zinc to exist. Zinc is involved in various enzymatic systems of bacteria and viruses (aldolases, proteases, polymerases, transcriptases, phosphatases, etc.). When zinc is added to culture media deficient in zinc, it increases the production of DNA, RNA and protein synthesis in bacteria. Zinc favors the stability and production of alkaline phosphatase in E. coli in part of the membrane and stabilizes many microorganisms. If there are high concentrations of zinc, then initial rhinovirus growth is inhibited in HeLa cells by inhibition of viral DNA synthesis in BS-1 cells infected with the herpes simplex virus. The alterations that produce infectious diseases in the host are very complex and include metabolic, biochemical and hormonal changes, with iron, zinc and copper involved in these changes. During the acute phase of infection, iron and zinc are moved to the liver, while copper leaves the liver and moves into the plasma, carried by ceruloplasmin. It is rapidly produced during inflammatory and infectious processes and it has been shown that zinc is stored in the liver and is thought to be bound to metallothioneins [72, 73]. Mobilization of plasma zinc to the liver is controlled by LEM (leukocyte endogenous mediator), which is produced in vitro by neutrophils and activated macrophages. This substance has been found in systemic infectious diseases, and has been observed in processes such as cellulite and shigellosis pielonefitis, at low concentrations as in viral infections. Zinc concentrations vary during different stages of infection. For example, during the acute phase, during the effervescence of the disease increases the amount of zinc, and during recovery, zinc concentration normalizes. Zinc is involved in a number of host defense mechanisms helping to maintain the stability of the cell membrane: at increased concentrations, zinc reduces or suppresses phagocytic function, and at low values T cells become dysfunctional. It is involved in the synthesis of nucleic acids and proteins and forms part of various metalloenzymes, like alkaline phosphatase, carbonic anhydrase and carboxypeptidases. Finally, zinc decreases concentrations favoring growth and reproduction of pathogenic germs. Zinc is very important in lymphocyte function. For example, in patients with *acrodermatitis enteropathica*, caused by a chronic deficiency of zinc, defects in cellular immunity result from dysfunctional T cells while B cells function normally. [74, 75]

*9.1.2.2. Recommendations and requirements*

Zinc requirements depend on the following criteria: the amount required to maintain balance in the body; the amount required to replace endogenous losses; and the amount needed to maintain the normal functions of zinc in the body. With an intake of 12.5 mg of zinc/day through a mixed diet, it should be possible to achieve a proper zinc balance. The RDA provides for at least 15 mg/day of zinc in adolescent and adult males, while in the case of women the amount recommended is 12 mg/day. The recommended amount during the first year of life is 5 mg/day [68, 70].

#### *9.1.2.3. Sources*

Food differs extensively in zinc content. The zinc concentration of oysters is 75 mg/100 g, while eggs only contain 0.02 mg/100 g. Chicken meat has 1 mg/100 g, and seafood, beef and other red meats are the best sources of zinc. Cereal grains are relatively rich in zinc; most bran contains zinc in the husk, and about 80% of the zinc is lost when the grains are processed or when milled. There is no uniform enrichment of cereal grains with zinc, although some cereal manufacturers do enrich as standard practice. Nuts and legumes are a relatively good source of zinc. The concentration of zinc in these foods can increase if they are grown in zinc rich soil, or if treated with fertilizers rich in zinc. Intake of zinc depends on the choice of food consumed. Animals, especially meat products, provide about 70% zinc. Cereals are the primary source of zinc. The zinc content of mixed diets varies between 10–15 mg/day and often depends on the amount of dietary protein. Diets based primarily on eggs, milk, poultry, and fish have a lower zinc content than those based on seafood, beef and other red meats. For a better view of the sources of zinc see Table 7, where sources of zinc are displayed in descending order of concentration, while Table 8 shows the sources of zinc in mg/100 g of food. These sources are representative of the most widely consumed foods in Mexico [57]. The variations are similar in vegetarian diets and water is low in zinc. During the first six months of life, zinc intake varies according to the form of milk supplied. Breastfed children ingest 1.9 mg per day, since the first month of age ingest up to 2.7 mg per day, while formula-fed continued children ingest 3.6–4.6 mg per day. Human colostrum contains 20 mg of zinc/L, decreasing rapidly so that after one month milk contains only 3 mg, and after two months, 2 mg/L. Children regularly take in 5–8 mg/day. Adolescent girls have a daily intake of 11 mg, while the daily intake of children between 8–13 years of age is approximately 8–10 mg. Most of the population consumes approximately 7–10 mg/day. Pregnant and lactating women have a similar zinc consumption. These differences are related to different energy intakes and not to the zinc content of diets, even if they are of poor nutritional quality or poorly balanced.

| **Food, (equivalent, service)** | **mg** |
|---|---|
| Oysters, eastern, ½ cup | 113.0 |
| Pacific oysters, ½ cup | 21.0 |
| Wheat germ, toasted ¼ cup | 4.7 |
| Ground beef, lean, 90 g | 4.6 |
| Beef liver, fried, 90 g | 4.6 |
| Turkey dark meat, baked, 90 g | 3.8 |
| Instant breakfast, one | 3.0 |
| Beef enchilada, one | 2.3 |
| Baked beans with pork, ½ cup | 1.9 |
| Skimmed ricotta cheese, ½ cup | 1.7 |
| Pecans, ¼ cup | 1.6 |
| Sesame butter, 1 cup | 1.6 |
| Dry peanuts, toasted ¼ cup | 1.4 |

| Food, (equivalent, service) | mg |
|---|---|
| Crab, canned, ¼ cup | 1.3 |
| Wild rice, cooked, ½ cup | 1.1 |
| Clams, canned ¼ cup | 1.1 |
| Lobster, cooked, ½ cup | 1.1 |
| Edam, 30 g | 1.1 |
| Milk, 2 % fat, 1 cup | 1.0 |
| Chicken breast, baked 1 | 1.0 |
| Nuts, English, ¼ cup | 0.8 |
| Bagel, 1 | 0.6 |
| Gingerbread, 1 piece | .6 |
| Egg, 1 | 0.6 |
| Baked Salmon, 30 g | 0.4 |

**Table 7.** Sources of zinc [76]

| Food | Content in mg /100 g edible portion |
|---|---|
| Rapeseed meal | 24.2 |
| Lyophilized hen egg | 12.5 |
| Ground beans | 8.4 |
| Cooked spinach | 7.2 |
| Breakfast cereal (Fiber One) | 6.7 |
| Barra type bun | 6.1 |
| Breakfast cereal (All-Bran) | 5.6 |
| Bisteck pulp, bovine | 5.4 |
| Pumpkin flowers, cooked | 5.3 |
| Beef, shoulder | 5.2 |
| Broccoli cooked | 5.1 |
| Lettuce, romaine | 4.7 |
| Coriander | 4.7 |
| Ground beef, fried | 4.3 |
| Beef, sirloin steak, boneless roast | 4.2 |
| Green beans, cooked | 4.2 |
| Huauzontle, cooked | 4.2 |
| Lentils, cooked | 4.1 |
| Peas, cooked | 4.0 |
| Cereal protein | 4.0 |

**Table 8.** Sources of zinc in Mexican food [77]

### *9.1.3. Iron*

#### *9.1.3.1. Features*

Iron is a metal considered by ancient civilizations (such as celestial being) sent by God for use as a metal or a pharmacological therapy or remedy. During Roman times, iron had therapeutic uses in treating diseases such as alopecia, acne, dermatitis, weakness, hemorrhoids, gout, pulmonary diseases, diarrhea, vomiting, edema, fever and cystitis. From the seventeenth century onwards the importance of treating iron chlorosis was known. The first descriptions of the importance of iron were made around 1871 and iron is now known to be one of the most abundant metals. Knowledge of the biological importance of the metal has grown, and interest in the iron deficiency known as anemia continues to this day. Iron is involved in various redox reactions such as hydrogen reduction, and is incorporated into carbohydrates during photosynthesis in the presence of ferredoxins. Aerobic metabolism depends on iron, and iron is an active component of the enzymes in the Krebs cycle [78]. Since 1966, over 30, 000 articles have been published on iron, on topics such as biology, medicine and nutrition. Between 1985 and 1991 a total of 9, 000 items was published describing iron metabolism, metabolic disorders caused by the deficiency or excess of iron, and iron ratio and nutrition. Despite the abundant information that exists on iron today, there are still many unresolved questions and uncertainties about the functions and metabolism of iron, mostly related to nutrition and immunity. An adult male has 3–5 g of iron in the blood, 30–40% of which is in the form of deposits [79].

#### *9.1.3.2. Function*

Iron is involved in the transport of oxygen and carbon dioxide, and functions as an active element in the related process of the cellular respiration of enzymes. The involvement of iron in the function of the immune and cognitive systems has not been fully clarified yet, reinforcing the importance of reducing the incidence of anemia worldwide [80].

Hemoglobin is a part of erythrocytes. The iron-containing protein heme is combined with oxygen in the lungs and carbon dioxide in the tissues. Myoglobin – a hemeprotein – also serves as an oxygen reservoir within the muscles. Oxidative ATP production in the mitochondria includes many enzymes containing heme and non-heme iron. Cytochromes, are involved with respiratory enzymes, several cofactors and the electron transfer chain, for energy production and storage, through oxidation-reduction processes ($Fe^{2+} \rightarrow Fe^{3+}$). It is also known that iron is important to the immune system, but the mechanisms by which it acts remain unknown. Neutrophils, white blood cells that engulf and destroy bacteria, are less effective when there is an iron deficiency. Transferrin and lactoferrin protect against infection by preventing microorganisms from using iron for growth. Another feature of iron is the effect it exerts on cognition, especially during childhood, where it improves psychomotor function and learning processes as well as attention and memory, especially in children with anemia. It is important to mention that in adults only 5–15% of iron from food is absorbed, only 2–10% of iron from plants is absorbed, and 10–30% from animal protein. This absorption also depends on several factors such as age, health status, iron status, the condition of the gastrointestinal tract, the amount and chemical form of the iron ingested, and the amount and proportion of various organic and inorganic components of the diet. Absorption occurs mainly in the duodenum in ferrous form. Iron absorption also depends on the chemical composition in which it is available. Amino acids, such as histidine and lysine, help in the absorption of iron [80, 81].

*9.1.3.3. Requirements and recommendations*

The requirement for iron and increased hemoglobin depends on the growth rate of the individual concerned. For example, growth is rapid during childhood and adolescence. There are various data that establish the average iron requirement. In children, the amount is 0.35–0.7 mg/day, and from 0.3–0.45 mg/day in pre-menstrual girls. An iron requirement of 38 mg/day for children of both sexes between 4- and 14-years-old, has been suggested.

Other iron requirements are based on the needs of the organism. For example, iron demands are increased in pregnant or menstruating women, and during the first two years of life due to the rapid growth into adolescence. Iron requirements during the first year of life are 0.4–1 mg/day rising to 1.5–2 mg/day during puberty. It is noteworthy that it has been difficult to accurately calculate iron requirements, since each individual varies in their capacity for absorption and the amount of iron available in food.

*9.1.3.4. Recommendations for iron [70]*

Because iron absorption is very limited, the recommendations are far superior to the requirements due to different situations such as pathological bleeding, and menstruation in the case of women. Adolescent girls who undertake strict diets are generally low in iron, pregnancy, lactation, which incidentally is necessary to supplement for 2 or 3 months before delivery.

*9.1.3.5. Sources of iron*

There are two types of iron in food: heme iron and non-heme or inorganic iron. Heme iron makes up 5% of iron in the diet and is found only in meat products (including poultry and fish). Non-heme iron is the iron in plant foods, and about 50% of the iron in animal products.

The distinction between the two types of iron is important because the body absorbs heme iron (the iron found in meat) much more efficiently. The main dietary sources of iron include organ meats, meats, legumes, cereals, eggs, seafood and nuts. (See Table 9). It has been found that the best source of iron is the liver, followed by oysters, kidney, heart, lean meat, poultry and fish. Dried beans and vegetables are also among the best sources. Some other iron-rich foods include egg yolk, dried fruits, dark molasses, whole grain breads and fortified wines and cereals. Milk and dairy products also contain a good amount. Iron in fortified cereals, flours and breads contributes significantly to the total intake, with fortified infant cereals being an important source prior to 12 months of age.

It is important to recognize that there are both stimulators and inhibitors to the absorption of iron. For example, besides being a source of heme iron, meat has better intestinal absorption which can be doubled in the presence of vitamin C. Other important inhibitors of iron absorption are tannins (found in tea, coffee and chocolate), phytates and oxalates (found in legumes, green vegetables and some grains). Table 10 shows further sources of Iron, as well as sources of vitamin C to enhance and improve the absorption of Iron. These data were obtained from tables of the nutritional value of the most widely consumed foods in Mexico.

| **Iron content (mg /100 g edible portion)** | **foods** |
|---|---|
| < 3 | single egg yolk, walnuts, peanuts, sesame, masa tortillas, whole grains, cooked legumes (beans, lentils, chickpeas) |
| 3–6 | poblano chili, turkey, dried fruit, beef |
| 6–9 | raw legumes (beans, lentils, chickpeas), pinole, Seafood |
| >9 | beef liver, chicken or pork, pumpkin seed, dried chilies, cheese, tuna, moringa |

**Table 9.** Guide to dietary iron sources

| **Food** | **Edible portion** | **Iron (mg)** |
|---|---|---|
| **Excellent** | | |
| Moronga / black pudding | 50 g | 22.5 |
| Pork liver | 100 g | 17.9 |
| Oysters | 100 g | 13.4 |
| Beef liver | 100 g | 6.3 |
| **Very good** | | |
| Beef | 100 g | 3.2 |
| Spinach, cooked | ½ cup | 3.2 |
| Beef kidney | 100 g | 3.0 |
| Shrimps | 100 g | 2.4 |
| Red beans | ½ cup | 2.3 |
| Veal | 100 g | 2.0 |
| Black beans | ½ g | 1.8 |
| Pork | 100 g | 1.6 |
| Fortified cereals | 30 g | 1.5–4.5 |
| **Good** | | |
| Artichoke | 1 unit | 1.6 |
| Nopales | 100 g | 1.6 |
| White fish | 100 g | 1.2 |

| Food | Edible portion | Iron (mg) |
|---|---|---|
| Chicken | 100 g | 1.2 |
| Peas (arvejas) | ½ cup | 1.2 |
| Rice | ½ cup | 1.1 |
| Avocado | ½ unit | 1.0 |
| **Moderate** | | |
| Bread | 1 slice | 0.9 |
| Raisin | ¼ cup | 0.8 |
| Tortilla (corn) | 1 small | 0.7 |
| Eggs | 1 | 0.7 |
| Pasta | ½ cup | 0.7 |
| Broccoli | ½ cup | 0.7 |
| Lettuce, dark green | 1 cup | 0.6 |
| Asparagus | ½ cup | 0.6 |
| Peanuts | ¼ cup | 0.6 |

**Table 10.** Iron sources and associated portion content [70]

There is a high diversity of Mexican foods that are rich in iron, and while it may be of great importance to know the amount of iron in our food, it is also important to increase the amount of it in the Mexican diet. Simply having a wider variety of foods will help us to improve our general nutritional status. This is why at the end of this paper, we analyze and discuss these sources of iron, as well as the importance of the relationship between iron intake and the need for iron supplementation when there is an iron deficiency, and/or the protective role of this element in the immune system in preventing long-term infections and/or malnutrition, which is also caused by the absence of other specific nutrients in the human diet [80].

The mechanism by which iron causes liver damage has been the subject of many studies. Iron overload produced under experimental conditions causes peroxidation of lipids in vivo. Peroxidation occurs at an iron concentration exceeding a specific threshold. Once this threshold is exceeded, the hepatocyte is unable to store iron in a nontoxic form, such as ferritin or hemosiderin, and an increase in iron storage occurs, allowing for the production of free radicals and consequent lipid peroxidation. Furthermore, in experimental models of iron overload, abnormalities of liver organelles (mitochondria, microsomes and lysosomes) are described in association with lipid peroxidation. However, studies have not yet confirmed the connection of these abnormalities the relationship to the development of liver fibrosis and cirrhosis. Research conducted at the beginning of the nineties showed that lipid peroxidation leads to increased transcription of the collagen gene and increased collagen production by human fibroblasts in culture [80, 81].

| **RAW VEGETABLES** | | **COOKED VEGETABLES** | | **CHILLI (PEPPER)** | | **OTHERS** | | **CEREALS** | |
|---|---|---|---|---|---|---|---|---|---|
| coriander | 39,92 | Spinach | 129.98 | cascabel | 30.60 | "axayacatl" worms | 48.00 | all- bran | 36.58 |
| Epazote | 36,54 | Huauzontle | 40.76 | chipotle | 28.00 | pork moronga | 36.63 | precooked (Nestum) | 34,87 |
| Yerbabuena | 29.10 | Quelite | 31.10 | Chipotle canned | 24.79 | roasted coffee | 21.40 | Raice Krispis | 22,23 |
| chard | 26.62 | Eggplant | 28.51 | pasilla | 24.57 | coconut oil | 19.60 | Crunchy Nut | 19.51 |
| parsley | 23.94 | Romeritos | 14.64 | mulato | 21.36 | dry mixed for beberages ( Nutrichoco) | 15.96 | Cereal proteínado | 18.60 |
| Coffee | 23.00 | Bróccoli | 13.41 | Chile piquin con semilla | 15.10 | almond capulín seed | 15.10 | Cerelac | 18.31 |
| Verdolaga | 20.20 | green tomato | 12.59 | guajillo | 18.81 | pingüica seeds | 14.70 | Corn Pops | 15.58 |
| mallow (Hidalgo) | 19.50 | Ejote | 12.70 | ancho black | 13.34 | cooked crab (Veracruz ) | 13.30 | bran | 14.60 |
| mallow (DF) | 19.40 | lentils | 12.37 | pickled jalapeño | 14.47 | ground beans | 12.40 | Soy | 13.70 |
| jundura plum | 18.60 | Chícharo | 12.25 | piquin with seeds | 14.28 | bean cócona | 11.50 | soybeans | 13.60 |
| bellow | 17.22 | Chilaca | 11.98 | guajillo dried cooked | 13.18 | lentils seeds | 11.20 | Fiber One | 13.50 |
| Chepilin | 16.78 | artichoke | 10.03 | mulato dried | 12.80 | pumkin seeds | 11.00 | wheat bran( all- bran ) | 13.10 |
| Calabacita (Morelos ) | 16.40 | green bean | 9.77 | moritta | 10.25 | sesame seeds | 10.68 | Corn Flakes | 12.21 |
| Alelón (Guerrero ) | 15.80 | | | guajillo | 10.10 | blue mass | 10.72 | **FLOURS OF** | |
| plum (Guerrero ) | 15.40 | **BREAD** | | | | flour for tamales | 10.45 | spinach | 88.00 |
| Tuna pressed red | 13.98 | Barra type bun | 76.90 | **ONLY SEEDS OF CHILLI** | | eelgrass seeds | 10.39 | potatoes | 72.40 |
| Hediondilla quelite | 13.80 | Enriched bread sliced | 39.40 | chile chipotle | 22.49 | corn for pozole | 10.27 | green bean | 18.20 |
| cucumber | 13.47 | deffated paste | 19.79 | chile mulato | 21.85 | yellow mass | 10.22 | black bean | 13.50 |
| mushroom "escobeta" | 13.20 | integral bread | 12.43 | chile pasilla | 17.60 | Tortilla of yellow corn | 9.87 | molturada Integral | 13,4 |
| Endive | 12.50 | | | chile guajillo | 15.30 | black bean cooked (Querétaro ) | 9.24 | carrots | 12.40 |
| Chaya dehydrated | 12.30 | | | guaje, dried | 15.20 | toast (tostiricas ) | 9.46 | | |
| Yurquilla curcuma | 12.21 | | | chile morita | 15.05 | bay bean big | 9.40 | | |

The name in parentheses indicates the origin of the food or brand. In some cases the original name of the food is maintained, as there is no adequate translation into the English language.

**Table 11.** Sources of Iron in Mexican Foods (mg Fe/100 g edible portion) [70, 77]

http://dx.doi.org/10.5772/61626

# 10. Lipids

## 10.1. n-3 Fatty acids

### *10.1.1. Oils, fatty acids and phospholipids*

Omega-3 fatty acids, DHA (docosahexaenoic acid), EPA (eicosapentaenoic acid), and polyunsaturated fats, especially the omega-3 type present in fish oil, have a protective effect because they decrease the viscosity of the blood, thus reducing the risk of thrombus formation. Monounsaturated fat, whose main representative is oleic acid present in olive oil, also has a protective effect in raising "good" cholesterol (HDL-c) and preventing the oxidation of "bad" cholesterol (LDL-c) [82]. Fish oils have large amounts of polyunsaturated omega-3 fatty acids (n-3 PUFA), DHA and EPA. The richest sources of these are oily fish (salmon, trout, sardines, and tuna). The high content of DHA and EPA in oily fish is the result of phytoplankton (rich in n-3 PUFAs), which contributes to the adaptation of fish to cold waters. In recent years, the ingestion of omega-3 fatty acids has been associated with a significant reduction in coronary heart disease, cardiac arrhythmias, and acute myocardial infarction (considered the leading cause of sudden cardiac death) due to an increased cardiac output, an almost complete inhibition of platelet aggregation, contribution to the prevention of thrombosis, relaxation of smooth muscle in blood vessels, and lower blood pressure. Polyunsaturated long-chain omega-3 fatty acids have been associated with antiarrhythmic, antithrombotic, antisclerotic, and antihypertensive effects, and prophylaxis of coronary disease [82, 83].

Other important sources include green vegetables like purslane, quelites and broccoli. The capacity of consumption of essential fatty acids such as omega-3 and omega-6 in a proper balance and quantity, helps to stabilize the metabolism of fats in the body and is involved in many other organic processes. As a result, cholesterol metabolism, quantity and transport are normalized, reducing the risk of cardiovascular disease. Omega-3 fatty acids are specifically involved in reducing cholesterol carried in low density lipoproteins (LDL), especially the higher risk, smaller and denser particles of LDL known as "bad" cholesterol, and in facilitating an increase in high-density lipoprotein (HDL) or "good cholesterol" that cleans the arteries instead of damaging them. They also have a role in the normal functioning of the endothelium (on the inside of the body's arteries) within which the lesions of atherosclerosis occur. A good balance in the supply of essential fatty acids and the significant contribution of polyunsaturated and monounsaturated fats delays the onset of atherosclerotic lesions. The intake of polyunsaturated fatty acids corrects delayed or adult onset diabetes, which also reduces, in turn, cardiovascular risk [84].

Many studies suggest that replacement of saturated fatty acids in the diet with n-3 PUFA results in enhanced susceptibility of LDL to oxidation. In humans, susceptibility to LDL oxidation is highly correlated with development of coronary stenosis. Adverse effects of n-3 PUFA in LDL oxidation in vitro do not occur in vivo, and thus may explain the protective effect of n-3 PUFAs against atherosclerosis, as has been demonstrated through reduced uptake by macrophages and a lower electrophoretic mobility of LDL group patients ingesting a daily supplement of 16 g of fish oil for three months compared to a control group of patients without dietary treatment [85, 86].

*10.1.2. Sources of Omega 3*

In Mexico, there is a real deficiency of these nutrients as consumption of fish (one of the main sources) and green plants is very low. The estimated per capita consumption is only 5–10g of fish per year, making it very difficult to obtain the minimum required amounts from this source. Green vegetable consumption is limited to only a few vegetables per day. Because these nutrients are part of membrane phospholipids, which are crucial for antioxidant defense against free radicals, the failure to include sufficient n-3 fatty acid promotes the lipoperoxidation observed in the development of slowly progressive or chronic degenerative diseases. When a nutrient is missing, it is likely that others are also deficient, and if within these are other antioxidants, the pattern becomes complicated. The development of chronic and degenerative diseases is likely to be promoted not only by the deficiency of antioxidant nutrients, but also by deficiency of other nutrients "working" with the same function. An imbalance in the intake of antioxidants can lead to free radicals having more of a role in the development of the diseases that cause major public health problems in developing countries. However, the solution will always be prevention through dietary education promoting an increased consumption of foods that are rich in antioxidants from the earliest stages of life, and even before birth [87, 88, 89].

In the table, 12 major sources of n-3 fatty acids are shown. The values of alpha linolenic acid (ALA), eicosapentaenoic (EPA) and docosahexaenoic acid (DHA) in edible plants are presented.

The most important sources of omega-3 fatty acids are fish and vegetables. Fish contains a good amount of alpha-linolenic acid, eicosapentaenoic acid (EPA) and docosahexaenoic acid (DHA). Vegetables are an excellent source of alpha-linolenic acid, as a precursor to EPA and DHA, but some other vegetables also contain these fatty acids as shown in Table 12.

| **a-LINOLENIC / CONCENTRATION mg/100g** | | **EPA / CONCENTRATION mg/100g** | | **DHA / CONCENTRATION mg/100g** | |
|---|---|---|---|---|---|
| Spinach | 36.764 | Broccoli | 4.987 | Romain lettuce | 6.240 |
| Watercress | 29.828 | Chard | 4.959 | Cauliflower | 5.743 |
| Yerbamora | 22.521 | Lettuce orejona | 4.792 | Coriander | 5.347 |
| Parsley | 20.139 | Romain lettuce | 4.375 | Nopal | 3.514 |
| Purslane | 19.774 | Papaloquelite | 3.817 | Orange pepper | 3.259 |
| Italian lettuce | 16.943 | Coriander | 2.848 | Huitlacoche | 3.183 |
| Lettuce heart | 13.303 | Serrano pepper (green) | 2.458 | Yerbabuena | 3.171 |
| Beetroot | 10.835 | Cabbage | 2.446 | Chayote spineless | 2.275 |

| a-LINOLENIC / CONCENTRATION mg/100g | | EPA / CONCENTRATION mg/100g | | DHA / CONCENTRATION mg/100g | |
|---|---|---|---|---|---|
| Serrano pepper (green) | 10.822 | Parsley | 2.144 | Serrano pepper (green) | 2.227 |
| Broccoli | 9.615 | Xoconostle | 2.009 | Red onion | 2.041 |
| Artichoke heart | 8.815 | Epazote | 1.764 | Mushroom | 1.857 |
| Poblano Chili | 8.279 | Milkweed | 1.752 | Cabbage | 1.696 |
| Coriander | 7.554 | Habanero pepper | 1.680 | Tomato | 1.338 |
| Lettuce sangría | 7.269 | Echalot | 1.393 | Round pumpkin (Creole) | 1.262 |
| White onion | 6.844 | Cebollines | 1.316 | Lettuce orejona | 1.213 |
| Lemon peel | 6.242 | Red onion | 1.299 | French lettuce | 1.183 |
| Peeled lemon | 6.229 | Pore | 1.280 | Papaloquelite | 1.175 |
| Red pepper | 6.193 | Huitlacoche | 1.242 | Chicory | 1.171 |
| Lentil seed | 6.096 | Green beans | 1.086 | Epazote | 1.083 |
| Chard | 5.267 | Cucumber with peel | 1.027 | Asparagus | 1.044 |
| Green bean | 5.236 | Mallow | 0.889 | Xoconostle | 1.034 |
| Red onion | 4.941 | Green beans | 0.875 | Radishes | 0.894 |
| Round pumpkin (Creole) | 4.779 | Beetroot | 0.859 | Green beans | 0.798 |
| Nopal | 4.562 | Nopal | 0.822 | Papalo | 0.679 |
| Chile apple | 4.550 | Poblano Chili | 0.784 | Beetroot | 0.546 |
| Papalo | 4.382 | Bean sprouts | 0.780 | Tomato ball | 0.538 |
| Chayote boneless | 4.234 | Round pumpkin (Creole) | 0.759 | Chayote boneless | 0.496 |
| Eggplant | 3.761 | Chayote boneless | 0.699 | Artichoke | 0.482 |
| Romeritos | 3.545 | Peeled cucumber | 0.691 | Swede | 0.463 |
| Celery | 3.385 | Cuaresmeño pepper (jalapeño) | 0.673 | Parsley | 0.450 |
| Cucumber with peel | 3.293 | Elongated gourd (Italian) | 0.657 | Lemon peel | 0.450 |
| Carrot | 3.249 | Tomato | 0.615 | Mushrooms | 0.442 |
| Cebollines | 3.099 | Swede | 0.584 | Habanero pepper | 0.414 |
| Yerbabuena | 3.084 | Carrot | 0.572 | Red cabbage | 0.355 |
| Cauliflower | 3.006 | Yerbabuena | 0.554 | White onion | 0.332 |
| Red cabbage | 2.974 | Xoconostle heart | 0.549 | Broccoli | 0.329 |
| Garlic | 2.878 | Orange pepper | 0.548 | Poblano Chili | 0.316 |
| Mushroom | 2.391 | Squash blossom | 0.507 | Yerbamora | 0.312 |
| Chilacayote | 2.239 | Mushroom | 0.497 | Cherry tomato | 0.248 |

| a-LINOLENIC / CONCENTRATION mg/100g | | EPA / CONCENTRATION mg/100g | | DHA / CONCENTRATION mg/100g | |
|---|---|---|---|---|---|
| Jícama | 2.189 | White onion | 0.469 | Huazontle | 0.243 |
| Tomato ball | 2.039 | Red pepper | 0.459 | Cebollines | 0.203 |
| French lettuce | 1.976 | Cob | 0.453 | Peas | 0.196 |
| Peeled cucumber | 1.971 | Chile apple | 0.354 | Cob | 0.168 |
| Cob | 1.813 | Papalo | 0.344 | Chilacayote | 0.150 |
| Orange pepper | 1.658 | Chicory | 0.322 | Echalot | 0.137 |
| Hongo escobetilla | 1.607 | Watercress | 0.281 | Hongo escobetilla | 0.075 |
| Pore | 1.590 | Asparagus | 0.263 | Eggplant | 0.069 |
| Habanero pepper | 1.577 | Chayote spineless | 0.253 | | |
| Hongo horn sheep | 1.483 | Peas | 0.241 | | |
| Asparagus | 1.393 | Huazontle | 0.239 | | |
| Chicory | 1.269 | Red cabbage | 0.229 | | |
| Swede | 1.265 | Tomato | 0.135 | | |
| Romain lettuce | 1.114 | Green pepper | 0.129 | | |
| Milkweed | 1.114 | Hongo yemita | 0.122 | | |
| Lettuce orejona | 1.093 | Radishes | 0.118 | | |
| Epazote | 1.045 | Chilacayote | 0.067 | | |
| Spring onion | 1.036 | Green tomato | 0.047 | | |

**Table 12.** Most important sources of omega 3 fatty acids in edible vegetables [94, 95]

In Mexico has over 12, 000 km of shoreline and as such is rich in different species of marine fish, and it also has a wide variety of freshwater fish. In the Gulf of Mexico alone, about 500 different species of fish rich in omega-3 have been identified along with many others in the Pacific Ocean, only some of which have been specifically identified and their nutrient composition with respect to omega 3 fatty acids reported. Our group has already published some of these results in international journals and forums [90, 91].

However, the per capita consumption of fish has diminished in recent years, and domestic production has been used mostly for export purposes. Current per capita consumption is approximately 10 g of fish annually, compared with countries such as Japan and China that have a per capita consumption of 70 kg annually. On the other hand, their consumption has been avoided by some groups due to the danger of allergic reactions, or in the case of pregnant women and children under one-year-old, on the recommendation of doctor. In addition, fish oils rich in omega 3 have been used in the preparation of food for pets and animals, rather than for humans [92, 93].

The consumption of green vegetables has also declined, and vegetables are only consumed daily in certain sectors of society, while in others they are essential not consumed at all — as discussed in the text.

There are currently few studies in Mexico showing deficiencies of omega 3 fatty acids, but the presence of several clinical signs of deficiency have already been observed in children. These include prematurity, malnutrition, retardation of intrauterine growth, recurrence of infections and allergies, psychomotor and mental neurodevelopment problems, delays in learning and damage to the vision of young children. In adults, there is an increase in retinopathy and a significant increase in neurodegenerative problems like Alzheimer's disease and other diseases associated with the central nervous system, at younger ages than in other countries.

Among the most serious issues is a significant increase in mortality from coronary heart disease, which is the primary cause of death in Mexico. It should be remembered that omega 3 fatty acids are considered to be cardio-protective. In pregnant women, both eclampsia and pre-eclampsia have been rising at the same time as premature rupturing of membranes and other problems associated with pregnancy. Our group has already documented a very serious deficiency of omega 3 fatty acids in the breast milk of clinically healthy women, and the same deficiency was observed in the children they fed. Even the serious problem of childhood obesity in Mexico is being analyzed from the clinical point of view of micronutrient deficiency, while that of the ingest excesses in others. Thus, increased consumption of foods with a high glycemic index and saturated fat content is combined with the consumption of very few foods with omega-3 fatty acids such as fish and plants. In addition, the high rate of physical inactivity and sedentary lifestyles, makes the problems somewhat more challenging to solve. If we add to this pattern the impending imbalance between consumption of antioxidant and pro-oxidant food, the result will be further disease and mortality — the main problems affecting public health in Mexico [94].

## 11. Antioxidant supplements and Mexican sportsmen

The consumption of vitamins and inorganic elements via supplements began back in 1930 following the chemical isolation of these compounds. In 1939, cyclists in the Tour de France reported an increase in performance level after taking vitamins. However, in 1940, further research did not significantly support this position (Applegate and Grivetti, 1997), and even in the current literature ergogenic capabilities of these micronutrients are not described. Among the vitamins and inorganic elements frequently used in sport are vitamin B12, antioxidants such as vitamin C, E, beta carotene, zinc, magnesium, copper, selenium and iron, carnitine, chromium, vanadium, and boron.

The significant increase in the demand for oxygen that physical activity involves — especially if it is intense and continuous — is responsible for a parallel rise in the formation of free radicals derived from oxygen (up to three times the resting value) and is considered to be one of the main initiators or amplifying mechanisms of muscle damage associated with exercise (Margaritis, 2003; Alonso, 2006). These highly reactive particles can activate a series of chain reactions and are able to damage collagen fibers, cell membranes and nuclear structures. The use of antioxidants in sport can be a useful weapon against oxidative stress, since it has been observed that athletes have better levels of vitamin C and glutathione peroxidase, and therefore higher levels of lipid peroxidation.

On this basis, it is easy to understand the importance of mechanisms enabling antioxidants to protect the body from the damage caused by exercise, especially when it is intense. Although it is noteworthy that the unjustified use of vitamins and inorganic elements as a special requirement for better physical performance was observed. Several studies have demonstrated that the use of these supplements can exceed daily recommended amounts in the blood by 10–15 times without causing any direct benefit in athletic performance [96].

In the study by Guzman in 2009 (Table 13 and 14) [97] the ingestion of nutrients by Mexican athletes in different sports was investigated. These sports were weightlifting, tae kwon do, swimming, zumba and spinning. In the same table, the average intake of vitamins and minerals calculated from records of the daily supplements taken by type of exercise, and in brackets the average percentage exceeding the daily recommendation, are shown. Not included in this section is a calculation of the intake of vitamins and minerals from the diet. We found that eating supplements alone met and even exceeded the recommendations for the daily intake of most vitamins in all types of exercise.

| Minerals | Weightlifting | Tae Kwon | Swimming | Zumba | Spinning |
|---|---|---|---|---|---|
| A (μg/d) | 1030.5±1140.5 (132%±169.4) | – | 2500 (625%) | 3300 (471%) | 3333 (476%) |
| D (μg/d) | 10 (200%) | – | 10±7.1 (200%±141.4) | 17 (340%) | 5 (100%) |
| E (mg/d) | 27.4± 21.2 (182%±141.5) | 25.83 (172%) | 3.3 (47%) | 162.1±127.4 (1080.7±849.4) | 99.9 (666%) |
| K (mg/d) | 28.3±10.4 (29%±3.2) | – | – | – | – |
| C (mg/d) | 43.1±61.1 (55%±82.6) | 600 (800%) | 410±445.3 (1667%±1747.4) | 1000 (1333% ) | 250 (333%) |
| $B_1$ (mg/d) | 8.4±23.4 (752%±2132.7) | – | 1 (167%) | 76 (6909% ) | 50 (4545%) |
| $B_2$ (mg/d) | 9.5±24.6 (848%±2239.9) | – | 1.2 (192%) | 76 (6909% ) | 50 (4545%) |
| $B_6$ (mg/d) | 9.9±24.5 (758%±18.8) | – | 2 (333%) | 72 (5538% ) | 50 (3846%) |
| $B_{12}$ (μg/d) | 14.1±25 (589%±1043.6) | 25 (1041%) | – | 6 (250%) | 75 (3125%) |
| Niacin | 21.7±24.5 | – | 6.7 | 82 | 50 |

| Minerals | Weightlifting | Tae Kwon | Swimming | Zumba | Spinning |
|---|---|---|---|---|---|
| (mg/d) | (144%±174) | | (83%) | (586%) | (357%) |
| Folic (μg/d) | 168±105 (42%±26.2) | 1000 (250%) | – | 400 (100%) | 200 (50%) |
| Pantotenic (μg/d) | 17.8±24.7 (356%±494.6) | – | – | 76 (1520% ) | 50 (1000%) |
| Biotina (μg/d) | 83.1±55 (284%±183.3) | – | – | 300 (1000% ) | 50 (167%) |

() Data brackets indicate percentage of the recommended amount

**Table 13.** Average daily intake and % RDI of vitamins from food supplements (Guzman, 2009) [97]

| Minerals | Weightlifting | Tae Kwon | Swimming | Zumba | Spinning |
|---|---|---|---|---|---|
| Calcium (mg/d) | 391.6±228.8 (39%±23.2 ) | – | 43.4 (5%) | 2000 (200%) | 25 (2%) |
| Iron (mg/d) | 5.8±6.1 (60%±61.2 ) | 115 (1045%) | – | – | 5 (28%) |
| Phosphorus (mg/d) | 421.7 ±339.5 (58%±49.6 ) | – | 66.7 (13%) | 228 (32% ) | – |
| Magnesium (mg/d) | 114.5±75.9 (29%±18.8 ) | – | – | 133 (41% ) | 25 (8%) |
| Zinc (mg/d) | 6.5±3.9 (66%±49.7 ) | – | – | 30 (375%) | 7.5 (94%) |
| Boron (μg/d) | – | – | – | – | 500 (50,000%) |
| Potasium (mg/d) | 99 (5%) | – | – | – | – |
| Chrome (μg/d) | 56.7±34.8 (164%±96.4 ) | – | – | 6 (24% ) | 50 (200%) |
| Selenium (μg/d) | 28.5±16.9 (52%±30.8 ) | – | – | 50 (91% ) | 75 (136%) |
| Cupper (μg/d) | 650±382.1 (72%±42.6 ) | – | – | 600 (67% ) | 1000 (111%) |

() Data brackets indicate percentage of the recommended amount

**Table 14.** Daily average and % RDI of inorganic elements from ingestion of food supplements [97]

## 12. Antioxidant capacity of foods and their use as nutraceuticals or functional foods

Nutraceuticals are natural biological substances extracted from natural sources that are characterized by their anti-denaturing biotechnological processes, which conserve all original properties without any chemical manipulation. Once extracted from the natural source, nutraceuticals are studied using processes similar to those used to identify the biological properties of drugs in animals and humans. Once their properties are documented, they can be used as food supplements for human consumption, not as substitutes for a normal diet. Sometimes, some of these natural substances may act as potential drugs due to their biological properties and can be prescribed as therapeutic adjuvants for preventive and / or curative purposes. As a concept, the nutraceutical is between a natural raw commodity and a xenobiotic chemical or a substance foreign to the body [98].

In most cases their functional properties are implicit to their use, and for this reason biological testing prior to use is not necessary. The concept of functional foods should not be confused with health foods, which often contain the same components. There are also combinations of essential nutrient with herbs and exotic fruits and and vegetables, somehow giving them any "extra" functional property which makes them very striking [99].

## 13. Nutraceutical as a therapeutic option

The novelty of these functional foods is the current scientific knowledge of the beneficial properties linked to disease prevention that they possess. Nutraceuticals have been associated with the prevention and treatment of at least four of the diseases that contribute to the high rate of mortality in developed countries (cancer, diabetes, cardiovascular disease and hypertension) and to the prevention of other diseases such as neural tube defects, osteoporosis and arthritis [100].

### 13.1. Global use of nutraceuticals

The growing expectation for good health has led to a large number of studies demonstrating the effective use of nutraceutical components. Although their indiscriminate use has been speculated, it is reversing the paradoxes of health in each country, which are shown as major public health problems, indicating a serious food deficiency or the mishandling of individual diets. Thus, it is as mentioned as serious public health problems, and in some cases even endemic in the population, diseases like cancer, diabetes mellitus, cardio and cerebrovascular disease, hypertension, obesity, cirrhosis and malnutrition. The concept of prevention is closely associated with the word nutraceutical, and this is a current that is gaining strength as prevention of disease is better than trying to find a cure. The costs to budgets of the latter impacts on virtually every hospital in the world [85, 86].

### 13.2. Convincing results

The cellular effects of nutraceuticals are easily understood as they are part of the phospholipid cell membrane. Their function is to protect the cells from virtually everything that can damage them such as natural, self or foreign oxidants which can cause lipid peroxidation.

Virtually all of these nutraceuticals act to protect against oxidants commonly called free radicals (FR), making them natural antioxidants (vitamins E, C, Selenium, CLA, etc). These antioxidants are found within — or are associated with — the lipid bilayer of the cell membrane, and many of them are even associated with membrane phospholipids.

Antioxidant protection is based on the union of the nutraceutical with FR, which is different with each nutraceutical, in that some take a shorter or longer time to join to FR, and it is in that space of time where the effectiveness of the nutraceutical reflects its protective biopotency. This may partly explain the anticarcinogenic property shown along with an antioxidant capacity. Most of these results have been obtained by in vivo and in vitro studies and have been confirmed by cell culture. Even where inhibition of carcinogenicity by nutraceuticals has been demonstrated (in conjugated linoleic acid (CLA) for example), it is in part accomplished by inducing the activity of antioxidant enzymes (such as SOD, catalase, GPx) and the generation of cytotoxic activities in certain tumor lines. When an antioxidant (an enzyme, vitamin or mineral) has failed to maintain protection of the cell membrane by itself, the combined activity of several nutraceuticals can achieve the expected protection [102, 103].

## 14. Conclusions

The use of healthy foods is increasing and is becoming more evidence-based with the use of functional foods and plants —or components thereof — with a nutritional content that can help prevent or treat disease. The conscious and optimal use of these nutraceuticals as antioxidants requires more knowledge of their properties and their experimentation probed epidemiologically.Regardless, higher consumption of vegetables and fruits that contain a lot of these antioxidant free radical scavengers is always recommended. It doesn't matter if the food is raw or cooked, it is more important that daily consumption increases. There is a discussion to be had regarding organic vegetables, and even organic animal products, but it is more important to include them as 50% of the diet, regardless of whether the source is organic or not. The appropriate consumption of fiber, prebiotics and probiotics is also necessary for adequate intestinal health and body processes.

It is controversial, but for energy and other substrates, cells need to oxidize nutrients such as glucose, amino acids or fatty acids. However, diet provides not only the body of nutritious elements, a high amount of consumed compounds have also an important oxidative capacity, and other compounds which are not oxidants per se, can create, once absorbed, an oxidative environment.

The most important advice for people below the poverty line is to recommend to them the high consumption of vegetables, which in Mexico are cheap and easily available, but there are still issues with directions for their use and handling, and further problems of proper living conditions and, of course, income. Both appropriate nutritional guidance directed at this group to address risk, and high-impact informative programs that influence diet, are lacking. A lack of knowledge of the nutritional composition of food is another factor which is not yet being addressed due to the lack of information and analysis of the same. However, efforts aimed at solving the problems are taking place, but very slowly.

Promotion of the national strategy for better health through the prevention of disease makes use of the available evidence. It includes universal, targeted and specific actions to reduce the effect of CNCD. All actions can only be conducted with the consensus of society and in conjunction with government agencies. In particular, attention should be focused on schools through programs directed at primary and secondary education, and the training of teachers and parents as well as regulation of the food sold on campuses and surrounding areas. The same actions can be applied to institutions that distribute food to their employees. These measures are complemented by demand for the development or adaptation of facilities enabling physical activity.

Agreements with the food industry and any organization related to the distribution of food are necessary, so that the population is allowed access to healthy food, or at least access to information of its composition. In this sense, measures corresponding to the composition of industrialized food labelling are already being taken according to Mexican standard (NOM-051-SCFISSA1-2010), where the nutritional values of the same should be limited, but specific information on industrialized or natural foods is still very limited and restricted to only certain groups of professionals. Information on healthy lifestyles must be disseminated by various means of communication in order to promote regular exercise and the correct choice of foods. Information about the risks of obesity and diabetes alone is not enough to change habits. The effect that this information has will be magnified if an environment conducive to putting it into practice can be created.

## Author details

José Luis Silencio Barrita[1*], Sara Montaño Benavides[2] and Santiago Sánchez[3]

*Address all correspondence to: silenciobarrita@live.com.mx

1 Instituto Nacional de Ciencias Medicas y Nutriciòn "Salvador Zubirán", Departamento de Ciencia y Tecnología de los Alimentos, México

2 Instituto Nacional de Ciencias Medicas y Nutriciòn "Salvador Zubirán", Departamento de Nutrición Animal, México

3 Instituto Mexicano del Seguro Social, IMSS, Hospital General de Zona No. 30, "Iztacalco", Departamento de Nutrición y Dietética, México

http://dx.doi.org/10.5772/61626

## References

[1] Clasificación Estadística Internacional de Enfermedades y Problemas Relacionados con la Salud (CIE), INEGI, 2013

[2] Rivera Dommarco, Juan, et al. Encuesta Nacional de Nutrición 1999. Estado nutricional de niños y mujeres en México. Cuernavaca, México, Instituto Nacional de Salud Pública, 2001. Available in www.insp.mx/enn

[3] Martínez Jasso I., Villezca Becerra PA. La alimentación en México: un estudio a partir de la Encuesta Nacional de Ingresos y Gastos de los Hogares. Revista de información y análisis, 2003; 21:26-37

[4] Silencio Barrita JL. Ácidos grasos poliinsaturados y selenio en leche humana. Nutrición Clínica, 2004; 7(4):227-39

[5] Córdova-Villalobos JA, Barriguete-Meléndez JA, Lara-Esqueda A, Barquera S, Rosas-Peralta M, Hernández-Ávila M, De León-May ME, Aguilar-Salinas CA. Chronic non-communicable diseases in Mexico: epidemiologic synopsis and integral prevention. Salud Pública Mex 2008; 50: 419-427.

[6] Aguilar-Salinas CA, Mehta R, Rojas R, Gomez-Perez FJ, Olaiz G, Rull JA. Management of the metabolic syndrome as a strategy for preventing the macrovascular complications of type 2 diabetes: controversial issues. Curr Diab Rev 2005; 1: 145-158.

[7] Gutiérrez-Delgado C, Reynoso-Noverón N, Hérnandez-Ávila M, MoharBetancourt A. Perfil epidemiológico de los cánceres en población adulta mexicana, periodo 2000-2010 [documento en revisión].

[8] Editorial. El cáncer en México: propuestas para su control. Salud púb Méx, 2014; 56(5): 418-420

[9] Aguilar-Salinas CA, Olaiz G, Valles V, Rios JM, Gomez-Perez FJ, et al. High prevalence of low HDL cholesterol concentrations and mixed hyperlipidemia in a Mexican nation wide survey. J Lip Res 2001; 42:1298- 1307

[10] Aguilar-Salinas CA, Rojas R, Gomez-Perez FJ, Valles V, Rios-Torres JM, Franco A, et al. High prevalence of the metabolic syndrome in Mexico. Arch Med Res 2004; 35:76-81.

[11] Aguilar-Salinas CA, Rojas R, Gomez-Perez FJ, Garcia E, Valles V, RíiosTorres JM, et al. Early onset type 2 diabetes in a Mexican, populationbased, nation-wide survey. Am J Med 2002; 113:569-574

[12] Estadísticas a propósito del... día mundial contra el cáncer (4 de febrero)" datos nacionales, 2015 disponible en: file:// EAP_del_dia_mundial_contra_el_cancer.pdf

[13] http://www.coneval.gob.mx/rw/resource/coneval/info_public/PDF_PUBLICACIONES/Evolucion_Historica_050411.pdf

[14] Olaiz G, Rojas R, Barquera S, Shamah T, Aguilar-Salinas C, Cravioto P, et al. Encuesta Nacional de Salud 2000. La salud de los adultos. Cuernavaca, Morelos: Instituto Nacional de Salud Pública, 2003.

[15] Oláiz-Fernández G, Rivera-Dommarco J, Shamah-Levy T, Rojas R, Villalpando-Hernández S, Hernández-Ávila M, et al. Encuesta Nacional de Salud y Nutrición 2006. Cuernavaca, Morelos: Instituto Nacional de Salud Pública, 2006.

[16] Bourges H. Glosario de términos nutriológicos. Cuadernos de Nutrición. Vol. 24. No. 1. Enero- Febrero 2001.

[17] Silencio Barrita JL. Vitaminas: conceptos generales, Nutrición Clínica 2006; 9(3):36-44

[18] Bourges Rodríguez H. Glosario de términos para la orientación alimentaria. Cuad Nutr 1988; 11(6):1-47.

[19] Ascencio Peralta C, Bourges Rodriguez H, Gomez Rodriguez FE, Silencio Barrita JL. "Conceptos basicos sobre la nutricion y los nutrimentos ", Fisiologia, Celulas, Organos y Sistemas, editores E. Julio Muñoz Martinez y Xaviera Garcia Gonzalez, Fondo de Cultura Economica, Tomo IV, Capitulo VII.1, Pag 23-45, 1997.

[20] Braunwald E., Isselbacher KJ. Petersdorf RG., Wilson JD., Martin JB., Fauci AS. Harrison´s Principles of Internal Medicine, Mc Graw Hill Co, 11a Ed, 1987, USA.

[21] Briggs MH. Vitamins in human Biology and Medicine. CRC press, 2nd edition 1987, USA.

[22] Hoffman-La Roche F. Compendio de vitaminas, 1970 Basilea, Suiza.

[23] Casanueva E, Valdés-Ramos R, Pfeffer F, Ricalde-Moreno A, García-Villegas E, Meza C. Serum retinol in urban Mexican women during the perinatal period. Salud Pública Mex 1999; 41:317-321.

[24] Mayor Oxilia R. Estrés Oxidativo y Sistema de Defensa Antioxidante. Rev. Inst. Med. Trop. 2010; 5(2):23-29

[25] Stephensen ChB. Vitamin A, infection, and immune function. Ann Rev of Nutr. 2001, 21: 167-192

[26] Huiming Y, Chaomin W, Meng M. Vitamin A for treating measles in children (Review), 2005, The Cochrane Collaboration. Published by John Wiley & Sons, Ltd

[27] Valkoa M, Leibfritzb D, Moncol J, Croninc MTD, Mazura M, Telserd J. Free radicals and antioxidants in normal physiological functions and human disease. Int J Biochem & Cell Biol, (2007) 39: 44–84

[28] Geil P, Shane-McWhorter L Dietary supplements in the management of diabetes: potential risks and benefits.

[29] Doldo E, Costanza G, Agostinelli S, Tarquini C, Ferlosio A, Arcuri G, Passeri D, Scioli MG, Orlandi A. Vitamin A, Cancer Treatment and Prevention: The New Role of Cel-

lular Retinol Binding Proteins. Biomed Res Int (2014), Article ID 624627, downloads.hindawi.com/journals/bmri/.../624627.pdf

[30] Bennasir H, Sridhar S, Abdel-Razek TT. Vitamin A... from physiology to disease prevention. Int J Pharmaceutical Sci Rev and Res (2010) 1(1):68-73

[31] Marks J. A guide to the vitamins, their role in health and disease. MTP Medical and Technical Pub. 1975, England.

[32] Brigelius-flohe R, Traber MG. Vitamin E: function and metabolism. (1999), FASEB J, 13:1145-1155

[33] Drevon CA. Absorption, Transport and Metabolism of Vitamin E. 1991, 14(4):229-246

[34] Gagné A, Wei SQ, Fraser WD, Julien P. Absorption, Transport, and Bioavailability of Vitamin E and its role in Pregnant Women. J Obstet Gynaecol Can 2008; 31(3):210–217

[35] Wu JH, Croft KD. Vitamin E metabolism. Mol Aspects Med 28 (2007) 437–452

[36] Dutta A, Dutta SK. Vitamin E and its Role in the Prevention of Atherosclerosis and Carcinogenesis: A Review. J Am College of Nutr, (2003); 22 (4):258–268

[37] González J, Pérez N, García G. Vitamina E: niveles séricos en hombres, mujeres y embarazadas. Bol Med Hosp Infant Mex. 1982; 39: 327-332.

[38] Vega-Franco L, Meza C, Meijenrik J, Alegret C. Concentración de vitamina E en niños con desnutrición proteico-energética. Bol Med Hosp Infant Mex. 1989; 46: 607-610.

[39] Garcıa-Closas R, Berenguer A, Tormo MJ, Sánchez MJ, Quiros JR, Navarro C, Arnaud R, Dorronsoro M, Chirlaque MD, Barricarte A, Ardanaz E, Amiano P, Martinez C, Agudo A, Gonzalez CA. Dietary sources of vitamin C, vitamin E and specific carotenoids in Spain. British J Nutr (2004), 91, 1005–1011

[40] Silencio Barrita JL.- Antioxidantes en el tratamiento de las enfermedades: Vitamina C. Capitulo 28.- En el libro: Los antioxidantes y las enfermedades crónico-degenerativas, tomo II, editores Morales González JA, Fernández Sánchez AM, Bautista Ávila M, Vargas Mendoza N, Madrigal Santillán EO. Universidad Autónoma del Estado de Hidalgo, 2009.

[41] Iqbal K, Khan A Ali Khan Khattak MM. Biological Significance of Ascorbic Acid (Vitamin C) in Human Health: A Review Pakistan J Nutr 2004, 3(1):5-13,

[42] Simon J.A Vitamin C and cardiovascular disease: a review J Ame Collage of Nutr, 1992, 11 (2):107-125

[43] Simon JA, Hudes ES, Tice JA. Relation of serum ascorbic acid to mortality among US adults. J Am Coll Nutr 2001; 20:255–63.

[44] Bourges H. Madrigal H. Chavéz A., Tablas del valor nutritivo de los alimentos mexicanos. Publicación L-12 INN, 1983, 13 ed. México

[45] Harris J.L. Vitamin C saturation test. Standarization meassurements at graded levels of intake. Lancet 1943, 1:515,

[46] Head KA, Ascorbic Acid in the Prevention and Treatment of Cancer. Altern Med Rev 1998; 3(3):174-186)

[47] Jacob RA, Pianalto FS, Agee RE. Cellular ascorbate depletion in healthy men. J Nutr. 1992; 122:1111-1118.

[48] Gomez E, Silencio JL, Bourges H. Vitamin B2, B6 and C status in patients with primary Sjögren's syndrome. 7th International congress of mucosal immunology, august 16-20, 1992, Prague Csechoslovaquia.

[49] Silencio Barrita JL, Santiago Sánchez MS. CHAPTER 18.-Antioxidant role of ascorbic acid and protective effects on chronic diseases. In the book Oxidative stress and chronic degenerative diseases- a role for antioxidants, edited By JA Morales-Gonzales, In Tech Editors, sept 2012, pages 449-484. ISBN 980-953-307-650-3

[50] Herrera, V. Más evidencia en contra del uso de vitaminas y antioxidantes en la prevención de enfermedades crónicas. Evidencia Actualización en la Práctica Ambulatoria-2002, 5(6): Nov-Dic

[51] Watt B.K. A.L. Composition of foods. Agriculture Handbook num. 9, United States, Departament of Agriculture, USA, 1975.

[52] The Merck indexa n enciclopedia of chemicals and drugs, eighth edition published by Merck &CO., Inc Rahway NJ USA, 1968

[53] Moreau RA, Whitaker BD, Hicks KB. Phytosterol, phytostanol and their conjugates in foods: structuraldiversity, quantitative analysis, and health promoting uses. Progress in Lipid Research. 2002; 41:457

[54] Duperon R. Thiersault M. Duperon P. Phytochemistry, 1984; 23:743

[55] Hicks KB, Moreau RA. Phytosterols and phytostanols; functional food colesterol busters. Food Technology, 2001; 55(1):63

[56] http://dcb-carot.unibe.ch/carotint.htm and http://www.carotenoidsociety.org/

[57] Sahyoun, R.N., 1996. Carotenoids, vitamins C and E and mortality in an elderly population. Am. J. epidemio, 144: 501-511

[58] Kumar S, Pandey AK, Rao MM, Razzaque WAA. Role of ß carotene/vitamin A in animal reproduction. Veterinary World, 2010; 3(5):236-237

[59] Andrews S, 2001. JANA, the role of lutein in human health. J Am Nutraceutical Assoc 4(2):1-30

[60] Gonzalez Cortes JH. Desgarro de epitelio pigmentario en la degeneración macular relacionada con la edad. XXV Congreso Mexicano de Oftalmología, Morelia Mich, 3-7 agosto 2002, disponible: http://dmremonterrey.blogspot.mx/

[61] Clinicopathological correlation of retinal pigment epithelial tears in exudative age related macular degeneration: pretear, tear, and scarred tear. Br J Ophthalmol. 2001 April; 85(4): 454–460

[62] Silencio Barrita JL. El selenio. Cuadr Nutr 2011, 34(6):230-235.

[63] Silencio Barrita JL, Bellido Villa A, Ritter Santiago T. Selenio. Nutrición Clínica 2004, 7(1):78-85

[64] Silencio Barrita JL. Las enfermedades renales y el selenio, Nutrición Clínica 2003, 6:291-292

[65] Ramiro Anaya M. Determinación de selenio por un método fluorométrico en 100 alimentos seleccionados. Tesis experimental, Facultad de Ciencias, UNAM, 2005.

[66] Rivero Silva OM. Determinación de selenio en 200 alimentos mexicanos. Tesis experimental, Instituto Nacional de Ciencias de la Salud, Universidad Autónoma del Estado de Hidalgo, 2010.

[67] Casanueva E, Kauffer Horwitz M, Pérez Lizaur AB, Arroyo P. Nutriología Médica, Editorial Panamericana, México, 2ª edición, capítulo III, 2001.

[68] Ortiz-Andrellucchi A. Nutrición e inmunidad. Rev Soc Med Quir Hosp Emerg Pérez de León 2007; 38(Suppl 1):12-18.

[69] Hunt J, Murphy M, Martner P, Faraji B, Swendseid M, Reynolds R, Sanchez A, Mejia A. Zinc, vitamin B6, and other nutrients in pregnant women attending prenatal clinics in Mexico. Am J Clin Nutr. 1987; 46: 563-9

[70] Bourges RH, Casanueva E, Rosado JL. Recomendaciones de ingestión de nutrimentos para la población mexicana: bases fisiológicas, tomo 1, Editorial Médica Panamericana, México, capítulos V, VI y VIII, 2005.

[71] Osredkar J, Sustar N. Copper and Zinc, Biological Role and Significance of Copper/Zinc Imbalance. J Clinic Toxicol 2011, S3:001

[72] Naithani M, Bharadwaj J, Darbari A. Review of Various Indicators for Assessment of Zinc Requirement and Effectiveness. Acta Medica International 2014; 1 (1):32-35

[73] Hambidge M (2000) Human zinc deficiency. Journal of Nutrition 130: 1344S–1349S

[74] Hambidge M (2003) Biomarkers of trace mineral intake and status. Journal of Nutrition 133: 948S–955S

[75] Nriagu J. Zinc Deficiency in Human Health in http://www.extranet.elsevier.com/homepage_about/mrwd/nvrn/Zinc%20Deficiency%20in%20Humans.pdf

[76] Raymond JL. Krause's Food, Nutrition and Diet Therapy, 13th ed., W.B. Saunders Co, 2010

[77] Hernańdez M, Chávez A, Bourges H. Valor nutritivo de los alimentos mexicanos: tablas de uso práctico. México: Instituto Nacional de la Nutrición, División de Nutrición, 1977

[78] Sarika Arora and Raj Kumar Kapoor (2012). Iron Metabolism in Humans: An Overview, Iron Metabolism, Dr. Sarika Arora (Ed.), ISBN: 978-953-51-0605-0 Intech on line Books.

[79] Lash A, Saleem A. Iron metabolism and its regulation. A review. Ann Clin Lab Sci, 1995; 25(1): 20-30

[80] Carocho M, Ferreira ICFR. A review on antioxidants, prooxidants and related controversy: Natural and synthetic compounds, screening and analysis methodologies and future perspectives. Food and Chem Toxicol, 2013; 51:15–25

[81] Brambilla D, Mancuso C, Scuderi MR, Bosco P, Cantarella G, Lempereur L, Di Benedetto G, Pezzino S, Bernardini R. The role of antioxidant supplement in immune system, neoplastic, and neurodegenerative disorders: a point of view for an assessment of the risk/benefit profile. Nutr J. 2008, 7:29

[82] Simopoulos AP. Human requirement for N-3 polyunsaturated fatty acids. Poult Sci. 2000 Jul; 79(7):961-70.

[83] Silencio Barrita JL, Guzmán Federle D, Molano Pérez F. implicación de los ácidos grasos n-3 en nutricion y diabetes, en el libro Nutrición y Diabetes, editores Solano Solano G, Del Castillo Arreola A, García Meras M, Guzmán Saldaña R, Romero Palencia A, compilado por la Universidad Autónoma del Estado de Hidalgo, 2011, paginas 283-316, ISBN 978-607-482-192-5

[84] Silencio Barrita JL. LCPUFAS y su importancia en la alimentación del recién nacido. diccionario de Especialidades Farmaceuticas en Pediatria. PLM, ed Mauricio Aguilera. MEXICO, 2010, ISBN 978-607-7767-20-6

[85] De Vriese SR, Dhont M, Christophe AB. Oxidative stability of low density lipoproteins and vitamin E levels increase in maternal blood during normal pregnancy. Lipids 2001; 36(4):3616.

[86] Agostoni C. The difficult balance between dietary polyunsaturated fatty acids. Acta Paediatr 2003; 92(12):13713

[87] Lands WEM. Biochemistry and physiology of n-3 fatty acids. FASEB J 1992; 6:2530-6.

[88] Ramírez-Silva et al.: Fatty acids intake in the Mexican population. Results of the National Nutrition Survey 2006. Nutrition & Metabolism, 2011 8:33.

[89] Silencio Barrita JL, Lara Flores G, Pérez Gil Romo F, Montaño Benavides S, Ortiz Huidobro RI, Castro González MI, Barrera Millán E, de Titto Carboni AM, López

Cabrera FT, Santiago Sánchez MS, Falcón A, Irisson R. Ácidos grasos en el calostro y en la leche madura de mujeres mexicanas, 2012, Rev Mex Ped, 79(1):5-12

[90] Silencio Barrita JL., Ortiz Ortega V., Castro González MA, Montaño S, Velasco S, Cárdenas B, Pérez-Gil Romo F., Bourges Rodríguez H. Trace element content of marine foods from the pacific mexican coast. J Trace Elem Exp Med 1998; 11(4): 471

[91] Castro González MI, Silencio Barrita JL, Juárez Silva ME, Montaño Benavides S, Pérez-GIL Romo F. Composición química proximal de la fauna de acompañamiento del camarón de Veracruz. (Golfo de Mexico). Biologia Tropical 46(2):249-256, 1998

[92] Castro González MI, Pérez Gil Romo F, Carranco Jáuregui ME, Montaño Benavides S, Silencio Barrita JL. Vitaminas y minerales de sardina en salsa de tomate, colectada en las zonas pesqueras del pacifico mexicano. Arch Latin Nutr 1999, 49(No. 4): 379-383

[93] Castro González MI, Ojeda A, Silencio Barrita JL, Cassis L, Ledesma H, Pérez-Gil F. Perfil lipídico de 25 pescados marinos mexicanos con especial énfasis en sus Ácidos grasos n-3 como componentes nutracéuticos, Archivos Latinoamericanos de Nutrición, 54 (3):328- 336, 2004

[94] Castro-González I, Maafs-Rodriguez AG, Silencio-Barrita JL, Galindo-Gómez C, Pérez-Gil F. Evaluation of the possible inclusion of certain fish species in chronic kidney disease diets based on their adverse and beneficial nutrient ratios. Int J Food Sci Nutr., 2012, 63 (4): 1-7

[95] Martínez Pérez D. Contenido de ácidos grasos en verduras producidas y consumidas en México, tesis licenciatura en nutrición Universidad SIGLO XXI –UAEM, febrero 2014, México.

[96] Williams MH. Dietary supplements and sports performance: introduction and vitamins. J Int Soc Sports Nutr. 2004; 1(2):1-6.

[97] Guzmán-Federle D. Evaluación del consumo de complementos alimenticios en individuos que realizan ejercicio físico en gimnasios y centros deportivos de Pachuca, Hidalgo, tesina 2009, Universidad Intercontinental, México

[98] Silencio Barrita JL. Investigación clinica con respecto a Nutricion y cancer. Rev Mex Ped 2009; 76 (No. 4), jul-ago:181-186.

[99] Biruete Guzman A, Juarez Hernandez E, Sieiro Ortega E, Romero Viruelas R. Silencio Barrita JL. Los Nutraceuticos. Lo que es conveniente saber. Rev Mex Ped 2009; 76(3): 136-145.

[100] Kalra EK. Nutraceutical- definition and introduction. AAPS Pharm Sci 2003; 5:1-2

[101] Nutracéuticos. En: http://www.maimonides.edu/gerontologia2007/2007/03/las_dianas_que_se_utilizan_con.html.

[102] Silencio J.L. Nutracéuticos, Nutri informate 2006, 11(4): 11-12

[103] Hawker N. Nutracéuticos, ¿alimentos o medicamentos? Junio 2003. www.vitafoods.co.uk/2002/nutrition/ffnut.htm.

# Chapter 7

# Rooibos (*Aspalathus linearis*) and its Major Flavonoids — Potential Against Oxidative Stress-Induced Conditions

Olawale R. Ajuwon, Jeanine L. Marnewick and Lester M. Davids

Additional information is available at the end of the chapter

http://dx.doi.org/10.5772/61614

**Abstract**

Reactive species are products of normal cellular metabolism and may be deleterious or beneficial. At low/moderate concentrations, reactive species are involved in physiological roles including cell signalling, defense against infectious agents and mitogenic responses. However, unbalanced defense mechanism of antioxidants, overproduction of reactive species or incorporation of free radicals into the living system from the environment may result in oxidative stress, a deleterious process that can lead to damage of important cell structures, including lipids and membranes, proteins and nucleic acids. The role of oxidative stress as a contributing factor in the pathophysiology of various diseases is increasingly being recognized, and augmenting the oxidative defense capacity of the cell through the intake of antioxidants as a way of preventing free radical-mediated cellular injuries is becoming a popular strategy. Much attention is being focused on the health beneficial role of phenolic phytochemicals derived from plants. They are considered to play an important role as physiologically functional foods and for the prevention of clinical conditions related to oxidative stress, even though their modes of action may still not be fully understood. Rooibos (*Aspalathus linearis*) is a popular South African tisane enjoyed for its taste and aroma. Rooibos has been made in the Cederberg mountain region of South Africa for generations and has been used medicinally for alleviation of allergies, asthma, infantile colic and skin problems. The potential antioxidative, immune-modulating, chemopreventive and chemotherapeutic actions of rooibos have been reported in several studies. This review provides a comprehensive data on the current knowledge of the biological and chemotherapeutic activity of rooibos and its major flavonoids. Most recent *in vitro* and *in vivo* (animal and human) studies were conducted with special attention paid to clinical conditions in which oxidative stress has been implicated. The conclusion described directions for future rooibos research to establish its activity and utility as a human chemopreventive and therapeutic agent.

**Keywords:** Antioxidant, Chemoprevention, Oxidative stress, reactive oxygen species, Rooibos

# 1. Introduction

Oxidative stress results when there is a disturbance in normal cellular and molecular function as a result of an imbalance between the production of various oxidizing chemical species (pro-oxidants) and natural antioxidant ability of cells in favour of the former. Pro-oxidants are made up primarily of free radicals, which are chemical species (molecule, ion or atom) that contains an unpaired or an odd electron in their outer orbit. As a result of the presence of unpaired electrons, free radicals are usually very reactive and highly unstable [1,2]. Free radicals target macromolecules in their proximity for their electrons, thereby oxidizing them and generating other free radicals. If the macromolecules targeted are important parts of the cellular structure such as nucleic acids, proteins and lipids, considerable oxidative injury can occur [3]. Free radicals can be derived from four elements, including oxygen, nitrogen, sulphur and chlorine, thereby creating reactive oxygen species (ROS), reactive nitrogen species (RNS), reactive sulphur species (RSS) and reactive chlorine species (RCS). The ROS and RNS are the two main families of relevant pro-oxidants in biology and medicine. Examples of reactive oxygen species are superoxide ($O_2^{\cdot-}$), hydroxyl radical ($OH^{\cdot}$), singlet oxygen ($^1O_2^{\cdot}$) and peroxyl radical ($ROO^{\cdot}$), while examples of reactive nitrogen species include nitric oxide (NO) and nitrogen dioxide ($NO_2$). Apart from the reactive-free radical species, there are also non-radical reactive species which, though do not contain unpaired electrons, are either oxidizing agents or easily converted to free radicals. Examples are hydrogen peroxide ($H_2O_2$), organic peroxides such as lipid hydroperoxides (ROOH), hypochlorous acid (HOCl) and peroxynitrite ($ONOO^{\cdot}$) [4–6]. ROS and RNS often act together to create a state of oxidative stress, as a result of either depletion of cellular antioxidant defense molecules or overproduction of the reactive species.

# 2. Generation and interactions of reactive oxygen species

Reactive oxygen species can be generated from many sources, which can be either endogenous or exogenous. Table 1 shows some major endogenous and exogenous sources of reactive species.

| Endogenous sources | Exogenous sources |
|---|---|
| Mitochondrial electron transport chain | Radiation (UV light, X-ray and $\gamma$-radiation) |
| Neutrophils and macrophages during inflammation | Environmental pollutants and toxins |
| Xanthine oxidoreductase, NADPH oxidase | Cigarette smoke, excessive alcohol, high-calorie diet |
| Microsomal oxidation in endoplasmic reticulum | Heavy metals |
| Myeloperoxidase (Phagocytes) | Infectious agent |
| Lipoxygenase, cyclooxygenases, prostaglandin synthase | Strenuous exercise |

**Table 1.** Endogenous and exogenous sources of reactive species

http://dx.doi.org/10.5772/61614

Most ROS are generated via endogenous sources as by-products of normal physiological and metabolic reactions such as energy generation through the mitochondrial electron transport chain reaction and protein assemblage by the endoplasmic reticulum [5,7]. The superoxide radical ($O_2^{\cdot -}$) is considered as the primary ROS and it can further interact with other molecules either directly or prevalently through enzyme- or metal-catalysed processes to generate secondary ROS [5,8]. The superoxide anion is formed mostly within the mitochondria of a cell during the electron transport chain reaction. Impairment of the electron transport mechanism because of damage to mitochondria either during pathophysiological conditions or during mitochondrial dysfunction results in ROS formation. The electron transport chain is made up of five multienzyme complexes, which are responsible for ATP generation and maintenance of mitochondrial membrane potential. The major site of ROS production is assumed to be complexes I and III [9]; however, ROS can also be generated by other electron complexes, as well as by other mitochondrial enzymes [10]. During energy transduction, there is a premature leakage of electrons to molecular oxygen at complexes I and III forming superoxide anion radical [5,6,11]. Xanthine oxidase is another source of superoxide radical. The enzyme catalyses a one-electron as well as a two-electron reduction of oxygen to form $O_2^{\cdot -}$ or $H_2O_2$, with itself being converted to uric acid [12]. Cyclooxygenase, lipoxygenase and cytochrome P450 are other enzymes that have been proposed to generate ROS. The primary function of these enzyme systems is not to generate ROS; however, as is the case with the mitochondria, a dysfunction of these systems in diseases or environmental toxicity generates ROS as a by-product. The superoxide anion radical produced as a result of a one-electron reduction of molecular oxygen in the mitochondria is a relatively stable intermediate and its dismutation by the enzyme manganese superoxide dismutase (MnSOD) will result in the formation of $H_2O_2$ [13]. Protein assemblage by the endoplasmic reticulum, mediated by the enzyme protein disulphide isomerase and oxidoreductin 1, as well as fatty acid degradation by peroxisomes, also accounts for $O_2^{\cdot -}$, which is dismutated to $H_2O_2$ by copper/zinc superoxide dismutase (Cu/ZnSOD) [14]. The $H_2O_2$ produced is a non-radical molecule and it is normally neutralized by $H_2O_2$-removing enzymes, catalase and the glutaredoxin enzyme system (consisting of glutathione peroxidase, glutathione reductase and glutathione (GSH)) to water and molecular oxygen [15]. Glutathione peroxidase uses reduced GSH as a substrate and it is converted to the oxidized form (GSSG). Reduced glutathione is regenerated in a reaction catalysed by glutathione reductase with subsequent oxidation of NADPH. However, during a period of iron overload (e.g. conditions of haemochromatosis, haemolytic anaemia and haemodialysis), in the presence of high amounts of $H_2O_2$, heavy metals such as iron (Fe) or copper (Cu) are freed from iron-containing molecules and there is a subsequent interaction of $O_2^{\cdot -}$ and $H_2O_2$ in a Haber–Weiss reaction or $Fe^{2+}$ ($Cu^{2+}$)-driven cleavage of $H_2O_2$ in a Fenton reaction to generate the highly reactive hydroxyl radical ($^{\cdot}OH$) as shown in Figure 1 [12,13,16]. The $^{\cdot}OH$ radical is highly reactive because of its very short half-life (<1 ns); thus, *in vivo*, it can react with macromolecules such as lipids, proteins and nucleic acids in the vicinity of its site of formation, resulting in oxidative damage to these important cellular components [16]. Other oxygen-derived free radicals that can be formed *in vivo* are the peroxyl radicals ($ROO^{\cdot}$) (Figure 1) which are involved in DNA damage and protein backbone modification and have also been found to synergistically enhance the induction of DNA damage by superoxide [7].

Reactive nitrogen species are nitrogen-centred free radicals and include the nitric oxide radical ($NO^{\cdot}$), nitrosonium cation ($^{\cdot}NO_2$), nitroxyl anion ($NO^-$), nitrous oxide radical ($NO_2^{\cdot}$) and peroxynitrite ($ONOO^-$). The $NO^{\cdot}$ radical is the precursor of other RNS and is produced in higher organisms by the oxidation of one of the terminal guanido-nitrogen atoms of Larginine in a reaction catalysed by nitric oxide synthase (NOS) [17,18]. The NOS family includes endothelial NOS (eNOS) and neuronal NOS (nNOS), both of which are constitutively expressed isoforms, as well as the transcriptionally regulated isoform, inducible NOS (iNOS), which has been shown to play an important role in host defense [19–21]. Overproduction of the RNS above the ability of the cell to neutralize them is known as nitrosative stress and may lead to nitrosylation reactions that can alter the structure of proteins and so inhibit their normal function [7]. Oxidative burst during inflammatory processes triggers the formation of both superoxide and nitric oxide radical from immune cells and they react together to produce the highly reactive peroxynitrite anion radical which is a potent oxidizing agent that can cause DNA fragmentation [5,7,22]. Other nitrating agents include the nitrosonium cation formed in a myeloperoxidase-catalysed reaction involving nitrite ($NO_2^-$) and hydrogen peroxide [23,24] and nitroso-peroxocarbonate ($ONO\text{-}OCO_2^-$) formed via the reaction of $CO_2$ with $ONOO^-$ [25,26]. A recent study has also shown that lipid peroxyl radicals are able to promote tyrosine nitration by inducing tyrosine oxidation and also by reacting with $NO_2^{\cdot}$ to produce nitrosonium cation ($^{\cdot}NO_2$) [27].

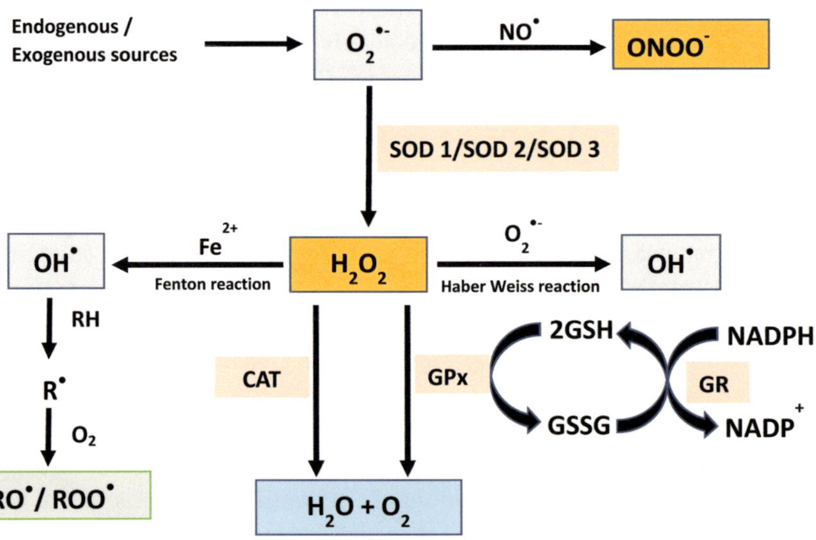


**Figure 1.** Sources and generation of different reactive oxygen species. CAT, catalase; GPx, glutathione peroxidase; GR, glutathione reductase; SOD, superoxide dismutase; RH, lipid membrane; R·, alkyl radical

Hypochlorous acid (HOCl) is a non-radical reactive chloride species involved in oxidation and chlorination reactions. It is usually generated via a myeloperoxidase-catalysed reaction involving $H^+$, chloride ion and $H_2O_2$ by the activated neutrophils [28,29]. Evidence has shown that HOCl can oxidize thiols and other biological molecules, and chlorinate amines, choles-

http://dx.doi.org/10.5772/61614

terol, unsaturated lipids, as well as DNA [28,30]. Although sulphur is considered a component of cellular antioxidant systems, emerging facts are indicating that reactive sulphur species (RSS) with similar stressor properties to ROS are generated under condition of oxidative stress when primary stressors react with cellular components. These reactive sulphur species are involved in protein activity modulation via thiolation reactions. An important example of an RSS implicated in physiological process is the thiyl radical ($RS^{\cdot}$) formed from a one-electron oxidation of a thiol group as an intermediate to the production of a disulphide, or via the reaction of sulphydryl groups with ROS such as superoxide radical [31,32]. Thiyl radicals participate in a series of radical chain reactions that generate other highly reactive sulphur-centred radicals. For example, a thiyl radical reacts with a thiolate to form the disulphide radical anion ($RSSR^{\cdot-}$) and with molecular oxygen to form the thiyl peroxyl radical ($RSOO^{\cdot}$). Furthermore, disulphide radical anions interact with molecular oxygen to generate the disulphide (RSSR) and the superoxide radical [31]. These reactions underline the close relationship between RSS and ROS.

## 3. Physiological importance of reactive oxygen species

Reactive oxygen and nitrogen species have generally been considered as being highly reactive and cytotoxic molecules; however, accumulating evidence has shown that besides their deleterious effects, ROS participates in physiological processes in a well-controlled manner. Physiological sources of ROS in the cell are many, but in a number of cells, ROS generated by the NADPH oxidase enzyme are those implicated in important physiological responses, such as defense against environmental pathogens or in cell signalling [33]. Reactive oxygen and nitrogen species are able to influence physiological processes because of their ability to modify the activity of key protein molecules containing domains sensitive to redox conditions. Reversible oxidation of the target domain results in the initiation of coupled events, such as the modification of the activity of downstream enzymes and/or transcription factors. Recent evidence has shown that at moderate concentrations, nitric oxide (NO), superoxide anion and related reactive species play an important role as regulatory mediators in cell-signalling processes and that many of the ROS-mediated responses actually protect the cells against oxidative stress and re-establish redox balance [4].

Accumulating evidence suggests that oxidants may be important regulators of the cellular response under hypoxic conditions (oxygen concentrations between 1% and 5%). Although the molecular basis for this is still obscure, nonetheless, there are suggestions that the release of ROS under these conditions functions as an important physiological regulator of hypoxia-inducible factor 1$\alpha$ (HIF-1$\alpha$), which, under low oxygen conditions, respond to a rise in ROS and then feedback and inhibit the production of ROS levels [10,34–36]. Other HIF-1$\alpha$ activities include the transcriptional regulation of metabolic enzymes such as lactate dehydrogenase A and pyruvate dehydrogenase kinase 1, which control the flow of carbon substrates into the mitochondria [37,38]. Other studies have also shown that HIF-1$\alpha$ regulates the expression of certain cytochrome components directly and also regulates specific microRNAs that in turn regulate the expression of components of the electron transport chain [10,39,40].

ROS mediate inflammatory responses in a number of ways. In an inflammatory environment, activated neutrophils and macrophages produce large quantities of superoxide radicals and other ROS via the phagocytic isoform of NADPH oxidase for the destruction of intracellular microbes [5,41]. Recent evidence has shown that mitochondrial ROS also contribute to the phagocytic response of the innate immune system as a result of the activation of a subset of Toll-like receptors (TLR1, TLR2 and TLR4) [42]. Mitochondrial ROS (mROS) have also been suggested to be involved in the activation of the inflammasome, thus acting as important signalling molecules to regulate inflammatory responses. Specifically, mROS have been implicated in the activation of NLRP3 (NOD-like receptor, pyrin domain-containing 3) receptor which senses a wide range of damage signals, including environmental irritants, microorganisms, as well as endogenous danger signals such as ATP and uric acid. When activated, NLRP3 forms a high-molecular-weight intracellular complex with a number of other protein partners to regulate the maturation and secretion of pro-inflammatory cytokines such as IL-1β [10,43–45].

Another physiological role of ROS is the regulation of vascular tone by cGMP. Soluble guanylate cyclase (sGC) is a heterodimeric protein which catalyses the formation of cGMP. Soluble guanylate cyclase can be activated by both $NO^{\cdot}$ and $H_2O_2$, with NO· activating sGC by binding to its $Fe^{2+}$-haem groups resulting in a conformational change at $Fe^{2+}$ that activates the enzyme. The cGMP formed is used as an intracellular amplifier and a second messenger to modulate the function of protein kinases, ion channels and other physiologically important targets, including smooth muscle tone regulation and platelet adhesion inhibition [5]. Reactive oxygen and nitrogen species have been implicated in the regulation of autophagy and apoptosis. Mitochondria-generated ROS regulate autophagy by the direct regulation of Atg4 (autophagy-related gene 4) activity [46,47]. ROS also played an important role in the release of cytochrome c and other pro-apoptotic proteins, which can trigger caspase activation and apoptosis. Therefore, small molecules like ROS can affect the complex networks of proteins mediating the induction and execution of cell death [13].

## 4. Consequences of oxidative stress

Oxidative attack by ROS is manifested as damage to nucleic acid bases, lipids and proteins, which can severely compromise cell health and viability or induce a variety of cellular responses through the generation of secondary reactive species, ultimately leading to cell death by necrosis or apoptosis [48]. Macromolecular damage via oxidative and/or nitrosative stress, if unchecked, can theoretically contribute to disease development, and an increasing amount of evidence suggests that oxidative and/or nitrosative stress is linked to the pathophysiologic mechanisms of a myriad of human diseases [3,6,48–50].

### 4.1. Lipid peroxidation

Lipid peroxidation refers to the oxidative degradation of lipids that occur when ROS attack lipids and extract a hydrogen atom from a methylene carbon in the side chain, and is one of

http://dx.doi.org/10.5772/61614

the major outcomes of free radical-mediated injury. The cell membrane is one of the most susceptible sites to ROS damage because the polyunsaturated fatty acid (PUFA) residues of phospholipids are very sensitive to oxidation. The process of lipid peroxidation occurs in three stages: initiation, propagation and termination. Reactive species such as $O_2^{\cdot-}$, $OH^{\cdot}$, $RCOO^{\cdot}$ and $ONOO^{\cdot}$ are all potent initiators of lipid peroxidation. During initiation, the fatty acid containing the double-bonded carbon chains is oxidized by a free radical to produce a lipid radical which stabilizes itself by reacting with oxygen to produce the peroxyl radical, which can then oxidize a neighbouring fatty acid to become a stable hydroperoxide. The oxidized fatty acid neighbour becomes a lipid radical, which propagates the oxidation process across the lipid membrane [7,16]. The lipid hydroperoxides formed may then be converted into conjugated dienes which are unstable and generate a variety of relatively stable end products, mainly aldehydic by-products, such as malondialdehyde (MDA) and more reactive $\alpha$-, $\beta$-unsaturated aldehydes, such as trans-4-hydroxy-2-nonenal (4-HNE) and 2-propenal (acrolein) [48,51–54]. Isoprostanes and neuroprostanes and more recently neurofurans are other products that have been derived from the endocyclization of lipid hyroperoxyl radicals [55–59], and the evaluation of these end products has been used as an index to determine the extent of oxidative damage in cells.

### 4.2. Oxidative protein damage

Proteins are primarily responsible for most functional processes within cells and are thus highly abundant in biological systems, making them important targets of ROS attack. The attack of ROS on the polypeptide backbone is initiated by an $\cdot OH$-dependent abstraction of the $\alpha$-hydrogen atom from an amino acid residue to form a carbon-centred radical. Under aerobic conditions, the carbon-centred radical readily interacts with molecular oxygen to form peroxyl radicals, which react with the protonated form of superoxide ($HO_2^{\cdot}$) and are converted to the alkyl peroxides [5,60]. Attack by ROS on proteins may lead to the oxidation of amino acid residue side chains, as well as oxidation of the protein backbone and formation of protein–protein cross-linkages. Consequently, protein fragmentation and generation of many protein oxidation products, which can cause damage to other biomolecules, may occur [53,61,62]. All amino acids within proteins may be attacked by reactive species (including ROS, RNS and electrophiles); however, the thiol group of cysteine are the most susceptible. Nevertheless, generally to a lesser extent than cysteine, sulphur-containing methionine and the aromatic amino acids tyrosine and tryptophan are also susceptible to oxidation [63].The susceptibility of the thiol group arises from the unique chemistry of cysteine, which confers on the thiol group specialized properties including nucleophilicity, high-affinity metal binding and/or ability to form disulphide bonds. The thiol group of cysteine is reported to be ionizable, generating a negatively charged thiolate group after deprotonation, boosting its reactivity [64]. The functional roles played by thiol groups are linked to this reactivity. Apart from having high affinities for metals, thiols may be subjected to alkylation by electrophiles (generated by xenobiotics), oxidation by reactive oxygen species and nitration by reactive nitrogen species [64–67]. These not only make the thiols versatile in their biological roles but may also lead to post-translational modifications that alter functions and subvert normal biology.

Every level of protein structure, from primary to quaternary (if multimeric proteins), may be altered when proteins are exposed to oxidative attack by ROS, causing major physical changes in protein structure [48]. Because most protein damage is irreparable, oxidative changes to the structure of protein can have a wide range of functional consequences including affecting the function of receptors, enzymes and transport proteins as well as generating new antigens that can provoke immune responses [68]. Secondary damage to other biomolecules can also result in the inactivation of DNA repair enzymes and loss of fidelity of damaged DNA polymerases in replicating DNA [68,69]. Although oxidized proteins are usually degraded by the proteosomal system and the lysosome–macroautophagy pathway, however, the degradation of damaged proteins may not be completely efficient, resulting in the formation of functionally inactive protein aggregates, which accumulate with age in separate compartments within cells or in the extracellular environment [63,70–72]. It is also possible that the proteolytic systems responsible for the removal of oxidized proteins may be impaired by oxidative stress itself, thereby accelerating the accumulation of damaged carbonylated-aggregating proteins [73]. Protein aggregates can be highly cytotoxic [74], and increased levels of carbonylated aggregates have been observed in patients with age-related disorders such as Parkinson disease, Alzheimer's disease and cancer [72].

### 4.3. Oxidative DNA damage

Reactive oxygen species, especially $OH^{\cdot}$ radical generated during oxidative stress, can react with and cause modifications in all the components of the DNA molecule (the purine and pyrimidine bases and deoxyribose sugar backbone), causing damage such as base or sugar lesions, single-strand breaks, double-strand breaks, abasic site formation and DNA–DNA or DNA–protein cross-links [3,53,75]. Reports have indicated that reactive oxygen and nitrogen species alone generate several kinds of single-strand breaks and more than 70 oxidative base and sugar products in DNA [76,77]. It has been estimated that the frequency of oxidative DNA damage in human cells is $10^4$ lesions/cell/day [78,79]. Although hydrogen peroxide is less reactive, it is more readily diffusible and thus more likely to be involved in the formation of oxidized bases through Fenton and Haber–Weiss reactions [60,80,81]. Oxidative damage not repaired before DNA replication may cause cell death, DNA mutation, replication errors and genomic instability [7,82–84]. Reports by Halliwell [85] and Valko et al. [8] indicated that oxidative DNA damage may be related to an increased risk of cancer development later in life. DNA can also undergo nitrative damage as a result of attack of reactive nitrogen species (RNS) on DNA bases to form 8-nitroguanine, a mutagenic DNA lesion that can preferentially lead to G–T transversions, and its formation has been observed in human samples [86]. Attack on DNA by aldehydic products of lipid peroxidation such as acrolein and 4-HNE can lead to the formation of bulky exocyclic adducts, which can promote DNA–DNA and DNA–protein cross-linking and impair transcription factors binding [87–89].

The most commonly used biomarker of DNA damage is the concentration of the nucleoside, 8-hydroxy-2′-deoxyguanosine (8-OHdG), which has been found to be mutagenic in bacterial and in mammalian cells [3,50,90,91]. Numerous reports have shown that urinary levels of 8-OHdG were elevated in humans with various malignancies [92–98] and also in experimental animal models of tumours [99–103].

## 5. Endogenous antioxidant defense system

Under normal circumstances, eukaryotic cells have evolved a defense mechanism to limit free radicals and the damage caused by them. These include systems based on the presence of antioxidant molecules, the repair of injured molecules and the removal of damaged molecules. The antioxidant defense system can be endogenous and/or exogenous. The endogenous system is made up of a network of antioxidant enzymes including superoxide dismutase (SOD), catalase (CAT), glutathione peroxidase (GPx) and the glutaredoxin and peroxiredoxin system as well as low-molecular-weight antioxidant molecules such as glutathione. Superoxide dismutase catalyses the dismutation of superoxide anion, converting it to molecular oxygen and $H_2O_2$. There are three isoforms of the SOD family and they all used a transition metal at their active site. There is a CuZnSOD form in the cytosol and the intermembrane mitochondrial compartment [104], a MnSOD in the mitochondrial matrix and another form in the extracellular compartment (e.g. blood) [105].

Although less reactive than the $O_2^{\cdot-}$ anion, the $H_2O_2$ must still be rapidly removed, and this can be accomplished by the enzymes CAT and GPx working coordinately. Catalase removes $H_2O_2$ at a high rate but shows low affinity for the peroxide; thus, it should be most useful during the peak of $H_2O_2$ production or accumulation [105]. Evidence has shown that this peak should occur *in vivo* since acatalasaemia (i.e. disorder caused by lack of catalase) increases oxidative stress and induces certain pathologies in humans [106]. Glutathione peroxidase may be present in the selenium- and non-selenium-dependent forms, and it has been shown that there are at least five selenium-containing GPxs in humans [107,108] whose activities may be manipulated by changing dietary selenium levels [109]. The isoforms are the ubiquitously expressed cytosolic GPx (GPx-1) and phospholipid hydroperoxide GPx (GPx-4), epithelium-specific gastrointestinal GPx (GPx-2), secreted plasma GPX (GPx-3) and GPx-6 found in the olfactory epithelium and embryonic tissue [108,110,111]. All the GPx isoforms (GPx 1-4 and GPx-6) can metabolize $H_2O_2$ and soluble fatty acid hydroperoxides; however, only GPx-4 can metabolize complex phospholipid hydroperoxides [107]. The phospholipid hydroperoxide GPx-4 also differs from other isoforms in that it is a monomer while others are tetrameric proteins [109]. Glutathione peroxidases use the reducing power of GSH (and other thiols, such as thioredoxin) to decompose $H_2O_2$, and it is the sulphydryl moiety of the cysteine residue that supplies the reducing equivalent for GPx activity. Two molecules of GSH are oxidized for every one molecule of $H_2O_2$ decomposed, resulting in the formation of GSSG, which can be re-reduced back to two molecules of GSH by glutathione reductase.

Various low-molecular-weight endogenous non-enzymatic antioxidants are found in animal and human tissues. These are usually depleted when they react with ROS, but are actually recycled back to the antioxidant form due to reduction by other molecules. Because of their low molecular weight, they are able to eliminate ROS at sites that much larger enzymes cannot access [105]. Reduced glutathione (GSH), thioredoxin and ascorbate are the main low-molecular-weight hydrophilic non-enzymatic antioxidant molecules in the cell. Glutathione (L-γ-glutamyl-L-cysteinylglycine) is the predominant intracellular non-protein thiol in eukaryotic cells. It possesses strong antioxidative properties and consequently plays a crucial role in intracellular protection against compounds such as ROS and other free radicals [112–114]. It functions as a nucleophile to form conjugates with many xenobiotics and/or their metabolites and can also serve as a reductant in the metabolism of hydrogen peroxides and other organic hydroperoxides [112,113,115–118]. During interaction with free radicals, the –SH group of GSH becomes oxidized, leading to the formation of corresponding disulphide compound (GSSG). Thus, a depletion of GSH is usually associated with an increase in GSSG concentration, and the ratio between reduced and oxidized forms of GSH serves as an important indicator of the redox environment [119–121]. The thioredoxin system is another major intracellular antioxidant system and comprises thioredoxin (Trx), thioredoxin reductase (TrxR) and NADPH. It is an oxidoreductase with a redox-active disulphide/dithiol at the active site [105,122]. Thioredoxin is cytoprotective against oxidative stress either directly by quenching singlet oxygen and scavenging hydroxyl radical or indirectly by reducing oxidized ROS-targeted proteins in cooperation with peroxiredoxin/thioredoxin-dependent peroxidases [123,124].

## 6. Exogenous antioxidant defenses: diet-derived compounds

Endogenous antioxidant defense system may not be sufficient when the formation of ROS and other free radicals is excessive, especially during pathophysiological conditions. Therefore, additional protective mechanism via exogenous/dietary antioxidants may be required. Well-known diet-derived antioxidants are vitamins C (ascorbate), E and A, as well as carotenoids. Next to glutathione, ascorbate is the other most abundant reduced non-enzymatic antioxidant in cells. It is endogenously synthesized and maintained at high levels in tissues (about 1 mM in rats), but in primates (including human), guinea pigs, fruit-eating bats and many bird species, it is obtained through diet [125]. Ascorbate can scavenge ROS and other free radicals by readily donating an electron to potentially damaging radicals such as hydroxyl ($OH^{\cdot}$), alkoxyl ($RO^{\cdot}$), peroxyl ($LOO^{\cdot}$), thiol radical ($GS^{\cdot}$) and tocopheroxyl radicals ($TO^{\cdot}$) becoming oxidized to the ascorbate radical ($Asc^{\cdot-}$) during the process [126,127]. Ascorbate is regenerated from the $Asc^{\cdot-}$ by the action of (i) NADH- and NADPH-dependent dehydroascorbate reductase [128–130], (ii) GSH-dependent dehydroascorbate reductase [131] or (iii) pH-dependent disproportionation reaction, which results in the formation of ascorbate and dehydroascorbate [132,133]. Reports have shown that GSH and ascorbate interact cooperatively *in vivo* to cope with ROS and that ascorbate serves as a co-antioxidant with vitamin E *in vivo* to protect LDL from detectable oxidative damage induced by aqueous peroxyl radicals [134].

Vitamin E and carotenoids are the most important antioxidants in the lipophilic environment of the cell. There are eight known isoforms of the vitamin E family, comprising $\alpha$-, $\beta$-, $\gamma$- and $\delta$-tocopherol and the corresponding tocotrienols [135–137]. Because of the presence of an –OH group on the chromanol ring of vitamin E, it is able to function as an antioxidant and reduces lipid peroxyl groups to hydroperoxides, thus terminating the propagation of lipid peroxidation [105,138]. Empirical evidence for carotenoids being important antioxidants *in vivo* is weak [105,139]. Recent reports on birds showed that carotenoids have a very low contribution to protection against oxidative damage under stressful conditions [140–144]. Many human trials where carotenoids and other dietary antioxidants (vitamin C and E) were administered singly or together did not show any positive effect and even contributed to mortality among study subjects [145–148].

In addition to the vitamins and carotenoids, plant-derived phenolics are other exogenous source of antioxidant defense compounds. The intake of exogenous antioxidants from teas, fruits, vegetables and spices is of nutritional interest since it has been associated with the modulation of oxidative stress and cellular damage, as well as with the provision of substantial health benefits. Diet has been known as a risk factor for chronic diseases for many years, and the upsurge in cases of chronic diseases worldwide in the past decades has been attributed in part to the change from traditional, largely plant-based diet to the high-fat, energy-dense diets with substantial content of animal fat [149]. In addition, the Mediterranean diet characterized by frequent consumption of fruits, vegetables, legumes, cereals and fish and low in meat and cheese is positively associated with a reduced risk of most chronic diseases including cardiovascular diseases (CVDs), diabetes and cancer [150]. Fruits, vegetables, spices and teas are excellent sources of fibre, vitamins and minerals, but they also contain components such as polyphenols, terpenes, alkaloids and phenolics, which may provide substantial health benefits beyond basic nutritional functions [151]. Disease prevention using whole plant extracts, or isolated compounds from plant (phytochemicals), is gaining more scientific attention worldwide, because it is a cost-effective alternative to orthodox treatment, and the phytochemicals are presumed from anecdotal evidence to be safe, with low toxicity and general acceptance [152,153]. Whole plant extracts as well as isolated molecules have been shown to possess antioxidant, anti-inflammatory, anticancer and tumour inhibitory effects. The health-promoting effects of these plants are suggested to be due to the polyphenolic antioxidant molecules that they contain. Plants often contain different polyphenolic phytochemicals, and it is the general assumption that the observed health effects of these plants may be due to the additive and/or synergistic effects of the complex mixture of these phytochemicals rather than the effect of a single molecule [154–156].

## 7. Plant phenolics and polyphenols

With an excess of 8000 phenolic structures being reported, plant phenolics are one of the most widely dispersed groups of phytochemicals in the plant kingdom. Structurally, they are characterized by having at least one aromatic ring with one or more hydroxyl groups and can range from low-molecular-weight, simple aromatic-ring compounds to large and complex tannins and derived polyphenols [157,158]. Phenolics are secondary metabolites produced during diverse physiological processes in the plants, such as growth, lignification, pigmentation, pollination and defense against pathogens, predators and environmental stress [159]. According to Crozier et al. [157], plant phenolics occurring naturally in healthy plant tissue can be classified into (i) flavonoids and (ii) non-flavonoids. The major non-flavonoids of dietary significance include phenolic acids, phenolic alcohols, stilbenes and the lignans. Flavonoids are widely distributed in food and beverages of plant origin, including teas, fruits, vegetables, spices, cocoa and wine that form part of human diet. In plants, they play key roles in growth and development [160], plant–insect interactions [161] and protection against harmful radiation by absorbing ultraviolet B (UVB) and scavenging ROS generated by UVB irradiation [162,163]. Flavonoids are known to provide health-promoting effects in humans who consume them. Reports have shown that flavonoids possess many useful health properties, including anti-inflammatory, oestrogenic, enzyme system modulation, antimicrobial, vascular and antitumour activity; however, the antioxidant activity is the most studied one that is attributed to flavonoids and has been found to be responsible for other biological activities in which the prevention of oxidative stress is beneficial [164–166]. Flavonoids also possess anticancer activity, which has been shown to go beyond the antioxidant, free radical-scavenging activity, but may involve mechanisms such as regulation of gene expression in cell proliferation, oncogenes and tumour suppressor genes, induction of cell cycle arrest and apoptosis, modulation of detoxification enzymes activity, stimulation of immune system and regulation of hormone metabolism [165]. The main subclasses of flavonoids are flavones, flavonols, flavanols, anthocyanins, isoflavones, flavanones, chalcones and dihydrochalcones [157,164–166]. Other classes include the coumarins, aurones, dihydroflavonols and flavan-3,4-diols.

## 8. Rooibos (Aspalathus linearis)

### 8.1. History, distribution of the species and morphology

Rooibos, *Aspalathus linearis* (Brum f) Dahlg. (Family Fabaceae; tribe Crotolarieae), is a hardy shrub that grows between 1.5 and 2 m in height with bright green, needle-shaped leaves and small, yellow flowers [167]. After harvesting, the needle-like leaves and stems can be either bruised and fermented prior to drying or dried immediately [168]. The unfermented product remains green in colour and is referred to as green rooibos. During fermentation, the colour changes from green to red as a result of the oxidation of the constituent polyphenols and the final product is often referred to as red tea or red bush tea. The genus *Aspalathus* comprises about 278 species and is endemic to South Africa, showing a high degree of polymorphism in terms of morphology, geographical distribution, ecology and phenolic constituents [169]. Rooibos has been consumed by locals for over 300 years, but it was not known outside South Africa until 1904 when Benjamin Ginsberg, a Russian immigrant to South Africa, recognized its potential and started trading with rooibos that he bought from the Khoi descendants and became the first exporter of rooibos. In the 1930s, Dr P.F. Le Fras Nortier, a local doctor and amateur botanist, realized the commercial potential of rooibos and initiated its cultivation in plantations by local farmers (http://www.sarooibos.co.za/content/view/31/79; retrieved on 08/10/2010). By 1999/2000, rooibos production was between 4500 and 6000 metric tonnes with the domestic market absorbing 70–75% of the annual production. Today, rooibos cultivation has reached a large scale of more than 12,000 metric tonnes, serving both local and increasing international market demand [170].

### 8.2. Phytochemical composition

The herbal beverage made from rooibos is naturally caffeine-free [171,172] and low in tannin when compared to *Camellia sinensis* teas [173]. The leaf tannin content of rooibos is reported to be about 3.2–4.4% [174,175]. Rooibos is unique in its monomeric flavonoid composition, containing two unique compounds, namely aspalathin and aspalalinin. Aspalathin is a C–C-linked dihydrochalcone glucoside [176,177] and aspalalinin is a cyclic dihydrochalcone [178], and both are only isolated from rooibos. Rooibos is also one of the only three known sources of nothofagin, a 3-dehydroxydihydrochalcone glucoside [179,180]. The other known sources of nothofagin are the heartwood of *Nothofagus fusca* [181] and the bark of a Chinese medicinal plant, *Schoepfia chinensis* [182]. Other flavonoids in rooibos include the C–C-linked β-D-glucopyranosides such as flavones orientin and iso-orientin [183] as well as vitexin and isovitexin, both flavone analogues of nothofagin [177]. The flavanones, dihydro-orientin and dihydroiso-orientin [184], as well as hemiphlorin [178] have also been isolated from rooibos. Other flavones isolated from rooibos include chrysoeriol, luteolin and luteolin-7-o-glucoside, while the flavonols quercetin, quercetin-3-orobinoside, hyperoside, isoquercitrin and rutin are also present [177,178,184–187]. The presence of phenolic acids, lignans and the coumarin, esculentin [177,178,187], as well as monomeric flavan-3-ol, (+)-catechin and oligomeric flavan-3-ol, procyanidin B3 and bis-fisetinidol-(4β,6:4β,8)-cathechin [187,188], has also been detected in rooibos. Figure 2 shows the major flavonoids that have been identified from rooibos.

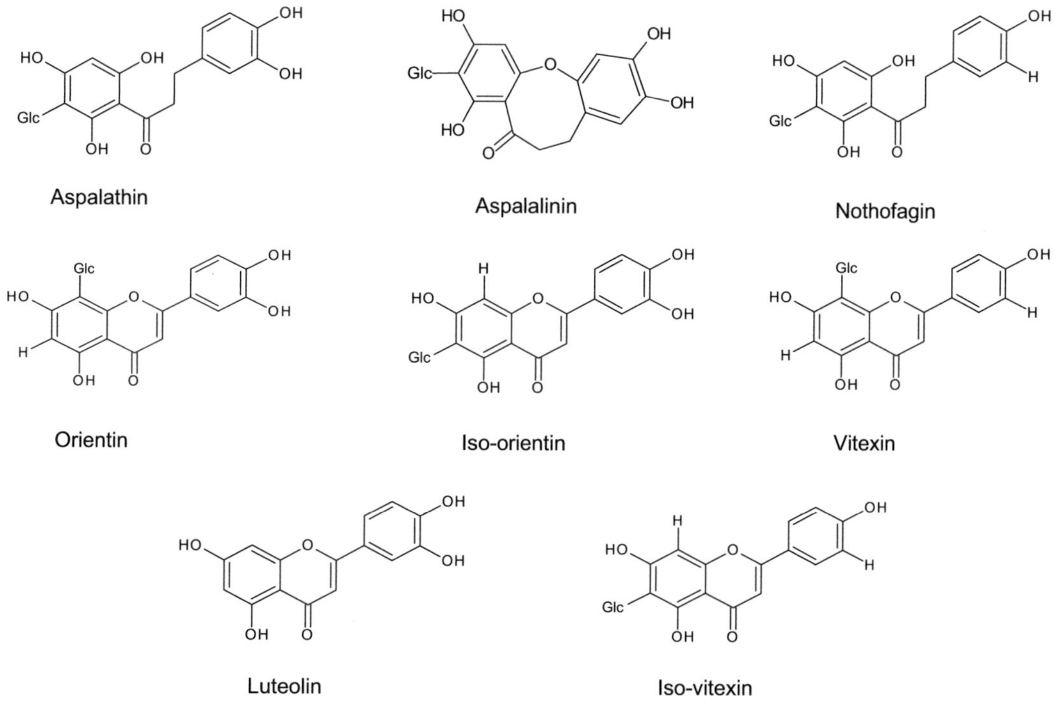


**Figure 2.** Some major flavonoids identified in rooibos (Glc = C-β-D-glucosyl group)

As a result of fermentation during the processing of rooibos, the content of polyphenols in rooibos is reduced. Studies have found a higher percentage of total polyphenols, flavonoids and non-flavonoids in unfermented/green rooibos compared to fermented/red rooibos [189–191]. These differences may be attributed to (i) the enzymatic and chemical modifications that occur during fermentation and (ii) the processing methods used [179,190]. During fermentation, aspalathin present in unfermented rooibos is extensively oxidized to dihydroiso-orientin. Nothofagin, iso-orientin, isovitexin and vitexin are degraded as well, but to a lesser extent [179,187,192,193]. *In vitro* studies revealed that fermentation and processing also affect the antioxidant activities of rooibos, with results showing that antioxidant activity (measured by different free radical-scavenging assays) decreases with fermentation [190,194–196], an effect attributed to a decreased total polyphenol content with fermentation.

## 9. Biological activities and potential of rooibos in oxidative stress-induced conditions

The attractiveness and use of naturally occurring compounds including those derived from fruits, vegetables, teas, various herbs and spices as potential chemopreventive and chemotherapeutic agents are gaining worldwide appeal. Evidence abounds that rooibos and/or its flavonoids hold great potential not only in the prevention but also in therapy of a wide variety of disease conditions. Since the first report on the biological activity of rooibos, an enormous body of work has revealed that rooibos extracts possess antioxidant, antimutagenic, anti-inflammatory, anti-diabetic, hepatoprotective, antimicrobial and, above all, cancer-preventive properties. The following sections will therefore take a detailed look at the various biological and health-promoting effects that have been shown for rooibos.

### 9.1. Antioxidant and lipid peroxidation inhibition activities

Although the controlled production of ROS has important physiological roles especially in energy production, phagocytosis, regulation of cell growth and cellular signalling, a high ROS production not counterbalanced by cellular antioxidant defense may result in oxidative stress. As previously mentioned, oxidative stress has been implicated in the pathogenesis of many disease conditions such as cancer, cardiovascular diseases, atherosclerosis, hypertension, ischaemia/reperfusion injury, diabetes mellitus, neurodegenerative disorders (Alzheimer's and Parkinson's disease), rheumatoid arthritis and ageing. Rooibos is a potent source of unique and beneficial phytochemicals, which are thought to contribute to its health benefits. Several studies have confirmed that rooibos showed antioxidant and lipid peroxidation inhibition activities both *in vitro* and *in vivo*. Extracts of fermented and unfermented rooibos as well as rooibos flavonoids showed *in vitro* antioxidant activity by scavenging free radicals [190,194–199]. Aqueous extracts of fermented rooibos showed *in vitro* inhibition of lipid peroxidation in cell membranes using rabbit erythrocyte membrane, rat liver microsome and rat liver homogenates [196,200], while methanolic extracts of fermented and unfermented rooibos inhibited microsomal lipid peroxidation [201]. Rooibos was also shown to scavenge alkyl peroxyl radicals formed during lipid peroxidation [202]. Aqueous and methanolic extracts of rooibos inhibited peroxyl radical-induced DNA strand scission in a dose-dependent manner,

http://dx.doi.org/10.5772/61614

with the aqueous extract being less effective compared with the methanolic extract [203]. Treatment with fermented rooibos increased the viability of Chinese hamster fibroblast (V79-4) cells following $H_2O_2$-induced oxidative stress by increasing the activity of antioxidant enzymes superoxide dismutase (SOD) and catalase (CAT) [204].

The antioxidant effects shown by rooibos *in vitro* has been confirmed in many experimental animal model studies. Aqueous rooibos extract administration suppressed age-related accumulation of lipid peroxides in several regions of the rat brain [205]. Gastric intubation of luteolin (a flavonoid found in rooibos, at 10-μmol/kg body weight) given 2 hours prior to $\gamma$-ray irradiation suppressed lipid peroxidation in mice bone marrow and spleen [206]. In a study in vitamin E-deficient rats, dietary supplementation of a freeze-dried hot water extract of rooibos decreased the level of lipid peroxidation in the liver, lung, small intestine and stomach of the animals, although the changes were not significant [207].

In the presence of reactive species, GSH is rapidly oxidized to GSSG and in its turn exported from cells. Intracellular levels of GSH, GSSG or the ratio of GSH:GSSG serve as good indicators of oxidative stress. The ability of rooibos extracts to inhibit lipid peroxidation and modulate antioxidant status in different organs and/or tissues has been widely reported. In 2003, Marnewick et al. [208] reported on the effect that chronic feeding of a 2% (w/v) aqueous extracts of both fermented and green rooibos had on the modulation of oxidative stress in male Fischer rats. Consumption of the fermented and green rooibos extract for 10 weeks did not have any effect on the antioxidant capacity (measured as ORAC) of the liver; however, both extracts significantly improved the glutathione (GSH) redox status of the liver by reducing the level of oxidized glutathione (GSSG) and thus increasing the GSH:GSSG ratio significantly. Simultaneously, the rooibos extracts also significantly enhanced the activities of the important hepatic phase II drug-metabolizing enzymes, glutathione S-transferase alpha and UDP-glucoronosyl transferase.

Using different models of induced oxidative stress, the effect of rooibos extracts supplementation has been reported widely. Fermented rooibos extract given for 10 weeks increased the antioxidant status of the $CCl_4$-treated rats by increasing $\alpha$-tocopherol and reducing the liver concentrations of coenzyme Q while inhibiting the formation of malondialdehyde [209]. A study by Awoniyi et al. [210] reported that fermented and green rooibos offer a measure of protection against tert-butyl hydroperoxide (t-BHP)-induced oxidative damage in rat sperm by increasing the antioxidant defense mechanism and thereby improving sperm motility and function. The authors concluded that a wide spectrum of polyphenolic compounds present in rooibos are effective as antioxidants; thus, rooibos could offer a therapeutic effect in the treatment of infertility. More recent studies in our laboratory have also reported on the ability of rooibos to inhibit lipid peroxidation and improve the glutathione redox status in the liver of rats. In one of the studies, male Wistar rats were supplemented with fermented rooibos for 10 weeks, while oxidative stress was induced by intraperitoneal injection of t-BHP in the last 2 weeks. Results from this study showed that rooibos supplementation was able to inhibit lipid peroxidation by decreasing the elevation in conjugated dienes and malondialdehyde levels induced by t-BHP. Simultaneously, the fermented rooibos extract also improves the glutathione redox status by increasing the reduced glutathione (GSH) concentration and invariably the GSH/GSSG ratio in the plasma and liver of the rats [211]. In another study, lipopolysaccharide (LPS) (*Escherichia coli*, serotype 0111:B4, 0.5 mg/kg b.w., i.p.) was used as an oxidative damage inducer in rats pre-fed fermented rooibos for 4 weeks. Results showed that pre-feeding rooibos inhibited plasma and hepatic formation of malondialdehyde (marker of lipid peroxidation) and also restored the impairment in glutathione redox status [212]. These studies showed that rooibos and its constituent polyphenols are effective antioxidants *in vivo*. Although the mechanisms by which rooibos offers protection in oxidative stress-induced conditions are not fully understood, however, we believed that the potential mechanisms by which this protection is achieved could involve one or more of several different antioxidant properties exhibited by flavonoids in rooibos. Rooibos polyphenols could be scavenging ROS directly by binding lipid peroxides or they may act as sacrificial antioxidants to inhibit the lipid peroxidation cascade. In improving the glutathione redox status, rooibos improved the antioxidant status of the cell. This effect may result from the ability of flavonoids present in rooibos to up-regulate the mRNA expression of $\gamma$-glutamylcysteine synthetase, the rate-limiting enzyme in the biosynthesis of glutathione.

### 9.2. Hepatoprotective effects

Chronic hepatic disease represents one of the foremost health problems worldwide, with liver cirrhosis and drug-induced liver injury accounting for the ninth leading cause of death in the western and developing countries [213]. *In vitro* and *in vivo* evidence suggests that free radical damage and oxidative stress contributing to lipid peroxidation is a critical mechanism implicated in the genesis and progression of different chronic liver diseases. It is a well-known fact that the available synthetic drug to treat liver disorders might also cause further damage to the liver [214]. Hence, medicinal plants have become increasingly popular and their use is widespread.

Hepatoprotective effects of rooibos has been reported in a number of studies. Consumption of rooibos protected against liver damage by suppressing the observed increase in plasma activities of aspartate aminotransferase (AST), alanine aminotransferase (ALT), alkaline phosphatase and bilirubin, and resulted in a histological regression of steatosis and cirrhosis in the liver of rats challenged with carbon tetrachloride (1 mL/kg, i.p.) twice a week for 10 weeks [215]. Another study from the same laboratory also reported that rooibos supported the regeneration of rat liver after intoxication with carbon tetrachloride, and this protective effect was ascribed to the ability of rooibos to inhibit lipid peroxidation in the liver [216].

In a t-BHP-induced hepatotoxicity model study, fermented rooibos consumption by rats reversed the elevation in serum aminotransferases and lactate dehydrogenase induced by t-BHP [211]. Histologically, the severe hepatic degeneration, hepatocyte vacuolation as well as massive lymphocyte and mononuclear cellular aggregation induced by t-BHP were also reversed. The hepatoprotective effect of rooibos was confirmed in another study by the same group when they reported on the amelioration of lipopolysaccharide-induced acute liver injury by a fermented rooibos extract [212]. Based on the result from these studies, the authors concluded that the hepatoprotective effects shown by fermented rooibos could be ascribed to a stabilizing effect of rooibos phenolics on the plasma membrane of hepatocytes, as well as the

http://dx.doi.org/10.5772/61614

repair of damaged hepatic tissues probably brought about by the stimulation of hepatocellular protein synthesis and accelerated regeneration of the hepatocytes, thus suggesting that rooibos may be of benefit in the prophylactic management of oxidant-induced liver injuries.

### 9.3. Anti-diabetic effects

Diabetes is one of the most common global diseases and affects approximately 200 million people worldwide [217,218], with the figure forecasted to rise to around 300 million people by 2025 [219]. Long-term effects of diabetes include progressive development of specific complements such as retinopathy, nephropathy and neuropathy. People with diabetes are also at risk for cardiovascular, peripheral vascular and cerebrovascular diseases. All these complications substantially increase the rates of morbidity and mortality associated with the disease and reduce the quality of life of the diabetic individuals. A chronic disease characterized by increased blood sugar, diabetes can result from lack of insulin production and/or resistance to insulin. Insulin resistance alone does not result in the development of type 2 diabetes, and reports have shown that there must be progressive dysfunction of pancreatic islets, which disrupt the secretion of glucagon from the $\alpha$ cells and insulin from $\beta$ cells, leading to uncontrolled hyperglycaemia, for the disease to develop [220].

Experimental evidence suggests the involvement of overproduction of free radicals in the onset of diabetes and diabetic complications, with oxidative stress proposed as a pathogenic mechanism linking insulin resistance with the dysfunction of beta cells and endothelium, impaired glucose tolerance and overt diabetes. $\beta$-Cell dysfunction in the diabetic state may result from a decrease in $\beta$-cell mass, and this has been attributed to oxidative stress-induced apoptosis [221]. The apoptotic state in the $\beta$-cells was reported to be triggered by high glucose concentration (glucotoxicity), possibly due to increased production of ROS. In addition, evidence has shown that $\beta$-cells have very low expression of antioxidant enzymes, making them more susceptible to oxidative damage [221–223]. Both pancreatic $\alpha$- and $\beta$-cells play vital roles in the maintenance of normal glucose homeostasis, and studies has shown that abnormal glucagon secretion is a hallmark of the diabetic state, suggesting that glucose sensing by the $\alpha$-cells is also dysfunctional [224,225]. The effect of oxidative stress on $\alpha$-cells in the diabetic state is less clear; however, evidence has shown that uncoupling protein 2 (UCP 2), a mitochondrial proton carrier protein, is highly expressed in pancreatic $\alpha$-cells [225]. A central role has been suggested for UCP 2 in ROS-mediated injury in the diabetic state, with several reports indicating that UCP 2 is negatively correlated with the level of ROS generation by respiring mitochondria [226]. Some authors have reported that in contrast to $\beta$-cells, islet $\alpha$-cells are resistant to diabetogenic toxins, an observation the authors adduced to the fact that $\alpha$-cells displayed high level of catalase expression and activity [227], suggesting that $\alpha$-cells may be more resistant to ROS damage. Biomarkers of oxidative damage including malondialdehyde, F2-isoprostane, protein carbonyls and advanced glycation end products (AGE) are reported to be elevated in either the plasma, erythrocytes or liver of streptozotocin (STZ)-induced diabetic rats [228–231], as well as in the plasma or urine of diabetic patients [232–234], giving further evidence of ROS contribution to the onset, progression and pathological consequence of diabetes. In addition, the level of cellular non-enzymatic antioxidants such as vitamin E, vitamin C and reduced GSH is depleted in different tissues of experimentally induced diabetes [235,236] and in diabetic patients [234,237–239]. The modulation of the antioxidant enzymes network in the diabetic state is another indirect evidence in support of the role of oxidative stress in diabetes and diabetic complications, an effect which has been reported to be highly variable, depending on the model of diabetes used and the tissue type assessed [240].

Several polyphenol-rich plants and/or plant extracts have been used for controlling diabetes [241,242]. Polyphenols may affect glycaemia through different mechanisms, including inhibition of intestinal glucose absorption, increasing glucose transport and metabolism in muscle and/or stimulating insulin secretion [217,219,243,244]. A report by Johnston et al. [245] demonstrated that glucose uptake into cells under sodium-dependent conditions was inhibited by flavonoid glycosides and non-glycosylated polyphenols in polarized Caco-2 intestinal cells. Under sodium-free conditions, aglycones and non-glycosylated polyphenols inhibited glucose uptake, whereas glycosides inhibited the active transport of glucose [246].

Investigation into the anti-diabetic properties of rooibos in streptozotocin-induced diabetic rats showed that the administration of aqueous and alkaline extracts of rooibos to diabetic rats did not affect markers of diabetic status such as glucose, glycated haemoglobin and fructosamine; however, biochemical markers characterizing hepatotoxic effects in plasma, advanced glycation end products (AGEs) and malondialdehyde in plasma and in different tissues of the diabetic rats were reduced [247]. Previously, an *in vitro* study by Kinae et al. [248] reported that a freeze-dried extract of fermented rooibos suppressed the formation of glycated albumin, including AGEs in a mixture of D-glucose and human serum albumin. Quercetin, a flavonol found in rooibos in small concentrations, has been shown to mediate the inhibition of the facilitated diffusion of glucose transporter 2 (GLUT2) in Chinese hamster ovary cells [249].

The hypoglycaemic and anti-diabetic activity of aspalathin (major polyphenol of rooibos) in type 2 diabetic model db/db mice was demonstrated for the first time in a study by Kawano et al. [250]. In the study, purified aspalathin from a fermented rooibos extract increased dose-dependently and significantly the glucose uptake by L6 myotubes at concentrations of 1-100 μM, irrespective of insulin absence and also increased insulin secretion from cultured RIN-5F cells at 100 μM. Dietary aspalathin (0.1–0.2%, g/kg diet) suppressed the increase in fasting blood glucose levels for 5 weeks and improved impaired glucose tolerance at 30, 60, 90 and 120 min in db/db mice. The results from this study presented the first evidence of the beneficial effects of aspalathin on glucose homeostasis in type 2 diabetes, through the stimulation of glucose uptake in muscle tissues, and insulin secretion from pancreatic $\beta$-cells.

In another study, a fermented aqueous extract and an aspalathin-enriched unfermented extract of rooibos were reported to lower the elevation in blood glucose observed, in an STZ-induced hyperglycaemic Wistar rat model [251]. The same study also reported that the elevated blood glucose observed in a diet-induced model of type 2 diabetes in vervet monkeys was lowered by a fermented aqueous extract of rooibos.

The hypoglycaemic potential of an aspalathin-enriched green rooibos extract, aspalathin, rutin and an aspalathin–rutin mixture was investigated by Muller et al. [252] in C2C12 myotubules and STZ-induced diabetic rats. Results from the study showed that the enriched green rooibos extract induced a dose–response increase in glucose uptake on C2C12 myotubules; aspalathin was effective at 1, 10 and 100 μM while rutin was effective at 100 μM. In STZ-induced diabetic rats, the aspalathin-enriched rooibos extract and the aspalathin–rutin mixture but not the single compounds separately reduced blood glucose concentrations, indicating that the hypoglycaemic effect may be due to synergistic interactions of polyphenols present in rooibos. Using a similar *in vitro* approach in C2C12 muscle cells, Mazibuko et al. [253] reported that an aspalathin-enriched green rooibos extract and an aqueous extract of fermented rooibos ameliorate palmitate-induced insulin resistance, with both extracts increasing glucose uptake, mitochondrial activity and ATP production. Further studies from the same group revealed that fermented rooibos extract has a potential for preventing obesity by inhibiting adipogenesis in differentiating 3T3-L1 adipocytes [254] and also showed cardioprotective effect in isolated cardiomyocytes derived from diabetic rats, subjected to experimentally-induced oxidative stress and ischaemia [255].

Vascular inflammation induced by high glucose is key in the initiation and progression of arthrosclerosis, a major diabetic complication. In another study investigating the benefit of rooibos in the treatment of diabetic complications, Ku et al. [256] assess whether aspalathin and nothofagin can suppress vascular inflammation induced by high glucose in human umbilical vein endothelial cells (HUVECs) and mice. Results from the study showed that aspalathin and nothofagin inhibited high glucose-mediated vascular permeability, adhesion of monocytes towards HUVEC and expression of cell adhesion molecules and suppressed ROS formation and nuclear factor-κβ (NF-κβ) activation *in vitro* and *in vivo*.

Although these studies strongly suggest that rooibos and its polyphenols effect their antidiabetic potential through multiple modes of action, some authors have suggested that because rooibos and/or aspalathin are capable of scavenging intracellular reactive oxygen species (ROS), the anti-diabetic potential of rooibos may be due to its antioxidative function, which may be involved in the activation of insulin-stimulated glucose uptake, and hence the modulation of glucose homeostasis [257].

### 9.4. Antimutagenic, antitumourigenic and anti-carcinogenic effects

The increasing mortality and morbidity arising from various cancers worldwide have made the search for an alternative strategy (such as the use of bioactive components from plants) in the prevention and management of cancers very imperative. Results from studies in recent decades have shown that bioactive compounds from plants have important roles in the prevention, and reducing the risks of chronic diseases, including cancer [258,259]. The beneficial effects shown by these compounds are attributed, among others, to their antioxidant and free radical-scavenging ability.

Rooibos is a rich and unique source of polyphenols, and the polyphenols present in rooibos being powerful antioxidants may play important roles in the prevention of cancer by reducing damage to DNA in the cell, and modulating the promotion of cancer. Several studies have demonstrated the *in vitro* antimutagenic properties of both fermented and unfermented rooibos extracts using various test systems. Fermented aqueous rooibos extract (3.33%, w/v) significantly suppressed the number of chromosomal aberrations in Chinese hamster ovary cells induced by benzo[a]pyrene (BaP) and mitomycin C (MMC) in the presence or absence of rat liver microsomal enzymes (S9). The clastogen-suppressing effect was obtained when cells were exposed to rooibos before and/or after mutagen treatment [260]. Aqueous extract of rooibos (2–10% of extract) suppressed oncogenic transformation of mouse fibroblast cells subjected to X-ray-induced transformation in a dose-dependent manner [261]. In another study, the effect of different concentrations of a fermented rooibos extract included in the culture medium, on growth and changes of growth parameters of cultured chick embryonic skeletal muscle cells, was investigated. It was discovered that the rooibos extract significantly inhibited cell proliferation as reflected by decreased DNA, RNA and protein contents in primary cell cultures of fibroblasts and myoblasts, in a dose-dependent manner [262]. The *in vitro* antimutagenic property of rooibos was further investigated by Marnewick et al. [189]. Both fermented and unfermented rooibos extracts (5% and 10%, w/v) were significantly effective against 2-acetylaminofluorene (2-AAF) and aflatoxin $B_1$ ($AFB_1$)-induced mutagenesis in the *Salmonella typhimurium* mutagenicity assay (tester strain TA98 and TA100) in the presence of metabolic activation with S9. However, poor inhibitory effects against the direct-acting mutagens, methyl methanesulphonate, cumene hydroperoxide and $H_2O_2$ using tester strain TA102, were observed. Further studies using the same *Salmonella* assay revealed that fermented rooibos exhibited a higher antimutagenic activity than unfermented rooibos against both $AFB_1$ and 2-AAF, although antimutagenic activity of both rooibos extracts was comparable to that of an extract prepared from *Camellia sinensis* [263]. Using genetically engineered V79 Chinese hamster fibroblasts that expressed human CYP1A2, N(O)-acetyltransferase (hNAT2*4) and sulphotransferase (hSULT1A1*1), Platt et al. [264] showed that rooibos moderately protects against the genotoxicity of 2-amino-3-methylimidazo(4,5-f)quinoline (IQ) and the protection shown was comparable to that shown by green and black teas. In another study, aqueous extracts of unfermented rooibos inhibited cell proliferation of human oesophageal cancer cells (WHCO5) by affecting energy (ATP) production [265]. The antimutagenic properties of the most prevalent flavonoids in rooibos (aspalathin and nothofagin) and their flavone derivatives were investigated by Snijman et al. [266]. The results showed that aspalathin, nothofagin and their structural flavonoid analogues displayed moderate antimutagenic properties, while luteolin and to some extent chrysoeriol showed activities comparable to those of the green tea flavonoids, EGCG. In the same study, quercetin and isoquercetin exhibited antimutagenic, pro-mutagenic and mutagenic effects in the presence or absence of metabolic activation.

The *in vitro* antimutagenic potential of rooibos extracts has been substantiated by several animal model experiments. Intraperitoneal injection of fermented rooibos extract (1 mL of 0.1%, w/v) 6 h prior to MMC treatment reduced the induction of micronucleated reticulocytes (MNRETs) in peripheral blood of ICR male mice [260]. The *ex vivo* antimutagenic potential of fermented and unfermented rooibos was established in a 10-week rooibos feeding study in male Fischer rats. It was discovered that liver cytosolic fractions from rats consuming fermented and unfermented rooibos protected against $AFB_1$-induced mutagenesis in the *Salmonella typhimurium* assay with tester strain TA100. However, only liver cytosolic fraction from rats fed the unfermented rooibos showed protection against 2-AAF-induced mutagenesis in the same assay with tester strain TA98 [267]. In addition, the activation potential of hepatic

http://dx.doi.org/10.5772/61614

microsomal preparations from rats consuming rooibos was evaluated in the study, with rooibos reducing the activation of $AFB_1$, but not that of 2-AAF.

The antitumourigenic activity of rooibos was established in a two-stage mouse skin carcinogenesis assay. Topical application of ethanol/acetone soluble fractions of fermented and unfermented rooibos prior to tumour promotion with 12-O-tetradecanoylphorbol-13-acetate (TPA) on ICR mouse skin initiated with 7,12-dimethylbenz(a)anthracene (DMBA) significantly reduced the number and size of tumours per mouse, as well as delayed tumour development [201]. More recently, a study in female SKH-1 hairless mice showed that the topical application of polyphenol-rich extracts of fermented and unfermented rooibos, prior to ultraviolet B-induced tumour promotion and after DMBA-initiated skin cancer, reduced the number of tumours per mouse by 91.39% and 75.37%, respectively, and also the tumour volume by 97.28% and 90.74%, respectively [268]. Another study related to tumour promotion in mouse skin showed that the expression of cyclooxygenase-2 (COX2) in ICR mouse skin by TPA was significantly reduced by a methanolic extract of fermented rooibos (60, 300, 600 μg, topically applied prior to TPA exposure) [269]. A recent study, which monitored the cancer-modulating properties of hot water extracts of fermented and unfermented rooibos in a liver carcinogenesis model against fumonisin $B_1$ ($FB_1$) promotion in male Fischer rat using diethylnitrosamine as cancer initiator, showed that unfermented rooibos was more protective against $FB_1$-induced cancer promotion, by significantly reducing the total number and size of pre-neoplastic foci staining positively for the placental form of $\gamma$-glutamyl transferase in the liver, presumably by arresting their growth [270]. A similar study investigated the protective effect of rooibos against methylbenzylnitrosamine-induced oesophageal cancer in male Fischer rats and reported that fermented and unfermented rooibos significantly reduced the number and size of papillomas, with rats drinking unfermented rooibos failing to develop larger papillomas [265,271]. Results from these studies suggest that rooibos could be developed as a nutraceutical for the chemoprevention of liver and oesophageal cancer.

### 9.5. Anti-inflammatory effects

Inflammation plays an important role in various diseases, such as rheumatoid arthritis, atherosclerosis and cardiovascular diseases, type 2 diabetes and cancers, which all show a high prevalence globally. Since ancient times, in various cultures worldwide, inflammatory disorders and related diseases have been treated with plants or plant-derived formulations, and the anti-inflammatory activity of several plant extracts and isolated compounds has been demonstrated scientifically [272]. A possible explanation for these anti-inflammatory effects of plant extracts may be found in the interplay between oxidative stress and inflammation. ROS are involved not only in the occurrence of oxidative stress but also in the promotion of inflammatory processes via activation of transcription factors such as NF-$\kappa\beta$ and activator protein (AP)-1, which induce the production of cytokines like TNF-$\alpha$ [273,274].

Scientific studies reporting on the possible anti-inflammatory effects of rooibos and/or its flavonoids are sparse. A study in Japan reported that unfermented rooibos extract (1.6%, w/v) administration as the only source of drinking fluid for 8 weeks reduced inflammation in dextran sodium sulphate (DSS)-induced colitis rats via an increased antioxidant activity [275].

There was a significant increase in serum SOD and urine 8-hydroxy-2′deoxyguanosine levels in rooibos rats compared to the control and the DSS rats. Based on these findings, the authors concluded that rooibos may prevent DNA damage and inflammation by its antioxidative activity *in vivo*. The anti-inflammatory activity of a rooibos tea extract in DMSO (0.5 mg/mL) and two of its flavonoids (luteolin and quercetin) was investigated, together with that of other herbal extracts, in a study using a lipopolysaccharide-stimulated macrophage model. The study results revealed that incubation with a rooibos extract (0.5 mg/mL) significantly reduced the secretion of the pro-inflammatory cytokine, IL-6, by at least 25% and marginally reduced the expression of inducible nitric oxide synthase (iNOS). However, the secretion of the anti-inflammatory cytokine, IL-10, was also reduced [272]. In the same study, rooibos flavonoids luteolin and quercetin at concentrations of 50 and 100 nM reduced the secretion of IL-6 and TNF-$\alpha$, while also inhibiting the expression of cyclooxygenase 2 (COX-2) and iNOS.

More recent studies in our laboratory has shown that fermented rooibos extract shows anti-inflammatory effect by suppressing lipopolysaccharide-triggered inflammatory response. In the liver, fermented rooibos consumption suppressed the elevation in pro-inflammatory cytokines, TNF-$\alpha$ and IL-6 [212], while increasing the plasma level of anti-inflammatory cytokine, IL-10 [276]. Lee and Bae [277] investigated the anti-inflammatory effects and underlying mechanisms of aspalathin and nothofagin against LPS-mediated vascular inflammatory responses in HUVECs and mice. Results from this study showed that both compounds possess anti-inflammatory functions *in vitro* by inhibiting LPS-induced barrier disruption, expression of cell adhesion molecules (CAMs) and adhesion/transendothelial migration of neutrophils to endothelial cells, as well as hyper-permeability and leukocyte migration *in vivo*. Furthermore, treatment with each compound ameliorated LPS-induced lethal endotoxaemia and suppressed the production of TNF-$\alpha$, IL-6 and activation of nuclear factor-$\kappa\beta$ (NF-$\kappa\beta$) or extracellular regulated kinases (ERK) 1/2. These studies demonstrated that rooibos and its major flavonoids may contribute to the reduction of inflammation and be preventive against related diseases.

### 9.6. Human studies

Reports examining the health benefits of rooibos in humans have been scarce and limited. However, various health-promoting effects, including antioxidant and oxidative stress modulation, inhibition of postprandial oxidative stress, inhibition of angiotensin-converting enzymes (ACEs), lipid profile modulation, promotion of hair growth and anti-wrinkle activity, among others, have been reported for rooibos in humans.

The first human study with rooibos was conducted by Hesseling et al. [278], when they studied the effects of rooibos compared with black tea and water on iron absorption in 30 healthy young men. The study revealed that rooibos consumption (200 mL for 14 days) did not have a deleterious effect on the iron status parameters, such as haemoglobin, ferritin, transferrin, serum iron and iron-binding capacity, compared with the control group taking water. A more recent parallel intervention study, involving 175 primary school children in South Africa, revealed that the consumption of 200-mL servings of rooibos twice daily for 16 weeks did not have any adverse effects on the iron status parameters including serum ferritin, transferrin, the total iron-binding capacity and the transferrin saturation [279].

A study to test the potential antihistaminic effects of rooibos was conducted in seven patients diagnosed with either asthma or hay fever. Ingestion of fermented rooibos or topical application of a rooibos poultice did not produce any anti-allergenic activities [280]. The dermatological potential of rooibos was demonstrated in a study that revealed the consumption of a diluted infusion of rooibos at least once a week is beneficial to patients with dermatological diseases. In this study, rooibos consumption decreased the incidence of recurrent *Herpes simplex* and human papilloma virus infection, while patients with atopic dermatitis were successfully treated, resulting in a decrease in itching sensation. In addition, rooibos consumption inhibited the infiltration of neutrophils associated with Behcet's disease, *Psoriasis vulgaris* and *Acne pustulosa* [281]. The anti-wrinkle activity of rooibos was demonstrated in a recent study, when a commercial extract of a rooibos blend with *Camelia sinensis* extract was topically applied to female skin. Results obtained from the study indicated that the extract reduced skin wrinkles by 10%; however, no effect was shown on skin smoothness, roughness and scaliness [282]. Also, Glynn [283] demonstrated that a blend of botanical extracts containing unfermented rooibos, when topically applied to male, promoted hair growth by increasing hair density, number of anagen follicles and hair growth rate.

Sauter [284] investigated the effect of an aspalathin-enriched extract of unfermented rooibos (15% aspalathin) on the plasma antioxidant status of 20 subjects who were given an oral dose (250 mg/tablet) twice daily for 14 days with diet restrictions to ensure low flavonoid intake during the study period. Rooibos consumption did not show any effect on the antioxidant status biomarkers monitored, except a slight decrease in the antioxidant status when the xanthine/xanthine oxidase assay was used. In another study, the effect of unfermented rooibos consumption on the antioxidant status of workers who were occupationally exposed to lead was investigated in a randomized placebo-controlled 8-week intervention trial. The study monitored the indices of lead exposure (blood lead (PbB), erythrocyte porphyrins (EPs) and delta-aminolevunilic acid dehydratase (ALAD)) and antioxidant status markers (SOD, GSH and MDA) in the erythrocytes and plasma. Rooibos consumption did not have any effect on the mean blood lead and erythrocyte porphyrin levels (lead exposure status), but there was a modulation of antioxidant status of the lead-exposed workers as shown by the decreased MDA levels and increased SOD activity and GSH levels in the plasma and erythrocytes of these factory workers [285]. More recently, a study by Francisco [286] showed that fermented rooibos consumption modulated postprandial glycaemia, lipaemia and oxidative stress in healthy subjects after an intake of a standardized fat meal. Consumption of rooibos by the subjects improved their redox status by increasing the total GSH and significantly lowering the level of conjugated dienes and TBARS after 6 hours. Further, the levels of total cholesterol, LDL cholesterol and triacylglycerol as well as high sensitive C-reactive proteins (hs-CRP) were all decreased in the same study. Guerreiro et al. [287] revealed that the consumption of a rooibos infusion with added glucose (25 g) did not change the glycaemic response when compared to the reference solution (water with 25 g of glucose) after 2 hours in a Caucasian population of young male and female adults aged 17–24 years, showing that rooibos is a healthy beverage for obese and diabetic patients. An acute, cross-over design intervention study involving 15 healthy volunteers consuming 500 mL of either water or fermented or unfermented rooibos showed that the plasma antioxidant capacity using the TRAP assay increased significantly with both rooibos extracts, reaching a peak 1 hour post consumption. No change in triacylglycerol, cholesterol or uric acid levels was observed with any of the treatments, while a transitory increase in glycaemia observed at 30 minutes was linked to a glucose upload [288]. Marnewick et al. [289] showed that the consumption of six cups of fermented rooibos herbal tea for 6 weeks improved the blood-lipid profile and reduced oxidative stress by decreasing lipid peroxidation and improving the redox status (GSH:GSSG) of adults at risk of CVDs. This study provided the first clinical evidence in humans of the ability of rooibos to modulate oxidative stress in adults at risk for developing heart disease.

A recent randomized, three-phase cross-over study investigated the effect of green tea, black tea and rooibos on the activity of the angiotensin-converting enzyme (ACE) and nitric oxide (NO) level in healthy volunteers. Study subjects received a single oral dose of 400 mL green tea, black tea or rooibos, with the activity of ACE and NO concentration measured at 0, 30, 60 and 180 minutes. The oral intake of a single dose (400 mL) of rooibos significantly inhibited ACE activity after 30 and 60 minutes, while no significant modulation was seen for green and black teas. When subjects were divided into subgroups according to ACE genotype, those with genotype II and ID showed significant inhibition of ACE activity after intake of rooibos. Rooibos consumption did not have any effect on NO concentration, blood pressure and heart rate [290]. The results of this study are highly significant, since ACE inhibitors are the first-line treatment of hypertension and thus common drugs used for cardiovascular diseases, suggesting that rooibos may have antihypertensive and cardiovascular effects through the inhibition of ACE activity.

### 9.7. Adverse reactions and/or toxicity

Although there is no evidence of any toxicological study that have been conducted for rooibos, however, a number of studies have addressed aspects of safety and toxicity. A widely used measure of safety of medicinal and plant extracts in literature is to determine their effect on serum levels of liver and kidney function markers. Increased serum levels of alanine aminotransferase (ALT), aspartate aminotransferase (AST), alkaline phosphatase (ALP) and lactate dehydrogenase (LDH) are used as surrogate markers for hepatic injury. Animal studies from our laboratory and elsewhere have shown that the consumption of unfermented and fermented rooibos as the sole source of drinking fluid over different lengths of time did not cause any adverse effect regarding body weight, liver weight or liver and kidney parameters including serum transaminases, alkaline phosphatase, creatinine, total and unconjugated bilirubin, total protein, total cholesterol or iron status [208,211,212,215,216,247]. Results from the first human intervention study also demonstrated the first scientific proof for human safety of rooibos consumption, as none of the participants reported any adverse effect and the clinical pathology results (ALT, AST, GGT, ALP, LDH, creatinine, bilirubin, glucose and protein) were all within the reference range [289]. The observations from these studies become all the more important, especially with recent cases of possible adverse hepatic effects of rooibos. Sinisalo et al. [291] reported a case of a 42-year-old patient diagnosed with a low-grade B-cell malignancy, Waldenstrom's macroglobulinaemia, 6 years earlier, who temporarily experienced elevated liver enzymes after consuming relatively large amounts of rooibos. The study reaffirms the excellent safety record of rooibos and concluded that the contamination of the rooibos herbal tea by some hepatotoxic compound, genetic predisposition and/or diseased state of the patient to react adversely to one of the many bioactive components in rooibos may be responsible for the effect. A very recent study also reported a case of acute hepatitis and liver failure associated with daily ingestion of rooibos and buchu herbal tea in a 52-year-old man with a history of hyperlipidaemia and stage III chronic kidney disease secondary to IgA nephropathy [292]. These two reports underlined the fact that further study still needs to be done on the safety of rooibos consumption in humans despite the excellent safety record of the tea to date, and also draw attention to the health status of the individual being very important when considering supplementation with dietary antioxidants.

## 10. Conclusion and future research directions

The attractiveness and use of naturally occurring compounds including those derived from fruits, vegetables and herbs as potential chemopreventive and chemotherapeutic agents have become a fascinating strategy, and from the various studies reviewed here, it is evident that rooibos and/or its polyphenols hold great potential not only in the prevention but also in therapy of a wide variety of oxidative stress-induced conditions. Since the first report on the biological activity of rooibos, an enormous body of work has revealed that rooibos extracts possess antioxidant, antimutagenic, anti-inflammatory, anti-diabetic, antimicrobial and, above all, cancer-preventive properties. However, most of the studies have been done *in vitro*, with *ex vivo*, *in vivo* and human trials still limited. This review strongly advocates a shift in focus from *in vitro* studies to animal and human studies, with future research aimed at defining the actual magnitude of health benefits, establishing a safe range of rooibos tea consumption associated with these benefits (if toxicity is discovered) and elucidating the mechanisms of action in order to predict its efficacy. In addition, definitive conclusions concerning the health benefits of rooibos consumption must come from well-designed observational epidemiological studies and clinical intervention trials. For rooibos and/or its flavonoids to have beneficial effects *in vivo*, they have to be absorbed and metabolized in such a way that they can reach and target specific sites. Therefore, more studies are needed to focus on rooibos bioavailability, as well as dose–response effects and pharmacokinetic profiling of the hypothetic active metabolites. Studies analysing the effect of rooibos extract in order to follow the synergistic bioactivity of the different flavonoids present in rooibos are also necessary.

## Acknowledgements

Olawale Ajuwon is supported by the postdoctoral research fellowships from the Claude Leon Foundation and the South African National Research Foundation. The authors acknowledge support from the University of Cape Town and the Cape Peninsula University of Technology, South Africa.

## Author details

Olawale R. Ajuwon[1*], Jeanine L. Marnewick[2] and Lester M. Davids[1]

*Address all correspondence to: Olawale.Ajuwon@uct.ac.za

1 Redox Laboratory, Department of Human Biology, Faculty of Health Sciences, University of Cape Town, Observatory 7925, Cape Town, South Africa

2 Oxidative Stress Research Centre, Institute of Biomedical and Microbial Biotechnology, Cape Peninsula University of Technology, Bellville, South Africa

## References

[1] Freeman, B.A., Campo, J.D. 1982. Biology of disease: free radicals and tissue injury. Laboratory Investigation, 47: 412–426.

[2] Halliwell, B. 2005. Free radicals and other reactive species in disease. Encyclopedia of Life Sciences (pp. 7): Wiley Online Library.

[3] Halliwell, B., Gutteridge, J.M.C. 2007. Free radicals in biology and medicine. Fourth Edition. Clarendon Press, Oxford, UK.

[4] Droge, W. 2002. Free radicals in the physiological control of cell function. Physiological Reviews, 82: 47–95.

[5] Valko, M., Leibfritz, D., Moncol, J., Cronin, M.T.D., Mazur, M., Telser, J. 2007. Free radicals and antioxidants in normal physiological functions and human disease. International Journal of Biochemistry & Cell Biology, 39: 44–84.

[6] Small, D.M., Coombes, J.S., Bennett, N., Johnson, D.W., Gobe, G.C. 2012. Oxidative stress, anti-oxidant therapies and chronic kidney disease. Nephrology, 17: 311–321.

[7] Valko, M., Rhodes, C., Moncol, J., Izakovic, M., Mazur, M. 2006. Free radicals, metals and antioxidants in oxidative stress-induced cancer. Chemico-Biological Interactions, 160: 1–40.

[8] Valko, M., Izakovic, M., Mazur, M., Rhodes, C.J., Telser, J. 2004. Role of oxygen radicals in DNA damage and cancer incidence. Molecular and Cellular Biochemistry, 266: 37–56.

[9] Murphy, M.P. 2009. How mitochondria produce reactive oxygen species. Biochemical Journal, 417: 1–13.

[10] Finkel, T. 2012. Signal transduction by mitochondrial oxidants. Journal of Biological Chemistry, 287: 4434–4440.

[11] Liu, Y., Fiskum, G., Schubert, D. 2002. Generation of reactive oxygen species by the mitochondrial electron transport chain. Journal of Neurochemistry, 80: 780–787.

[12] Kalyanaraman, B. 2013. Teaching the basics of redox biology to medical and graduate students: oxidants, antioxidants and disease mechanisms. Redox Biology, 1: 244–257.

[13] Ott, M., Gogvadze, V., Orrenius, S., Zhivotovsky, B. 2007. Mitochondria, oxidative stress and cell death. Apoptosis, 12: 913–922.

[14] Lugrin, J., Rosenblatt-Velin, N., Parapanov, R., Liaudet, L. 2014. The role of oxidative stress during inflammatory processes. Biological Chemistry, 395: 203–230.

[15] Michiels, C., Raes, M., Toussaint, O., Remacle, J. 1994. Importance of Se-glutathione peroxidase, catalase, and Cu/Zn-SOD for cell survival against oxidative stress. Free Radical Biology and Medicine, 17: 235–248.

[16] Jomova, K., Jenisova, Z., Feszterova, M., Baros, S., Liska, J., Hudecova, D., Rhodes, C.J., Valko, M. 2011. Arsenic: toxicity, oxidative stress and human disease. Journal of Applied Toxicology, 31: 95–107.

[17] Palmer, R., Rees, D.D., Ashton, D.S., Moncada, S. 1988. L-arginine is the physiological precursor for the formation of nitric oxide in endothelium-dependent relaxation. Biochemical and Biophysical Research Communications, 153: 1251–1256.

[18] Bryan, N.S., Bian, K., Murad, F. 2009. Discovery of the nitric oxide signaling pathway and targets for drug development. Frontiers in Bioscience, 14: 1–18.

[19] Nathan, C. 1992. Nitric oxide as a secretory product of mammalian cells. FASEB Journal, 6: 3051–3064.

[20] Geller, D.A., Billiar, T.R. 1998. Molecular biology of nitric oxide synthases. Cancer and Metastasis Reviews, 17: 7–23.

[21] White, P.J., Charbonneau, A., Cooney, G.J., Marette, A. 2010. Nitrosative modifications of protein and lipid signaling molecules by reactive nitrogen species. American Journal of Physiology, Endocrinology and Metabolism, 299: E868–E878.

[22] Carr, A.C., McCall, M.R., Frei, B. 2000. Oxidation of LDL by myeloperoxidase and reactive nitrogen species reaction pathways and antioxidant protection. Arteriosclerosis, Thrombosis, and Vascular Biology, 20: 1716–1723.

[23] Eiserich, J.P., Hristova, M., Cross, C.E., Jones, A.D., Freeman, B.A., Halliwell, B., van der Vliet, A. 1998. Formation of nitric oxide-derived inflammatory oxidants by myeloperoxidase in neutrophils. Nature, 391: 393–397.

[24] Sampson, J.B., Ye, Y.Z., Rosen, H., Beckman, J.S. 1998. Myeloperoxidase and horseradish peroxidase catalyze tyrosine nitration in proteins from nitrite and hydrogen peroxide. Archives of Biochemistry and Biophysics, 356: 207–213.

[25] Denicola, A., Freeman, B.A., Trujillo, M., Radi, R. 1996. Peroxynitrite reaction with carbon dioxide/bicarbonate: kinetics and influence on peroxynitrite-mediated oxidations. Archives of Biochemistry and Biophysics, 333: 49–58.

[26] Lang Jr, J.D., Chumley, P., Eiserich, J.P., Estevez, A., Bamberg, T., Adhami, A., Crow, J., Freeman, B.A. 2000. Hypercapnia induces injury to alveolar epithelial cells via a nitric oxide dependent pathway. American Journal of Physiology, 279: L994–L1002.

[27] Bartesaghi, S., Wenzel, J., Trujillo, M., L pez, M., Joseph, J., Kalyanaraman, B., Radi, R. 2010. Lipid peroxyl radicals mediate tyrosine dimerization and nitration in membranes. Chemical Research in Toxicology, 23: 821–835.

[28] Phaniendra, A., Jestadi, D.B., Periyasamy, L. 2015. Free radicals: properties, sources, targets and their implication in various diseases. Indian Journal of Clinical Biochemistry, 30: 11–26.

[29] Ye, Z.W., Zhang, J., Townsend, D.M., Tew, K.D. 2015. Oxidative stress, redox regulation and diseases of cellular differentiation. Biochimica et Biophysica Acta, 1850: 1607–1621.

[30] Prutz, W.A. 1996. Hypochlorous acid interactions with thiols, nucleotides, DNA, and other biological substrates. Archives of Biochemistry and Biophysics, 332: 110–120.

[31] Giles, G.I., Jacob, C. 2002. Reactive sulfur species: an emerging concept in oxidative stress. Biological Chemistry, 383: 375–388.

[32] Gruhlke, M.C.H., Slusarenko, A.J. 2012. The biology of reactive sulfur species. Plant Physiology and Biochemistry, 59: 98–107.

[33] Pourova, J., Kottova, M., Voprsalova, M., Pour, M. 2010. Reactive oxygen and nitrogen species in normal physiological processes. Acta Physiologica, 198: 15–35.

[34] Brunelle, J.K., Bell, E.L., Quesada, N.M., Vercauteren, K., Tiranti, V., Zeviani, M., Scarpulla, R.C., Chandel, N.S. 2005. Oxygen sensing requires mitochondrial ROS but not oxidative phosphorylation. Cell Metabolism, 1: 409–414.

[35] Guzy, R.D., Hoyos, B., Robin, E., Chen, H., Liu, L., Mansfield, K.D., Simon, M.C., Hammerling, U., Schumacker, P.T. 2005. Mitochondrial complex III is required for hypoxia-induced ROS production and cellular oxygen sensing. Cell Metabolism, 1: 401–408.

[36] Mansfield, K.D., Guzy, R.D., Pan, Y., Young, R.M., Cash, T.P., Schumacker, P.T., Simon, M.C. 2005. Mitochondrial dysfunction resulting from loss of cytochrome c impairs cellular oxygen sensing and hypoxic HIF-$\alpha$ activation. Cell metabolism, 1: 393–399.

[37] Semenza, G.L., Jiang, B.H., Leung, S.W., Passantino, R., Concordet, J.P., Maire, P., Giallongo, A. 1996. Hypoxia response elements in the aldolase A, enolase 1, and lac-

tate dehydrogenase A gene promoters contain essential binding sites for hypoxia-inducible factor 1. Journal of Biological Chemistry, 271: 32,529–32,537.

[38] Kim, J., Tchernyshyov, I., Semenza, G.L., Dang, C.V. 2006. HIF-1-mediated expression of pyruvate dehydrogenase kinase: a metabolic switch required for cellular adaptation to hypoxia. Cell Metabolism, 3: 177–185.

[39] Fukuda, R., Zhang, H., Kim, J., Shimoda, L., Dang, C.V., Semenza, G.L. 2007. HIF-1 regulates cytochrome oxidase subunits to optimize efficiency of respiration in hypoxic cells. Cell, 129: 111–122.

[40] Chan, S.Y., Zhang, Y.Y., Hemann, C., Mahoney, C.E., Zweier, J.L., Loscalzo, J. 2009. MicroRNA-210 controls mitochondrial metabolism during hypoxia by repressing the ironsulfur cluster assembly proteins ISCU1/2. Cell Metabolism, 10: 273–284.

[41] Keisari, Y., Braun, L., Flescher, E., 1983. The oxidative burst and related phenomena in mouse macrophages elicited by different sterile inflammatory stimuli. Immunobiology, 165: 78–89.

[42] West, A.P., Brodsky, I.E., Rahner, C., Woo, D.K., Erdjument-Bromage, H., Tempst, P., Walsh, M.C., Choi, Y., Shadel, G.S., Ghosh, S. 2011. TLR signalling augments macrophage bactericidal activity through mitochondrial ROS. Nature, 472: 476–480.

[43] Nakahira, K., Haspel, J.A., Rathinam, V.A.K., Lee, S.J., Dolinay, T., Lam, H.C., Englert, J.A., Rabinovitch, M., Cernadas, M., Kim, H.P. 2011. Autophagy proteins regulate innate immune responses by inhibiting the release of mitochondrial DNA mediated by the NALP3 inflammasome. Nature Immunology, 12: 222–230.

[44] Zhou, R., Yazdi, A.S., Menu, P., Tschopp, J. 2011. A role for mitochondria in NLRP3 inflammasome activation. Nature, 469: 221–225.

[45] Bulua, A.C., Simon, A., Maddipati, R., Pelletier, M., Park, H., Kim, K.Y., Sack, M.N., Kastner, D.L., Siegel, R.M. 2011. Mitochondrial reactive oxygen species promote production of proinflammatory cytokines and are elevated in TNFR1-associated periodic syndrome (TRAPS). Journal of Experimental Medicine, 208: 519–533.

[46] Scherz-Shouval, R., Shvets, E., Fass, E., Shorer, H., Gil, L., Elazar, Z. 2007. Reactive oxygen species are essential for autophagy and specifically regulate the activity of Atg4. EMBO Journal, 26: 1749–1760.

[47] Scherz-Shouval, R., Elazar, Z. 2011. Regulation of autophagy by ROS: physiology and pathology. Trends in Biochemical Sciences, 36: 30–38.

[48] Dalle-Donne, I., Rossi, R., Colombo, R., Giustarini, D., Milzani, A. 2006. Biomarkers of oxidative damage in human disease. Clinical Chemistry, 52: 601–623.

[49] Dalle-Donne, I., Giustarini, D., Colombo, R., Rossi, R., Milzani, A. 2003. Protein carbonylation in human diseases. Trends in Molecular Medicine, 9: 169–176.

[50] Klaunig, J.E., Kamendulis, L.M., Hocevar, B.A. 2010. Oxidative stress and oxidative damage in carcinogenesis. Toxicologic Pathology, 38: 96–109.

[51] Uchida, K. 2003. 4-Hydroxy-2-nonenal: a product and mediator of oxidative stress. Progress in Lipid Research, 42(4): 318–343.

[52] Carini, M., Aldini, G., Facino, R.M. 2004. Mass spectrometry for detection of 4-hydroxy-trans2-nonenal (HNE) adducts with peptides and proteins. Mass Spectrometry Reviews, 23: 281–305.

[53] Mangialasche, F., Polidori, M.C., Monastero, R., Ercolani, S., Camarda, C., Cecchetti, R., Mecocci, P. 2009. Biomarkers of oxidative and nitrosative damage in Alzheimer's disease and mild cognitive impairment. Ageing Research Reviews, 8: 285–305.

[54] Vignini, A. 2011. Stroke and oxidative stress. Oxidative Stress and Free Radical Damage in Neurology, 137–152.

[55] Roberts, L.J., Morrow, J. 2002. Products of the isoprostane pathway: unique bioactive compounds and markers of lipid peroxidation. Cellular and Molecular Life Sciences, 59: 808–820.

[56] Cracowski, J.L., Durand, T., Bessard, G. 2002. Isoprostanes as a biomarker of lipid peroxidation in humans: physiology, pharmacology and clinical implications. Trends in Pharmacological Sciences, 23: 360–366.

[57] Song, W.L., Lawson, J.A., Reilly, D., Rokach, J., Chang, C.T., Giasson, B., FitzGerald, G.A. 2008. Neurofurans, novel indices of oxidant stress derived from docosahexaenoic acid. Journal of Biological Chemistry, 283: 6–16.

[58] Halliwell, B., Lee, C.Y.J. 2010. Using isoprostanes as biomarkers of oxidative stress: some rarely considered issues. Antioxidants & Redox Signaling, 13; 145–156.

[59] Niki, E. 2010. Free radicals and lipid peroxidation. Free Radical Biology in Digestive Diseases, 29: 1–11.

[60] Stadtman, E.R. 1992. Protein oxidation and aging. Science, 257: 1220–1224

[61] Pandey, K.B., Rizvi, S.I. 2010. Resveratrol may protect plasma proteins from oxidation under conditions of oxidative stress in vitro. Journal of the Brazilian Chemical Society, 21: 909–913.

[62] Pandey, K.B., Rizvi, S.I. 2011. Biomarkers of oxidative stress in red blood cells. Biomedical Papers, 155: 131–136.

[63] Butterfield, D.A., Dalle-Donne, I. 2014. Redox proteomics: from protein modifications to cellular dysfunction and disease. Mass Spectrometry Reviews, 33: 1–6.

[64] Poole, L.B. 2015. The basics of thiols and cysteines in redox biology and chemistry. Free Radical Biology and Medicine, 80: 148–157.

[65] Winterbourn, C.C., Hampton, M.B. 2008. Thiol chemistry and specificity in redox signaling. Free Radical Biology and Medicine, 45: 549–561.

[66] Higdon, A., Diers, A.R., Oh, J.Y., Landar, A., Darley-Usmar, V.M. 2012. Cell signalling by reactive lipid species: new concepts and molecular mechanisms. Biochemical Journal, 442: 453–464.

[67] Winther, J.R., Thorpe, C. 2014. Quantification of thiols and disulfides. Biochimica et Biophysica Acta, 1840: 838–846.

[68] Halliwell, B., Whiteman, M. 2004. Measuring reactive species and oxidative damage in vivo and in cell culture: how should you do it and what do the results mean? British Journal of Pharmacology, 142: 231–255.

[69] Wiseman, H., Halliwell, B. 1996. Damage to DNA by reactive oxygen and nitrogen species: role in inflammatory disease and progression to cancer. Biochemical Journal, 313: 17–29.

[70] Dunlop, R.A., Brunk, U.T., Rodgers, K.J. 2009. Oxidized proteins: mechanisms of removal and consequences of accumulation. IUBMB Life, 61: 522–527.

[71] Seifert, U., Bialy, L.P., Ebstein, F., Bech-Otschir, D., Voigt, A., Schröter, F., Prozorovsky, T., Lange, N., Steffen, J., Rieger, M. 2010. Immunoproteasomes preserve protein homeostasis upon interferon-induced oxidative stress. Cell, 142: 613–624.

[72] Avery, S. 2011. Molecular targets of oxidative stress. Biochemical Journal, 434: 201–210.

[73] Cecarini, V., Gee, J., Fioretti, E., Amici, M., Angeletti, M., Eleuteri, A.M., & Keller, J.N. 2007. Protein oxidation and cellular homeostasis: emphasis on metabolism. Biochimica et Biophysica Acta, 1773: 93–104.

[74] Campioni, S., Mannini, B., Zampagni, M., Pensalfini, A., Parrini, C., Evangelisti, E., Relini, A., Stefani, M., Dobson, C.M., Cecchi, C. 2010. A causative link between the structure of aberrant protein oligomers and their toxicity. Nature Chemical Biology, 6: 140–147.

[75] Dizdaroglu, M., Jaruga, P., Birincioglu, M., Rodriguez, H. 2002. Free radical-induced damage to DNA: mechanisms and measurement. Free Radical Biology and Medicine, 32: 1102–1115.

[76] Sander, M., Cadet, J., Casciano, D.A., Galloway, S.M., Marnett, L.J., Novak, R.F., Pettit, S.D., Prestoa, R.J., Skare, J.A., Williams, G.M. 2005. Proceedings of a workshop on DNA adducts: biological significance and applications to risk assessment. Toxicology and Applied Pharmacology, 208: 1–20.

[77] Hoeijmakers, J.H.J. 2009. DNA damage, aging, and cancer. New England Journal of Medicine, 361: 1475–1485.

[78] Fraga, C.G., Shigenaga, M.K., Park, J.W., Degan, P., Ames, B.N. 1990. Oxidative damage to DNA during aging: 8-hydroxy-2'-deoxyguanosine in rat organ DNA and urine. Proceedings of the National Academy of Sciences, 87: 4533–4537.

[79] Lu, A.L., Li, X., Gu, Y., Wright, P.M., Chang, D.Y. 2001. Repair of oxidative DNA damage. Cell Biochemistry and Biophysics, 35; 141–170.

[80] Guyton, K., Kensler, T. 1993. Oxidative mechanisms in carcinogenesis. British Medical Bulletin, 49: 523–544.

[81] Barber, D.A., Harris, S.R. 1994. Oxygen free radicals and antioxidants: a review. American Pharmacy, NS34: 26–35.

[82] Marnett, L.J. 2000. Oxyradicals and DNA damage. Carcinogenesis, 21: 361–370.

[83] Cooke, J.P. 2003. NO and angiogenesis. Atherosclerosis Supplements, 4: 53–60.

[84] Klaunig, J.E., Kamendulis, L.M. 2004. The role of oxidative stress in carcinogenesis. Annual Review of Pharmacology and Toxicology, 44: 239–267.

[85] Halliwell, B. 2002. Effect of diet on cancer development: is oxidative DNA damage a biomarker. Free Radical Biology and Medicine, 32: 968–974.

[86] Kawanishi, S., Hiraku, Y. 2006. Oxidative and nitrative DNA damage as biomarker for carcinogenesis with special reference to inflammation. Antioxidants and Redox Signaling 8: 1047–1058.

[87] Kozekov, I.D., Nechev, L.V., Moseley, M.S., Harris, C.M., Rizzo, C.J., Stone, M.P., Harris, T.M. 2003. DNA interchain cross-links formed by acrolein and crotonaldehyde. Journal of the American Chemical Society, 125: 50–61.

[88] Liu, X., Lovell, M.A., Lynn, B.C. 2006. Detection and quantification of endogenous cyclic DNA adducts derived from trans-4-hydroxy-2-nonenal in human brain tissue by isotope dilution capillary liquid chromatography nano-electrospray tandem mass spectrometry. Chemical Research in Toxicology, 19: 710–718.

[89] Lovell, M.A., Markesbery, W.R. 2007. Oxidative DNA damage in mild cognitive impairment and late-stage Alzheimer's disease. Nucleic Acids Research, 35: 7497–7504.

[90] Guetens, G., Boeck, G.D., Highley, M., van Oosterom, A.T., de Bruijn, E.A. 2002. Oxidative DNA damage: biological significance and methods of analysis. Critical Reviews in Clinical Laboratory Sciences, 39: 331–457.

[91] Mateos, R., Bravo, L. 2007. Chromatographic and electrophoretic methods for the analysis of biomarkers of oxidative damage to macromolecules (DNA, lipids, and proteins). Journal of Separation Science, 30: 175–191.

[92] Kuo, H.W., Chou, S.Y., Hu, T.W., Wu, F.Y., Chen, D.J. 2007. Urinary 8-hydroxy-2'deoxyguanosine (8-OHdG) and genetic polymorphisms in breast cancer patients. Mutation Research, 631: 62–68.

[93] Rasanen, J., Sihvo, E., Ahotupa, M., Färkkilä, M., Salo, J. 2007. The expression of 8hydroxydeoxyguanosine in oesophageal tissues and tumors. European Journal of Surgical Oncology, 33(10), 1164–1168.

[94] Chang, D., Wang, F., Zhao, Y.S., Pan, H.Z. 2008. Evaluation of oxidative stress in colorectal cancer patients. Biomedical and Environmental Sciences, 21: 286–289.

[95] Yang, Y., Tian, Y., Yan, C., Jin, X., Tang, J., Shen, X. 2009. Determinants of urinary 8-hydroxy-2′-deoxyguanosine in Chinese children with acute leukemia. Environmental Toxicology, 24: 446–452.

[96] Valavanidis, A., Vlachogianni, T., Fiotakis, C. 2009. 8-hydroxy-2′-deoxyguanosine (8-OHdG): a critical biomarker of oxidative stress and carcinogenesis. Journal of Environmental Science and Health Part C, 27: 120–139.

[97] Agha-Hosseini, F., Mirzaii-Dizgah, I., Farmanbar, N., Abdollahi, M. 2012. Oxidative stress status and DNA damage in saliva of human subjects with oral lichen planus and oral squamous cell carcinoma. Journal of Oral Pathology and Medicine, 41: 736–740.

[98] Gonenç, A., Hac şevki, A., Aslan, S., Torun, M., Şimşek, B. 2012. Increased oxidative DNA damage and impaired antioxidant defense system in patients with gastrointestinal cancer. European Journal of Internal Medicine, 23: 350–354.

[99] Gottschling, B.C., Maronpot, R.R., Hailey, J.R., Peddada, S., Moomaw, C.R., Klaunig, J.E., Nyska, A. 2001. The role of oxidative stress in indium phosphide-induced lung carcinogenesis in rats. Toxicological Sciences, 64: 28–40.

[100] Muguruma, M., Unami, A., Kanki, M., Kuroiwa, Y., Nishimura, J., Dewa, Y., Umemura, T., Oishi, Y., Mitsumori, K. 2007. Possible involvement of oxidative stress in piperonyl butoxide induced hepatocarcinogenesis in rats. Toxicology, 236: 61–75.

[101] Iqbal, M., Okazaki, Y., Okada, S. 2009. Curcumin attenuates oxidative damage in animals treated with a renal carcinogen, ferric nitrilotriacetate (Fe-NTA): implications for cancer prevention. Molecular and Cellular Biochemistry, 324: 157–164.

[102] Kumar, M., Sharma, V.L., Sehgal, A., Jain, M. 2012. Protective effects of green and white tea against benzo(a)pyrene induced oxidative stress and DNA damage in murine model. Nutrition and Cancer, 64: 300–306.

[103] Sintara, K., Thong-Ngam, D., Patumraj, S., Klaikeaw, N. 2012. Curcumin attenuates gastric cancer induced by N-methyl-N-nitrosourea and saturated sodium chloride in rats. Journal of Biomedicine and Biotechnology, 2012, Article ID 915380, 8 pages.

[104] Okado-Matsumoto, A., Fridovich, I. 2001. Subcellular distribution of superoxide dismutases (SOD) in rat liver. Journal of Biological Chemistry, 276: 38,388–38,393.

[105] Pamplona, R., Costantini, D. 2011. Molecular and structural antioxidant defenses against oxidative stress in animals. American Journal of Physiology, 301: R843–R863.

[106] Góth, L. 2000. Lipid and carbohydrate metabolism in acatalasemia. Clinical Chemistry, 46: 560–576.

[107] Brigelius-Flohé, R., Kipp, A. 2009. Glutathione peroxidases in different stages of carcinogenesis. Biochimica et Biophysica Acta, 1790: 1555–1568.

[108] Brigelius-Flohé, R., Maiorino, M. 2013. Glutathione peroxidases. Biochimica et Biophysica Acta, 1830: 3289–3303.

[109] Arthur, J. 2000. The glutathione peroxidases. Cellular and Molecular Life Sciences, 57: 1825–1835.

[110] Papp, L.V., Lu, J., Holmgren, A., Khanna, K.K. 2007. From selenium to selenoproteins: synthesis, identity, and their role in human health. Antioxidants & Redox Signaling, 9: 775–806.

[111] Steinbrenner, H., Sies, H. 2009. Protection against reactive oxygen species by selenoproteins. Biochimica et Biophysica Acta, 1790: 1478–1485.

[112] Anderson, M.E. 1996. Glutathione and glutathione delivery compounds. Advances in Pharmacology, 38: 65–78.

[113] Anderson, M.E., Luo, J.L. 1998. Glutathione therapy: from prodrugs to genes. Seminars in Liver diseases, 18: 415–424.

[114] Nordberg, J., Arnér, E.S.J. 2001. Reactive oxygen species, antioxidants, and the mammalian thioredoxin system1. Free Radical Biology and Medicine, 31: 1287–1312.

[115] Rahman, Q., Abidi, P., Afaq, F., Schiffmann, D., Mossman, B.T., Kamp, D.W., Athar, M. 1999. Glutathione redox system in oxidative lung injury. Critical Reviews in Toxicology, 29: 543–568.

[116] Deneke, S.M. 2001. Thiol-based antioxidants. Current Topics in Cellular Regulation, 36: 151–180.

[117] Suntres, Z.E. 2002. Role of antioxidants in paraquat toxicity. Toxicology, 180: 65–77.

[118] Suntres, Z.E. 2011. Liposomal antioxidants for protection against oxidant-induced damage. Journal of Toxicology, 2011, Article ID 152474, 16 pages.

[119] Dickinson, D.A., Forman, H.J. 2002. Glutathione in defense and signaling. Annal of New York Academy of Science, 973: 488–504.

[120] Marnewick, J.L., Van Der Westhuizen, F.H., Joubert, E., Swanevelder, S., Swart, P., et al. 2009. Chemoprotective properties of rooibos (*Aspalathus linearis*), honeybush (*Cyclopia intermedia*) herbal and green and black (*Camellia sinensis*) teas against cancer promotion induced by fumonisin B1 in rat liver. Food Chem Toxicol, 47: 220–229.

[121] Aquilano, K., Baldeli, S., Ciriolo, M.R. 2014. Glutathione: new roles in redox signaling for an old antioxidant. Frontiers in Pharmacology, 5: 196.

[122] Hashemy, S.I. 2011. The human thioredoxin system: modifications and clinical applications. Iranian Journal of Basic Medical Sciences, 14: 191–204.

[123] Nakamura, H. 2005. Thioredoxin and its related molecules: update 2005. Antioxidants & Redox Signaling, 7: 823–828.

[124] Karlenius, T.C., Tonissen, K.F. 2010. Thioredoxin and cancer: a role for thioredoxin in all states of tumor oxygenation. Cancers, 2: 209–232.

[125] Barja, G. 1996. Ascorbic acid and aging. Sub-cellular Biochemistry, 25: 157–188.

[126] Buettner, G.R. 1993. The Pecking order of free radicals and antioxidants: lipid peroxidation, tocopherol, and ascorbate. Archives of Biochemistry and Biophysics, 300: 535–543.

[127] Du, J., Cullen, J.J., Buettner, G.R. 2012. Ascorbic acid: chemistry, biology and the treatment of cancer. Biochimica et Biophysica Acta, 1826: 443–457.

[128] Rose, R., Bode, A. 1993. Biology of free radical scavengers: an evaluation of ascorbate. FASEB Journal, 7: 1135–1142.

[129] Navas, P., Villalba, J.M., Córdoba, F. 1994. Ascorbate function at the plasma membrane. Biochimica et Biophysica Acta, 1197: 1–13.

[130] Linster, C.L., Van Schaftingen, E. 2007. Vitamin C. FEBS Journal, 274: 1–22.

[131] Wells, W.W., Xu, D.P. 1994. Dehydroascorbate reduction. Journal of Bioenergetics and Biomembranes, 26: 369–377.

[132] Bielski, B.H.J., Allen, A.O., Schwarz, H.A. 1981. Mechanism of the disproportionation of ascorbate radicals. Journal of the American Chemical Society, 103: 3516–3518.

[133] Corti, A., Casini, A.F., Pompella, A. 2010. Cellular pathways for transport and efflux of ascorbate and dehydroascorbate. Archives of Biochemistry and Biophysics, 500: 107–115.

[134] Frei, B., England, L., Ames, B.N. 1989. Ascorbate is an outstanding antioxidant in human blood plasma. Proceedings of the National Academy of Sciences, 86: 6377–6381.

[135] Wang, X., Quinn, P.J. 1999. Vitamin E and its function in membranes. Progress in Lipid Research, 38: 309–336.

[136] Bron, D., Asmis, R. 2001. Vitamin E and the prevention of atherosclerosis. International Journal for Vitamin and Nutrition Research, 71: 18–24.

[137] Surai, P.F. 2002. Natural antioxidants in avian nutrition and reproduction. Nottingham University Press.

[138] Esterbauer, H., Schaur, R.J., Zollner, H. 1991. Chemistry and biochemistry of 4-hydroxynonenal, malonaldehyde and related aldehydes. Free Radical Biology and Medicine, 11: 81–128.

[139] Halliwell, B. 2012. Free radicals and antioxidants: updating a personal view. Nutrition Reviews, 70: 257–265.

[140] Costantini, D., Møller, A. 2008. Carotenoids are minor antioxidants for birds. Functional Ecology, 22: 367–370.

[141] Costantini, D., Fanfani, A., Dell'Omo, G. 2008. Effects of corticosteroids on oxidative damage and circulating carotenoids in captive adult kestrels (*Falco tinnunculus*). Journal of Comparative Physiology B, 178: 829–835.

[142] Isaksson, C., Andersson, S. 2008. Oxidative stress does not influence carotenoid mobilization and plumage pigmentation. Proceedings of the Royal Society B, 275: 309–314.

[143] Horak, P., Sild, E., Soomets, U., Sepp, T., Kilk, K. 2010. Oxidative stress and information content of black and yellow plumage coloration: an experiment with greenfinches. Journal of Experimental Biology, 213: 2225–2233.

[144] Larcombe, S.D., Mullen, W., Alexander, L., Arnold, K.E. 2010. Dietary antioxidants, lipid peroxidation and plumage colouration in nestling blue tits *Cyanistes caeruleus*. Naturwissenschaften, 97: 903–913.

[145] Heinonen, O.P., Albanes, D. 1994. The effect of vitamin E and beta carotene on the incidence of lung cancer and other cancers in male smokers. New England Journal of Medicine, 330: 1029–1035.

[146] Albanes, D., Heinonen, O.P., Taylor, P.R., Virtamo, J., Edwards, B.K., Rautalahti, M., et al. 1996. $\alpha$-Tocopherol and $\beta$-carotene supplements and lung cancer incidence in the $\alpha$-tocopherol, $\beta$-carotene cancer prevention study: effects of base-line characteristics and study compliance. Journal of the National Cancer Institute, 88: 1560–1570.

[147] Lee, I.M., Cook, N.R., Manson, J.A.E., Buring, J.E., Hennekens, C.H. 1999. $\beta$-carotene supplementation and incidence of cancer and cardiovascular disease: the Women's Health Study. Journal of the National Cancer Institute, 91: 2102–2106.

[148] Muntwyler, J., Hennekens, C.H., Manson, J.A.E., Buring, J.E., Gaziano, J.M. 2002. Vitamin supplement use in a low-risk population of US male physicians and subsequent cardiovascular mortality. Archives of Internal Medicine, 162: 1472–1476.

[149] Roberts, C.K., Barnard, R.J., 2005. Effects of exercise and diet on chronic disease. Journal of Applied Physiology, 98: 3–30.

[150] Martinez-Gonzalez, M.A., Bes-Rastrollo, M., Serra-Majem, L., Lairon, D., Estruch, R., Trichopoulou, A. 2009. Mediterranean food pattern and the primary prevention of chronic disease: recent developments. Nutrition Reviews, 67: S111–S116.

[151] Aggarwal, B.B., Shishodia, S., Sandur, S.K., Pandey, M.K., Sethi, G. 2006. Inflammation and cancer: how hot is the link? Biochemical Pharmacology, 72: 1605–1621.

[152] Surh, Y.J. 2003. Cancer chemoprevention with dietary phytochemicals. Nature Reviews Cancer, 3: 768–780.

[153] Weng, C.J., Yen, G.C. 2012. Chemopreventive effects of dietary phytochemicals against cancer invasion and metastasis: phenolic acids, monophenol, polyphenol, and their derivatives. Cancer Treatment Reviews, 38: 76–87.

[154] Liu, R.H. 2003. Health benefits of fruit and vegetables are from additive and synergistic combinations of phytochemicals. American Journal of Clinical Nutrition, 78: 517S–520S.

[155] Liu, R.H. 2004. Potential synergy of phytochemicals in cancer prevention: mechanism of action. Journal of Nutrition, 134: 3479S–3485S.

[156] De Kok, T.M., Van Breda, S.G., Manson, M.M. 2008. Mechanisms of combined action of different chemopreventive dietary compounds. European Journal of Nutrition, 47: 51–59.

[157] Crozier, A., Jaganath, I.B., Clifford, M.N. 2009. Dietary phenolics: chemistry, bioavailability and effects on health. Natural Product Reports, 26: 1001–1043.

[158] Tsao, R. 2010. Chemistry and biochemistry of dietary polyphenols. Nutrients, 2: 1231–1246.

[159] Balasundram, N., Sundram, K., Samman, S. 2006. Phenolic compounds in plants and agri-industrial by-products: antioxidant activity, occurrence, and potential uses. Food Chemistry, 99: 191–203.

[160] Taylor, L.P., Grotewold, E. 2005. Flavonoids as developmental regulators. Current Opinion in Plant Biology, 8: 317–323.

[161] Simmonds, M.S.J. 2001. Importance of flavonoids in insect-plant interactions: feeding and oviposition. Phytochemistry, 56: 245–252.

[162] Xu, C., Natarajan, S., Sullivan, J.H. 2008a. Impact of solar ultraviolet-B radiation on the antioxidant defense system in soybean lines differing in flavonoid contents. Environmental and Experimental Botany, 63: 39–48.

[163] Xu, C., Sullivan, J.H., Garrett, W.M., Caperna, T.J., Natarajan, S. 2008b. Impact of solar ultraviolet-B on the proteome in soybean lines differing in flavonoid contents. Phytochemistry, 69: 38–48.

[164] Yao, L.H., Jiang, Y.M., Shi, J., Tomas-Barberan, F.A., Datta, N., Singanusong, R., Chen, S.S. 2004. Flavonoids in food and their health benefits. Plant Foods for Human Nutrition, 59: 113–122.

[165] Lin, J., Weng, M.S. 2006. Flavonoids as nutraceuticals. The Science of Flavonoids, 213–238.

[166] Pereira, D.M., Valentão, P., Pereira, J.A., Andrade, P.B. 2009. Phenolics: from chemistry to biology. Molecules, 14: 2202–2211.

[167] Van Wyk, B.E., Gericke, M. 2000. People's plants. A Guide to Useful Plants of Southern Africa. Briza Publications, Pretoria, South Africa, pp. 102–103.

[168] McKay, D.L., Blumberg, J.B. 2007. A review of the bioactivity of South African herbal teas: rooibos (*Aspalathus linearis*) and honeybush (*Cyclopia intermedia*). Phytotherapy Research, 21: 1–16.

[169] Van Heerden, F., Van Wyk, B.E., Viljoen, A., Steenkamp, P. 2003. Phenolic variation in wild populations of *Aspalathus linearis* (rooibos tea). Biochemical Systematics and Ecology, 31: 885–895.

[170] Wilson, N.L.W. 2005. Cape natural tea products and the US market: rooibos rebels ready to raid. Applied Economic Perspectives and Policy, 27: 139–148.

[171] Cheney, R.H., Scholtz, E. 1963. Rooibos tea, a South African contribution to world beverages. Economic Botany, 17: 186–194.

[172] Morton, J.F. 1983. Rooibos tea, *Aspalathus linearis*, a caffeineless, low-tannin beverage. Economic Botany, 37: 164–173.

[173] Joubert, E., Gelderblom, W., Louw, A., De Beer, D. 2008a. South African herbal teas: *Aspalathus linearis, Cyclopia spp.* and *Athrixia phylicoides* – a review. Journal of Ethnopharmacology, 119: 376–412.

[174] Reynecke, J., Coetzee, W., Bester, J. 1949. Rooibos tea. A preliminary report on the composition. Farming in South Africa, 24: 397–398.

[175] Blommaert, K.L.J., Steenkamp, J. 1978. Tannien- en moontlike kafeieninhoud van rooibostee, *Aspalathus* (Subgen. Nortiera) *linearis* (Burm. Fil) R. Dahlgr. Agroplantae, 10: 49.

[176] Koeppen, B., Roux, D. 1966. C-glycosylflavonoids. The chemistry of aspalathin. Biochemical Journal, 99: 604–609.

[177] Rabe, C., Steenkamp, J.A., Joubert, E., Burger, J.F.W., Ferreira, D. 1994. Phenolic metabolites from rooibos tea (*Aspalathus linearis*). Phytochemistry, 35: 1559–1565.

[178] Shimamura, N., Miyase, T., Umehara, K., Warashina, T., Fujii, S. 2006. Phytoestrogens from *Aspalathus linearis*. Biological and Pharmaceutical Bulletin, 29: 1271–1274.

[179] Joubert, E. 1996. HPLC quantification of the dihydrochalcones, aspalathin and nothofagin in rooibos tea (*Aspalathus linearis*) as affected by processing. Food Chemistry, 55: 403–411.

[180] Joubert, E., Gelderblom, W., De Beer, D. 2009. Phenolic contribution of South African herbal teas to a healthy diet. Natural Product Communications, 4: 701–718.

[181] Hillis, W., Inoue, T. 1967. The polyphenols of *Nothofagus* species II: the heartwood of *Nothofagus fusca*. Phytochemistry, 6: 59–67.

[182] Huang, C., Gan, X., Bai, H., Ma, L., Hu, L. 2008. Schoepfin A, B, C: three new chalcone C-glycosides from *Schoepfia chinensis*. Natural Product Research, 22: 623–627.

[183] Koeppen, B., Roux, D. 1965. C-glycosylflavonoids. The chemistry of orientin and isoorientin. Biochemical Journal, 97: 444–448.

[184] Bramati, L., Minoggio, M., Gardana, C., Simonetti, P., Mauri, P., Pietta, P. 2002. Quantitative characterization of flavonoid compounds in rooibos tea (*Aspalathus linearis*) by LC-UV/DAD. Journal of Agricultural and Food Chemistry, 50: 5513–5519.

[185] Snyckers, F., Salemi, G. 1974. Studies of South African medicinal plants. Part 1. Quercetin as the major in vitro active component of rooibos tea. Journal of South African Chemical Institute, 27: 5–7.

[186] Kazuno, S., Yanagida, M., Shindo, N., Murayama, K. 2005. Mass spectrometric identification and quantification of glycosyl flavonoids, including dihydrochalcones with neutral loss scan mode. Analytical Biochemistry, 347: 182–192.

[187] Krafczyk, N., Glomb, M.A. 2008. Characterization of phenolic compounds in rooibos tea. Journal of Agricultural and Food Chemistry, 56: 3368–3376.

[188] Ferreira, D., Marais, C., Steenkamp, J.A., Joubert, E. 1995. Rooibos tea as a likely health food supplement. Proceedings of Recent Development of Technologies on Fundamental Foods for Health, 73–88.

[189] Marnewick, J.L., Gelderblom, W.C.A., Joubert, E. 2000. An investigation on the antimutagenic properties of South African herbal teas. Mutation Research, 471: 157–166.

[190] Standley, L., Winterton, P., Marnewick, J.L., Gelderblom, W.C.A., Joubert, E., Britz, T.J. 2001. Influence of processing stages on antimutagenic and antioxidant potentials of rooibos tea. Journal of Agricultural and Food Chemistry, 49: 114–117.

[191] Joubert, E., Winterton, P., Britz, T.J., Gelderblom, W.C.A. 2005. Antioxidant and prooxidant activities of aqueous extracts and crude polyphenolic fractions of rooibos (*Aspalathus linearis*). Journal of Agricultural and Food Chemistry, 53: 10,260–10,267.

[192] Joubert, E., Ferreira, D. 1996. Antioxidants of rooibos tea – a possible explanation for its health promoting properties? South African Journal of Food Science and Nutrition, 8: 79–84.

[193] Bramati, L., Aquilano, F., Pietta, P. 2003. Unfermented rooibos tea: quantitative characterization of flavonoids by HPLC-UV and determination of the total antioxidant activity. Journal of Agricultural and Food Chemistry, 51: 7472–7474.

[194] Von Gadow, A., Joubert, E., Hansmann, C. 1997a. Comparison of the antioxidant activity of rooibos tea (*Aspalathus linearis*) with green, oolong and black tea. Food Chemistry, 60: 73–77.

[195] Joubert, E., Winterton, P., Britz, T.J., Ferreira, D. 2004. Superoxide anion and α, αdi-phenyl-β-picrylhydrazyl radical scavenging capacity of rooibos (*Aspalathus linearis*) aqueous extracts, crude phenolic fractions, tannin and flavonoids. Food Research International, 37: 133–138.

[196] Joubert, E., Richards, E.S., Merwe, J.D.V., De Beer, D., Manley, M., Gelderblom, W. 2008b. Effect of species variation and processing on phenolic composition and in vitro antioxidant activity of aqueous extracts of *Cyclopia* spp. (honeybush tea). Journal of Agricultural and Food Chemistry, 56: 954–963.

[197] Von Gadow, A., Joubert, E., Hansmann, C. 1997b. Effect of extraction time and additional heating on the antioxidant activity of rooibos tea (*Aspalathus linearis*) extracts. Journal of Agricultural and Food Chemistry, 45: 1370–1374.

[198] Snijman, P.W., Joubert, E., Ferreira, D., Li, X. C., Ding, Y., Green, I.R., Gelderblom, W.C.A. 2009. Antioxidant activity of the dihydrochalcones aspalathin and nothofagin and their corresponding flavones in relation to other rooibos (*Aspalathus linearis*) flavonoids, epigallocatechin gallate, and Trolox. Journal of Agricultural and Food Chemistry, 57: 6678–6684.

[199] Krafczyk, N., Heinrich, T., Porzel, A., Glomb, M.A. 2009. Oxidation of the dihydrochalcone aspalathin leads to dimerization. Journal of Agricultural and Food Chemistry, 57; 6838–6843.

[200] Hitomi, E., Tamura, S., Tsurumoto, Y., Tsuda, T., Nakano, M. 1999. Antioxidative activity of rooibos tea (*Aspalathuslinearis*). Journal of the Japanese Society of Food Science and Technology, 46: 779–785.

[201] Marnewick, J., Joubert, E., Joseph, S., Swanevelder, S., Swart, P., Gelderblom, W. 2005. Inhibition of tumour promotion in mouse skin by extracts of rooibos (*Aspalathus linearis*) and honeybush (*Cyclopia intermedia*), unique South African herbal teas. Cancer Letters, 224: 193–202.

[202] Akaike, T., Ijiri, S., Sato, K., Maeda, H. 1995. Determination of peroxyl radical-scavenging activity in food by using bactericidal action of alkyl peroxyl radical. Journal of Agricultural and Food Chemistry, 43: 1864–1870.

[203] Lee, E.J., Jang, H.D. 2004. Antioxidant activity and protective effect on DNA strand scission of rooibos tea (*Aspalathus linearis*). Biofactors, 21: 285–292.

[204] Yoo, K.M., Lee, C.H., Lee, H., Moon, B.K., Lee, C.Y. 2008. Relative antioxidant and cytoprotective activities of common herbs. Food Chemistry, 106: 929–936.

[205] Inanami, O., Asanuma, T., Inukai, N., Jin, T., Shimokawa, S., Kasai, N., Nakano, M., Sato, F., Kuwabara, M. 1995. The suppression of age-related accumulation of lipid peroxides in rat brain by administration of rooibos tea (*Aspalathus linearis*). Neuroscience Letters, 196: 8588.

[206] Shimoi, K., Masuda, S., Shen, B., Furugori, M., Kinae, N. 1996. Radioprotective effects of antioxidative plant flavonoids in mice. Mutation Research, 350: 153–161.

[207] Hitomi, E., Nishikiori, A., Matsumoto, A., Moriguchi, S., Kojo, S., Tokumaru, S., Nakano, M. 2004. Effect of rooibos tea (*Aspalathus linearis*) extract on lipid peroxidation in vitamin E deficient rats. ITE Letters on Batteries New Technologies and Medicine, 5: 30–38.

[208] Marnewick, J.L., Joubert, E., Swart, P., van der Westhuizen, F., Gelderblom, W.C. 2003. Modulation of hepatic drug metabolizing enzymes and oxidative status by rooibos (*Aspalathus linearis*) and honeybush (*Cyclopia intermedia*), green and black (*Camellia sinensis*) teas in rats. Journal of Agricultural and Food Chemistry, 51: 8113–8119.

[209] Kucharska, J., Ulicna, O., Gvozdjakova, A., Sumbalova, Z., Vancova, O., Bozek, P., Nakano, M., Greksak, M. 2004. Regeneration of coenzyme Q9 redox state and inhibition of oxidative stress by rooibos tea (*Aspalathus linearis*) administration in carbon tetrachloride liver damage. Physiological Research, 53: 515–522.

[210] Awoniyi, D.O., Aboua, Y.G., Marnewick, J., Brooks, N. 2012. The effects of rooibos (*Aspalathus linearis*), green tea (*Camellia sinensis*) and commercial rooibos and green tea supplements on epididymal sperm in oxidative stress-induced rats. Phytotherapy Research, 26: 1231–1239.

[211] Ajuwon, O.R., Katengua-Thamahane, E., Van Rooyen, J., Oguntibeju, O.O., Marnewick, J. 2013. Protective effects of rooibos (*Aspalathus linearis*) and/or red palm oil (*Elaeis guineensis*) supplementation on tert-butyl hydroperoxide-induced oxidative hepatotoxicity in Wistar rats. Evidence-Based Complementary and Alternative Medicine, 2013, Article ID 984273, 19 pages.

[212] Ajuwon, O.R., Oguntibeju O.O., Marnewick J.L. 2014. Amelioration of lipopolysaccharide-induced liver injury by aqueous rooibos (*Aspalathus linearis*) extract via inhibition of pro-inflammatory cytokines and oxidative stress. BMC Complementary and Alternative Medicine, 14: 392.

[213] Saleem, T.S.M., Chetty, C.M., Ramkanth, S., Rajan, V., Kumar, K.M., Gauthaman, K. 2010. Hepatoprotective herbs – a review. International Journal of Research in Pharmaceutical Sciences, 1: 1–5.

[214] Chen, C.J., Deng, A.J., Liu, C., Shi, R., Qin, H.L., Wang, A.P. 2011. Hepatoprotective activity of *Cichorium endivia* L. extract and its chemical constituents. Molecules, 16: 9049–9066.

[215] Ulicna, O., Greksák, M., Vancova, O., Zlatos, L., Galbavy, S., Bosek, P., Nakano, M. 2003. Hepatoprotective effect of rooibos tea (*Aspalathus linearis*) on $CCl_4$-induced liver damage in rats. Physiological Research, 52: 461–466.

[216] Ulicna, O., Vancova, O., Waczulikova, I., Bozek, P., Janega, P., Babal, P., Liskova, S., Greksak, M. 2008. Does rooibos tea (*Aspalathus linearis*) support regeneration of rat

liver after intoxication by carbon tetrachloride? General Physiology and Biophysics, 27: 179–186.

[217] McCune, L.M., Johns, T. 2002. Antioxidant activity in medicinal plants associated with the symptoms of diabetes mellitus used by the indigenous peoples of the North American boreal forest. Journal of Ethnopharmacology, 82: 197–205.

[218] Büyükbalci, A., El, S.N. 2008. Determination of *in vitro* antidiabetic effects, antioxidant activities and phenol contents of some herbal teas. Plant Foods for Human Nutrition, 63: 27–33.

[219] Pandey, K.B., Rizvi, S.I. 2010. Resveratrol may protect plasma proteins from oxidation under conditions of oxidative stress in vitro. Journal of the Brazilian Chemical Society, 21: 909–913.

[220] Meece, J. 2007. Pancreatic islet dysfunction in type 2 diabetes: a rational target for incretin-based therapies. Current Medicinal Research and Opinions, 23: 933–944.

[221] Wali, J.A., Rondas, D., McKenzie, M.D., Zhao, Y., Elkerbout, L., Fynch, S., Gurzov, E.N., Akira, S., Mathieu, C., Kay, T.W.H., Overbergh, L., Strasser, A., Thomas, H.E. 2014. The proapoptotic BH3-only proteins Bim and Puma are downstream of endoplasmic reticulum and mitochondrial oxidative stress in pancreatic islets in response to glucotoxicity. Cell Death and Disease, 5: e1124.

[222] Lenzen, S., Drinkgern, J., Tiedge, M. 1996. Low antioxidant enzyme gene expression in pancreatic islets compared with various other mouse tissues. Free Radical Biology and Medicine, 20: 463–466.

[223] Jonas, J.C., Bensellam, M., Duprez, J., Elouil, H., Guiot, Y., Pascal, S.M. 2009. Glucose regulation of islet stress responses and beta-cell failure in type 2 diabetes. Diabetes Obesity and Metabolism, 11(Suppl 4): 65–81.

[224] Burcelin, R., Knauf, C., Cani, P.D. 2008. Pancreatic $\alpha$-cell dysfunction in diabetes. Diabetes and Metabolism, 34: S49–S55.

[225] Diao, J., Allister, E.M., Koshkin, V., Lee, S.C., Bhattacharjee, A., Tang, C., Giacca, A., Chan, C.B., Wheeler, M.B. 2008. UCP2 is highly expressed in pancreatic $\alpha$-cells and influences secretion and survival. Proceedings of the National Academy of Sciences, 105: 12,057–12,062.

[226] Rolo, A.P., Palmeira, C.M. 2006. Diabetes and mitochondrial function: role of hyperglycemia and oxidative stress. Toxicology and Applied Pharmacology, 212: 167–178.

[227] Bloch, K., Shichman, E., Vorobeychik, M., Bloch, D., Vardi, P. 2006. Catalase expression in pancreatic alpha cells of diabetic and non-diabetic mice. Histochemistry and Cell Biology, 127: 227–232.

[228] Van Dam, P.S., Van Asbeck B.S., Van Oirschot J.F., Biessels G.J., Hamers F.P., Marx J.J. 2001. Glutathione and $\alpha$-lipoate in diabetic rats: nerve function, blood flow and oxidative state. European Journal of Clinical Investigation, 31: 417–424.

[229] Palmer, A., Thomas, C.R., Gopaul, N., Dhir, S., Änggård, E.E., Poston, L.., Tribe, R.M. 1998. Dietary antioxidant supplementation reduces lipid peroxidation but impairs vascular function in small mesenteric arteries of the streptozotocin-diabetic rat. Diabetologia, 41: 148–156.

[230] Portero-Otín, M., Pamplona, R., Ruiz, M.C., Cabiscol, E., Prat, J., Bellmunt, M.J. 1999. Diabetes induces an impairment in the proteolytic activity against oxidized proteins and a heterogeneous effect in non-enzymatic protein modifications in the cytosol of rat liver and kidney. Diabetes 48: 2215–2220.

[231] Ling, X., Nagai, R., Sakashita, N., Takeya, M., Horiuchi, S., Takahashi, K. 2001. Immunohistochemical distribution and quantitative biochemical detection of advanced glycation end products in fetal to adult rats and in rats with streptozotocin-induced diabetes. Laboratory Investigation, 81: 845–861.

[232] Gopaul, N.K., Angglard, E.E., Mallet, A.I., Betteridge, D.J., Wolff, S.P., Nourooz-Zadeh, J. 1995. Plasma 8-epi-$PGF_{2alpha}$ levels are elevated in individuals with non-insulin dependent diabetes mellitus. FEBS Letters, 368: 225–229.

[233] Davi, G., Ciabattoni, G., Consoli, A., Mezzetti, A., Falco, A., Santarone, S., et al. 1999. In vivo formation of 8-iso-prostaglandin $F_{2alpha}$ and platelet activation in diabetes mellitus. Circulation, 99: 224–229.

[234] Martin-Gallan, P., Carrascosa, A., Gussinye, M., Dominguez, C. 2003. Biomarkers of diabetes-associated oxidative stress and antioxidant status in young diabetic patients with or without subclinical complications. Free Radical Biology and Medicine, 34: 1563–1574.

[235] Kashiba, M., Oka, J., Ichikawa, R., Kasahara, E., Inayama, T., Kageyama, A., et al. 2002. Impaired ascorbic acid metabolism in streptozotocin-induced diabetic rats. Free Radical Biology and Medicine, 33: 1221–1230.

[236] Obrosova, I.G., Fathallah, L., Liu, E., Nourooz-Zadeh, J. 2003. Early oxidative stress in the diabetic kidney: effect of DL-alpha-lipoic acid. Free Radical Biology and Medicine, 34: 186–195.

[237] Sundaram, R.K., Bhaskar, A., Vijayalingam, S., Viswanathan, M., Mohan, R., Shanmugasundaram, K.R. 1996. Antioxidant status and lipid peroxidation in type II diabetes mellitus with and without complications. Clinical Science, 90: 255–260.

[238] Nourooz-Zadeh, J., Rahimi, A., Tajaddini-Sarmadi, J., Tritschler, H., Rosen, P., Halliwell, B., et al. 1997. Relationships between plasma measures of oxidative stress and metabolic control in NIDDM. Diabetologia, 40: 647–653.

[239] Davison, G.W., George, L., Jackson, S.K., Young, I.S., Davies, B., Bailey, D.M., et al. 2002. Exercise, free radicals, and lipid peroxidation in type 1 diabetes mellitus. Free Radical Biology and Medicine, 33: 1543–1551.

[240] Niedowicz, D.M., Daleke, D.L. 2005. The role of oxidative stress in diabetic complications. Cell Biochemistry and Biophysics, 43: 289–330.

[241] Erasto, P., Adebola, P., Grierson, D., Afolayan, A. 2005. An ethnobotanical study of plants used for the treatment of diabetes in the Eastern Cape Province, South Africa. African Journal of Biotechnology, 4: 1458–1460.

[242] Jung, M., Park, M., Lee, H.C., Kang, Y.H., Kang, E.S., Kim, S.K. 2006. Antidiabetic agents from medicinal plants. Current Medicinal Chemistry, 13: 1203–1218.

[243] Gallagher, A., Flatt, P., Duffy, G., Abdel-Wahab, Y. 2003. The effects of traditional antidiabetic plants on *in vitro* glucose diffusion. Nutrition Research, 23: 413–424.

[244] Kelble, A. 2005. Spices and type 2 diabetes. Nutrition and Food Science, 35: 81–87.

[245] Johnston, K., Sharp, P., Clifford, M., Morgan, L. 2005. Dietary polyphenols decrease glucose uptake by human intestinal Caco-2 cells. FEBS Letters, 579: 1653–1657.

[246] Han, X., Shen, T., Lou, H. 2007. Dietary polyphenols and their biological significance. International Journal of Molecular Sciences, 8: 950–988.

[247] Ulicna, O., Vancova, O., Bozek, P., Carsky, J., Sebekova, K., Boor, P., Nakano, M., Greksák, M. 2006. Rooibos tea (*Aspalathus linearis*) partially prevents oxidative stress in streptozotocin-induced diabetic rats. Physiological Research, 55: 157–164.

[248] Kinae, N., Shimoi, K., Masumori, S., Harasawa, M., Furugori, M. 1994. Suppression of the formation of advanced glycosylation products by tea extracts. In: Ho, C.-T., Osawa, T., Huang, M.-T., Rosen, R.T. (Eds.), Food Phytochemicals for Cancer Prevention II, ACS Symposium Series, vol. 547. American Chemical Society, Washington, DC, USA, pp. 68–75.

[249] Song, J., Kwon, O., Chen, S., Daruwala, R., Eck, P., Park, J.B., Levine, M. 2002. Flavonoid inhibition of SVCT1 and GLUT2, intestinal transporters for vitamin C and glucose. Journal of Biological Chemistry, 277: 15,252–15,260.

[250] Kawano, A., Nakamura, H., Hata, S., Minakawa, M., Miura, Y., Yagasaki, K. 2009. Hypoglycemic effect of aspalathin, a rooibos tea component from *Aspalathus linearis*, in type 2 diabetic model db/db mice. Phytomedicine, 16: 437–443.

[251] Joubert, E., De Beer, D., Malherbe, C., Louw, J., Muller, C., Gelderblom, W.C.A. 2010. Aspalathin, a rooibos dihydrochalcone C-glucoside – occurrence, bioavailability and potential health benefits. Abstract Book of the 64th Annual Meeting of the Japanese Society of Nutrition and Food Science, 22–23 May, Tokushima, Japan, p. 31.

[252] Muller, C., Joubert, E., De Beer, D., Sanderson, M., Malherbe, C.J., Fey, S.J., Louw, J. 2012. Acute assessment of an aspalathin-enriched green rooibos (*Aspalathus linearis*) extract with hypoglycemic potential. Phytomedicine, 20: 32–39.

[253] Mazibuko, S.E., Muller, C.J.F., Joubert, E., De Beer, D., Johnson, R., Opoku, A.R., Louw, J. 2013. Amelioration of palmitate-induced insulin resistance in C2C12 muscle cells by rooibos (*Aspalathus linearis*). Phytomedicine, 20: 813–819.

[254] Sanderson, M., Mazibuko, S.E., Joubert, E., De Beer, D., Johnson, R., Pheiffer, C., Louw, J., Muller, C.J.F. 2014. Effects of fermented rooibos (*Aspalathus linearis*) on adipocyte differentiation. Phytomedicine, 21: 109–117.

[255] Dludla, P., Muller, C.J.F., Louw, J., Joubert, E., Salie, R., Opoku, A.R., Johnson, R. 2014. The cardioprotective effect of an aqueous extract of fermented rooibos (*Aspalathus linearis*) on cultured cardiomyocytes derived from diabetic rats. Phytomedicine, 21: 595–601.

[256] Ku, S.K., Kwak, S., Kim, Y., Bae, J.S. 2014. Aspalathin and nothofagin from rooibos (*Aspalathus linearis*) inhibits high glucose-induced inflammation *in vitro* and *in vivo*. Inflammation, 1–11.

[257] Hanhineva, K., Törrönen, R., Bondia-Pons, I., Pekkinen, J., Kolehmainen, M., Mykkänen, H., Poutanen, K. 2010. Impact of dietary polyphenols on carbohydrate metabolism. International Journal of Molecular Sciences, 11: 1365–1402.

[258] Colic, M., Pavelic, K. 2002. Molecular, cellular and medical aspects of the action of nutraceuticals and small molecules therapeutics: from chemoprevention to new drug development. Drugs under Experimental and Clinical Research, 28: 169–176.

[259] Mehta, R.G., Murillo, G., Naithani, R., Peng, X. 2010. Cancer chemoprevention by natural products: how far have we come? Pharmaceutical Research, 27: 950–961.

[260] Sasaki, Y.F., Yamada, H., Shimoi, K., Kator, K., Kinae, N. 1993. The clastogen-suppressing effects of green tea, Po-lei tea and rooibos tea in CHO cells and mice. Mutation Research, 286: 221–232.

[261] Komatsu, K., Kator, K., Mitsuda, Y., Mine, M., Okumura, Y. 1994. Inhibitory effects of rooibos tea (*Aspalathus linearis*) on X-ray-induced C3H10T12 cell transformation. Cancer Letters, 77: 33–38.

[262] Lamosova, D., Juráni, M., Greksák, M., Nakano, M., Vaneková, M. 1997. Effect of rooibos tea (*Aspalathus linearis*) on chick skeletal muscle cell growth in culture. Comparative Biochemistry and Physiology C, 116: 39–45.

[263] Van der Merwe, J., Joubert, E., Richards, E., Manley, M., Snijman, P., Marnewick, J., Gelderblom, W. 2006. A comparative study on the antimutagenic properties of aqueous extracts of *Aspalathus linearis* (rooibos), different *Cyclopia* spp. (honeybush) and *Camellia sinensis* teas. Mutation Research, 611: 42–53.

[264] Platt, K.L., Edenharder, R., Aderhold, S., Muckel, E., Glatt, H. 2010. Fruits and vegetables protects against the genotoxicity of heterocyclic aromatic amines activated by human xenobiotic-metabolizing enzymes expressed in mammalian cells. Mutation Research, 703: 90–98.

[265] Sissing, L. 2008. Investigations into the cancer modulating properties of *Aspalathus linearis* (rooibos), *Cyclopia intermedia* (honeybush) and *Sutherlandia frutescens* (cancer bush) in oesophageal carcinogenesis. M.Sc. (Physiology) Thesis. University of the Western Cape, Bellville, South Africa.

[266] Snijman, P.W., Swanevelder, S., Joubert, E., Green, I.R., Gelderblom, W.C.A. 2007. The antimutagenic activity of the major flavonoids of rooibos (*Aspalathus linearis*): some dose–response effects on mutagen activation–flavonoid interactions. Mutation Research, 631: 111–123.

[267] Marnewick, J., Batenburg, W., Swart, P., Joubert, E., Swanevelder, S., Gelderblom, W. 2004. Ex vivo modulation of chemical-induced mutagenesis by subcellular liver fractions of rats treated with rooibos (*Aspalathus linearis*) tea, honeybush (*Cyclopia intermedia*) tea, as well as green and black (*Camellia sinensis*) teas. Mutation Research, 558: 145–154.

[268] Petrova, A. 2009. Modulation of ultraviolet light-induced skin carcinogenesis by extracts of rooibos and honeybush using a mouse model: elucidating possible protective mechanisms. MTech (Biomedical Technology) Thesis. Cape Peninsula University of Technology, South Africa.

[269] Na, H.K., Mossanda, K.S., Lee, J.Y., Surh, Y.J. 2004. Inhibition of phorbol ester-induced COX-2 expression by some edible African plants. Biofactors, 21: 149–153.

[270] Marnewick, J.L., Van Der Westhuizen, F.H., Joubert, E., Swanevelder, S., Swart, P., Gelderblom, W.C.A. 2009. Chemoprotective properties of rooibos (*Aspalathus linearis*), honeybush (*Cyclopia intermedia*) herbal and green and black (*Camellia sinensis*) teas against cancer promotion induced by fumonisin B1 in rat liver. Food and Chemical Toxicology, 47: 220–229.

[271] Sissing, L., Marnewick, J., de Kock, M., Swanevelder, S., Joubert, E., Gelderblom, W. 2011. Modulating effects of rooibos and honeybush herbal teas on the development of esophageal papillomas in rats. Nutrition and Cancer, 63: 600–610.

[272] Mueller, M., Hobiger, S., Jungbauer, A. 2010. Anti-inflammatory activity of extracts from fruits, herbs and spices. Food Chemistry, 122: 987–996.

[273] MacNee, W. 2001. Oxidative stress and lung inflammation in airways disease. European Journal of Pharmacology, 429: 195–207.

[274] Rahman, I. 2002. Oxidative stress, transcription factors and chromatin remodelling in lung inflammation. Biochemical Pharmacology, 64: 935–942.

[275] Baba, H., Ohtsuka, Y., Haruna, H., Lee, T., Nagata, S., Maeda, M., Yamashiro, Y., Shimizu, T. 2009. Studies of anti-inflammatory effects of rooibos tea in rats. Pediatrics International, 51: 700–704.

[276] Katengua-Thamahane, E., Marnewick, J.L., Ajuwon, O.R., Chegou, N.N., Szűcs, G., Ferdinandy, P., Csont, T., Csonka, C., Van Rooyen, J. 2014. The combination of red palm oil and rooibos show anti-inflammatory effects in rats. Journal of Inflammation, 11: 41.

[277] Lee, W., Bae, J.S. 2015. Anti-inflammatory effects of aspalathin and nothofagin from rooibos (*Aspalathus linearis*) *in vitro* and *in vivo*. Inflammation, 1–15.

[278] Hesseling, P., Klopper, J., Van Heerden, P. 1979. The effect of rooibos tea on iron absorption. South African Medical Journal, 55: 631–632.

[279] Breet, P., Kruger, H.S., Jerling, J.C., Oosthuizen, W. 2005. Actions of black tea and rooibos on iron status of primary school children. Nutrition Research, 25: 983–994.

[280] Hesseling, P., Joubert, J. 1982. The effect of rooibos tea on the type I allergic reaction. South African Medical Journal, 62: 1037–1038.

[281] Shindo, Y., Kato, K. 1991. Effect of rooibos tea on some dermatological diseases. In: Proceedings of the International Symposium on Tea Science, 385–389, Shizuoka, Japan.

[282] Chuarienthong, P., Lourith, N., Leelapornpisid, P. 2010. Clinical efficacy comparison of antiwrinkle cosmetics containing herbal flavonoids. International Journal of Cosmetic Science, 32: 99–106.

[283] Glynn, K.M. 2010. From in-vitro to in-vivo: corporate development and efficacy of a topical hair growth agent derived from natural extracts. MSc (Biology) Thesis. Grand Valley State University, MI, USA.

[284] Sauter, W. 2005. Vergleich antioxidativer Eigenschaften von Extrakten aus *Camellia sinensis, Rosmarinus officinalis, Cyclopia genistoides, Cyclopia sessiliflora* und *Aspalathus linearis*. PhD Desertation. Technische Universität München, Munchen, Germany.

[285] Nikolova, V., Petrova, S., Petkova, V., Pavlova, S., Michailova, A., Georgieva, T. 2007. Antioxidative effects of rooibos tea on workers occupationally exposed to lead. Toxicology Letters, 172: S120–S121.

[286] Francisco, N.M. 2010. Modulation of postprandial oxidative stress by rooibos (*Aspalathus linearis*) in normolipidaemic individuals. MTech (Biomedical Technology) Thesis. Cape Peninsula University of Technology, South Africa.

[287] Guerreiro, S., Alcada, M., Azevedo, I. 2010. Bebidas acucaradas e glicemia. Acta Medica Portuguesa, 23, 567–578 (In Portuguese).

[288] Villaño, D., Pecorari, M., Testa, M.F., Raguzzini, A., Stalmach, A., Crozier, A., Tubili, C., Serafini, M. 2010. Unfermented and fermented rooibos teas (*Aspalathus linearis*) in-

crease plasma total antioxidant capacity in healthy humans. Food Chemistry, 123: 679–683.

[289] Marnewick, J.L., Rautenbach, F., Venter, I., Neethling, H., Blackhurst, D.M., Wolmarans, P., Macharia, M. 2011. Effects of rooibos (*Aspalathus linearis*) on oxidative stress and biochemical parameters in adults at risk for cardiovascular disease. Journal of Ethnopharmacology, 133: 46–52.

[290] Persson, I.A.L., Persson, K., Hagg, S., Andersson, R.G.G. 2010. Effects of green tea, black tea and rooibos tea on angiotensin-converting enzyme and nitric oxide in healthy volunteers. Public Health Nutrition, 13: 730–737.

[291] Sinisalo, M., Enkovaara, A-L., Kivisto, K.T. 2010. Possible hepatotoxic effect of rooibos tea: a case report. European Journal of Clinical Pharmacology, 66: 427–428.

[292] Engels, M., Wang, C., Matoso, A., Maidan, E., Wands, J. 2013. Tea not tincture: hepatotoxicity associated with rooibos herbal tea. ACG Case Reports Journal, 1: 58–60.

Section 3

# Antioxidants – Health and Disease

Chapter 8

# Oxidative Stress and Antioxidant Defenses Induced by Physical Exercise

Juana M. Morillas-Ruiz and Pilar Hernández-Sánchez

Additional information is available at the end of the chapter

http://dx.doi.org/10.5772/61547

**Abstract**

This chapter intends to present the physiological and biochemical mechanisms by which exercise induces the appearance of oxidative stress, as well as the characteristics of the physical exercise that involve the appearance of oxidative stress in the human organism.

On the other hand, in this chapter are also presented antioxidant defenses that are magnified with physical exercise and the mechanisms of action, as well as other types of antioxidant defenses that can be incorporated to the body to increase the total antioxidant capacity.

The understanding of possible mechanisms associated with physiological responses that explain how exercise increases oxygen toxicity, and the design of appropriate measures to minimize this toxicity, are critical to:

1) Increase the effectiveness of the exercise as a preventive and therapeutic tool in clinical practice.

2) Control the exercise-induced tissue damage.

Oxidative stress is a situation in which the cells are exposed to pro-oxidant agent and antioxidant defense mechanisms are exceeded, affecting the cellular redox state. This fact occurs during and/or at a high intensity exercise phase.

The knowledge of how the antioxidants interact provides rational bases to develop nutritional strategies to confront the progress in exercise activities and the health maintaining in amateur and professional subjects.

**Keywords:** oxidative stress, physical exercise, antioxidant defense, nutritional strategies

## 1. Introduction

The beneficial effects of regular non-exhaustive physical exercise have been known for a long time. Exercise is part of the treatment of common diseases such as diabetes mellitus or coronary heart disease. It improves plasma lipid profile, increases bone density, and helps to lose weight. However, the benecial effects of exercise are lost with exhaustive exercise and with a lack of training. Some of this damage is due to the production of free radicals. Exhaustive exercise causes muscle damage and inflammation, producing free radicals (Fig 1). Phagocytes, such as neuthrophyls, located in the inflamed area may all contribute to free radical production.

As a consecuence, oxidative stress occurs. Oxidative stress is a disturbance in the prooxidant-antioxidant balance in favor of the former, giving an overproduction of free radicals, causing insuffucient antioxidant defenses.

Exercise increases energy demands, increasing the oxygen consumption 10–20 times (200 times in active muscles). It is well known that one of the forms of radical production during exercise is due to a leak in the mitochondrial electron transport chain (ETC). Molecular oxygen typically reacts with species generated by single electron transfer. In such reactions, molecular oxygen is reduced by one electron to form a superoxide radical ($O_2^{\cdot-}$).

Free radicals are chemical compounds produced in cells in a wide range of processes by reactive species oxygen (ROS) (superoxide, hidroxile, alcoxile, peroxile, hydrogen peroxide) and nitrogen (nitric oxide, nitrogen dioxide, peroxinitrile). Free radicals are widely thought to be of cardinal importance in effecting both the damage and the adaptation that accompany significant physical activity or exercise. They are reactive prooxidant agents to carbohydrates, proteins, and lipids

Submaximal long-duration exercise training may augment the physiological antioxidant defenses in several tissues. Antioxidant defenses can be enzimatic (SOD, GPx, and catalase) and non-enzimatic (glutation, vitamin C and E, carotenoids, lipoic acid, transferrine, and polyphenols).

The effect of antioxidant suplementation (vitamin C, vitamin E, carotenoids, polyphenols, etc.) in the oxidative stress during the exercise has been demonstrated. Within antioxidants, flavonoids are the most important polyphenolic compounds found in rich abundance in all land plants. Flavonoids often exhibit strong antioxidant properties due to its ability to trap free radicals. Furthermore, it is currently unclear whether regular vigorous exercise increases the need for dietary intake of antioxidants. Clearly, additional research that analyzes the antioxidant requirements of individual athletes is needed. This controversy may be due to different reasons: age and sex, training methods, differences in biomaker analysis methods, etc.

Chronic exercise also leads to the upregulation of the body's antioxidant defence mechanism, which helps minimize the oxidative stress that may occur after an acute bout of exercise. Recent studies show a beneficial role of the reactive species, produced during a bout of exercise, that lead to important training adaptations: angiogenesis, mitochondria biogenesis, and muscle hypertrophy. The adaptations occur depending on the mechanic, and consequently biochemical, stimulus within the muscle. This is a new area of study that promises important findings in the sphere of molecular and cellular mechanisms involved in the relationship between oxidative stress and exercise.

Healthy exercise is being done on a regular basis (several days a week) at a moderate intensity so that the human body in its capacity for homeostatic adaptation (with this type of exercise) increases the physiological antioxidant defenses (enzyme systems such as glutathione peroxidase, catalase, superoxide dismutase), and will offset the appearance of oxidizing species upon exercise (radical and non-radical) with this increased enzyme activity; however, the exercise leads to an increase in the oxidative body state. When more oxidizing species are generated, the body can counteract the so-called oxidative stress, which appears to be unhealthy. If the body does not sufficiently increase the physiological antioxidant defense, it is necessary to provide these through dietary antioxidants such as those included in fruits and vegetables.

Brites et al. (1999) observed an increase in plasma levels of low molecular weight antioxidants (ascorbic acid, uric acid, and $\alpha$-tocopherol) in a group of trained players to sedentary controls. This increase can be attributed to a mobilization of these antioxidants from tissues into plasma, which would justify the improvement of the total plasma antioxidant status with training.

It would actually be very convenient for researchers, clinicians, coaches, etc. if optimal levels for relevant parameters would be decided and are available, but to date, and for a number of reasons, there are no clear data, such as definite reference intervals. The best practice seems to be thefrequent monitoring, comparison of the individualized values, and relative assessment of the training settings.

The optimal response to training would be the achievement of the relevant adaptations, such as the enhancement of the antioxidant capacity and the subsequent health improvement. Well-designed exercise training is regarded as a preferred way to attain these benefits.

## 2. Free radicals and exercise

Free radicals during exercise can be formed by different sources [1]:

1. It may be due to an electron leak in the mitochondrial ETC on the ubiquinone-citocrome b, giving superoxide radical anion ($O_2^{\bullet -}$). In this reaction, hydrogen ions and electrons are transported by the electron transporter to the oxygen giving water as final product.

$$4H^{+} + 4e^{-} + O_2 \rightarrow 2H_2O$$

The ETC releases energy and resynthesized ATP in associated reactions. The transport of one electron pair produces enough energy to resynthesize an average of three ATP moles. A total of 12 electron pairs are produced by 180 g of glycogen degradation, producing 36 ATP moles.

Therefore, during aerobic metabolism, the major part of the 36 ATP moles are resynthesized in the ETC.

Due to the increasing $O_2$ consumed during the exercise (10–20 times) and considering the flow of $O_2$ in the muscle is 10 times higher [2], the production of superoxide radical anions ($O_2^{\bullet -}$) is equally increased. The superoxide ions' consequent reactions produce other ROS, hydrogen peroxide ($H_2O_2$), and hidroxile radical ($OH^-$) [1].

The muscle contraction is connected with the generation of ROS [3] and the consequent muscle fatigue [4].

The prolonged reduction of the strength muscle is probable due to the muscle damage produced by the superoxide radical after the exercise. After exercising, neuthrophils produce ROS, causing the inflammatory response [5]. Neutrophils are the predominant phagocytes of circulating blood, and they are the first cells to arrive at sites of infection. ROS produced during the exercise favor the neutrophils' muscle infiltration, promoting the increase of vascular permeability.

The interaction with the vascular endothelium is produced through membrane receptors: adhesion molecule interleukocyte and leukocyte endothelial adhesion molecule. It has been demonstrated that body temperature increase the leukocytes adhesion to the endothelial cells causing cellular damage. Moderate exercise increases cellular respiration and high intensity exercise tends to suppress cellular respiration [6].

It has been recently demonstrated that the neuthrophils activation factor is induced by the gram-negative lipopolysaccharide. The activation of the leukocytes have toxic effects such as proteinases release, ROS, and ecosanoids.

The stimulation of xantine-oxidase located in endothelial cells during the ischemia reperfusion also produces superoxide radicals. In this reaction, oxygen penetrates the cells producing urate and superoxide radicals with high toxicity [7].

Superoxide radicals stimulate the neuthrophils activation, increasing leukocyte activity in different organs, causing damaged tissues.

Another endogenic form to obtain superoxide radicals is the peroxidation of araquidonic acid, which activates lipoxigenase and ciclooxigenase rutes [8]. It is important to consider that ROS also have beneficial biological effects [9].

The production of septic shock produces nitric oxide derived from L-arginine. This compound is generate in nervous cells and hepatocytes stimulated by citoquines and leukocytes giving a vasodilating effect.

Nitric oxide has several functions: inmunosupression, neurotoxicity, and alteration of the sensorial transmission. Human studies reveal high levels of this compound during sepsis [10].

**2.** Ischemia-reperfusion. Periods of intensive exercise can cause temporary ischemia or hypoxia in certain regions of the body (kidney, splanic region). Hipoxia is higher as the intensity of the activity increases. After the intensive exercise, the damaged regions are reoxigenated, and then ischemia-reperfusion producing free radicals occurs [11].

http://dx.doi.org/10.5772/61547

Ischemia-reperfusion can occur with intensive exercise, such as rowing, using more oxygen when the resistance of the shoulder, arms, back, and legs is tested [12].

3. Catecholamines autoxidation. The level of catecholamines increases when exercise intensity increases.

4. Xanthine oxidase. Free radical production during exhaustive exercise may also be caused by the enzyme xanthine oxidase [7]. Periods of intensive exercise can cause temporary ischemia or hypoxia, causing ATP to be converted to ADP, AMP, inosine, and finally hypoxanthine. Under such ischemic conditions, intracellular xanthine dehydrogenase (XD) can be converted to xanthine oxidase (XO) by cysteine residue.

Under normal physiological conditions, XD is the dominant form of the enzyme and oxidizes both hypo-xanthine and xanthine (to uric acid) in a process that, concomitantly, reduces NAD $^+$ to NADH. Xanthine oxidase, on the other hand, can no longer utilize $NAD^+$ as the electron acceptor, and instead, preferentially reduces oxygen directly to superoxide and hydrogen peroxide. During ischemia, oxygen concentrations are low and intracellular concentrations of XO and hypoxanthine can rise. When oxygen is finally reintroduced (reperfusion) a burst of $O_2^-$ and $H_2O_2$ can occur. It should be noted that XO generates both $O_2^-$ and $H_2O_2$ directly.

5. Lipid peroxidation of arachidonic acid produces superoxide radicals [8].

6. The alteration of calcium homeostasis produced when the muscle is stressed during exhaustive exercise, particulary anaerobic/excentric exercise leads to ischemia (hipoxantine formation), creating an excess of contracted activity and muscle damage (with proteases activation) [13].

7. As a result of the respiratory reaction due to the activation of leukocytes after muscle damaged induced during exercise.

## 3. Oxidative stress induced by exercise

### 3.1. Oxidative stress induced by extenuant exercise

The increase in energy consumed during exercise increases the oxygen demands of the active tissues, increasing up to 20 times in comparison with basal state [14]. The oxygen flow in the peripheral skeletal muscle tissue can increase up to 200 times, increasing 30 times the blood flow, and the oxygen difference in the arteriovenous blow increases 3 times. As a result, the oxidative metabolism is increased, maximizing the energy produced by unit of substrate and avoiding lactate accumulation [15].

Dillard et al. (1978) [16] first described that extenuant exercise induced lipid damage in tissues. After that, many other investigations focused on the effects of exercise and training in oxygen toxicity and the body defense response. It is accepted that oxygen toxicity can be implicated in some pathologic situations.

The understanding of the mechanisms associated with physiological responses that explain how exercise increases the oxygen toxicity and the design of appropiate measures to minimize toxicity are indispensable to:

1. Increase exercise efficacy as a preventive and therapeutic instrument in clinical practise

2. Control the damaged tissue induced by exercise

Oxidative stress induced by extenuant exercise is a situation by which cells are exposed to a prooxidant environment and defense mechanisms are not enough, affecting the redox estate of the cells. Due to this, nutritional supplements of antioxidants such as vitamin C, vitamin E, carotenids, and polyphenols in the diet are important [13].

In humans, antioxidant defenses in the skeletal muscle and heart are limited. Basal metabolism in the heart is 100% higher than in the liver. This involves a higher risk of oxidative damage in the heart [17]. In basal state, the oxygen consumed of a kilogram of heart and liver is 94 and 44 mL/min, respectively [18]. In adults, superoxide dismutase (SOD) and catalase (CAT) activities are 40 and 16 times smaller in muscles compared with the liver activity of these enzimes [19].

Davies et al. (1982) [20] showed using spin electronic resonance (SER) that exhaustive exercise increases free radical concentration in the liver and muscles and induce oxidative damage in these tissues. Futhermore, they verify the decrease of antioxidant levels and free radical damage could be implicated in the mitochondrial biosynthesis.

Sakellariou et al. (2014) [3] observed the effect of intense muscular contraction activity during 30 min of exercisse and showed 70% increment of free radicals in the active muscles compared with the muscles in basal state.

The theory explains that muscle damage, particularry after eccentric muscle exercise, is responsible for the inflammatory stress after the exercise.

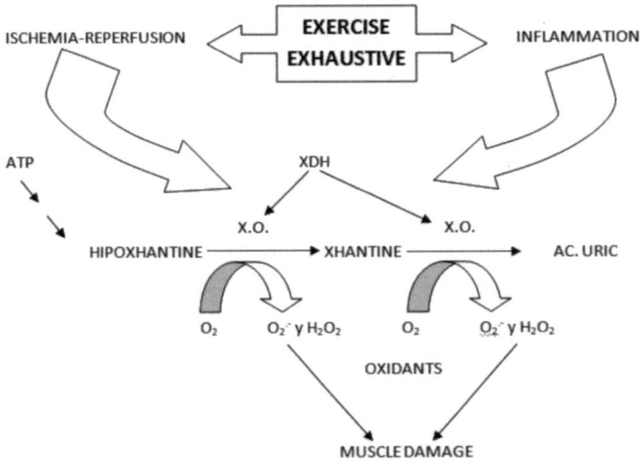


**Figure 1.** Scheme of the relationship between exhaustive exercise and muscle damage.

http://dx.doi.org/10.5772/61547

After exercise, neuthrophils, monocytes and macrofagues go to the damaged area and provoke the elimination of degraded proteins and cellular remainders. These cells are able to produce ROS and proinflammatory cytokines such as IL-1 TNF-$\alpha$ or IL-8, producing oxidative stress and eventually inflammation. Concentric exercise is associated with an increase in inflammation markers (IL-6) but not in muscle damage parameters (CK). However, excentric muscular exercise shows a typical increase in CK after 72 h. In this case, there is no increase of IL-6 [13].

Barclay et al. (1991) [21] suggest that oxidative stress is implicated in muscle fatigue. There is no evidence of the effect of superoxide radicals in the presence of the free radical hidroxil trapper, blocking the xanthine oxidase activity. Powers et al. (2008) [22] observed a relation between oxidative stress and the mscle strengh.

The factor that triggers the muscle contraction is an electric signal based on the $Na^+$ entry and followed by a $K^+$ release of the cell. Animal and human studies have demonstrated a rise in the $K^+$ plasma concentrations after muscle concentration. As a result of this increase of $K^+$ that escape to the extracellular medium and water introduction to the cell, intensive muscle contraction decreases by 6–20% of the intracellular $K^+$ contraction. This can promote the fatigue [23].

Glutathione (GSH) oxidation in different tissues is a valid parameter to appreciate oxidative stress. In this situation, intracellular GSH rapidily oxidizes to GSSG. Intracellular GSSG can be reduced to GSH in the presence of a reductase glutathione and NADPH as cofactor. When the oxidative stress is high, the relation between GSSG/GSH can be higher than the reduction ability of the cells. In this situation, the heart and skeletal muscle cells pour GSSG out of the cells [24].

In extenuant exercise, an increase of GSSG and a decrease of total glutation (GSSG + GSH) in the skeletal muscle tissues such as the liver and heart has been observed [25, 26].

This increasing production of GSSG exceeds the reductase glutathione's ability to reduce disulfide group, thus explaining that the GSSG spill from the tissue to the plasma [27]. The increasing oxidized glutathione plasma concentration as a result of the exercise has been demonstrated in many studies [28, 26, 29].

Gohil et al., (1988) [30] showed that submaximal exercises at 65% $VO_{2max}$ increased the oxidation of blood glutathione during the first 15 min of the exercise. In another study, the level of GSSG in blood increased significantly after 14 min during a maximal test in the cycle ergometer or after pedalling for 30 min in an aerobic threshold or after pedalling 30 min in an anaerobic threshold [26]. In contrast, they did not find significant changes in GSSG in the blood after 60 min and 120 min of the exercise [25]. Sen et al. (1995) [26] demonstrated that 24 hours of recuperation is enough to establish GSSG values in the blood before the exercise.

The glutathione synthesis ability in the liver is high and exercise induces a decrease of glutathione, promoting a protective response of the liver [26].

Studies in hepatectomized (HX) rats reveal that the GSH level in the heart muscle depends on its supply in the liver; however, this fact does not apply to skeletal muscles [31]. These cells are very active in glutathione production. It has been estimated that muscle cells are able to produce 3 mM concentrations of glutathione [27].

The use of gluthatione oxidation as a parameter to detect free radical damage in exercise has demontrated that the damage only appears in exercise exhaustion, meaning that the effect of free radicals only occurs when the subject do exercise above the anaerobic threshold [32].

ROS synthesis induced by neutrophils in exercise has been demostrated by many authors [33, 34].

In mammals, oxidative DNA damage is related to the metabolism rate [35]. After racing for 10 hours, the relationship between oxidase nucleosides/creatine is 1.3 higher with respect to basal state [36]. However, Viguie et al., (1993) [37] did not observe significant changes in 8-hidroxiguanosine after 90 min of racing.

Oxidants as hydroxyle radicals and peroxide radicals can react with proteins. Oxidase proteins rapidily break down into amino acids. Some of these, such as methionine, tryptophan, histidine, and sulfhydryl residue are very sensitive to oxidative damage. The protein oxidation include receptor modification, alteration in translated signals, and other processes (Aoi et al., 2014) [38].

Reznick et al. (1992) [39] observed that exercise increases the protein oxidation of skeletal muscles in rats. Rajguru et al. (1994) [40] showed that after exercise, there is a decrease of sulfhydryl groups in the skeletal muscle. This fact is important in protein crosslinking.

### 3.2. Exercise as an oxidative stress protector

Up to now, the work has been focused in the damaging effect of exhaustive exercise. However, moderate exercise results in a healthy and beneficial practise that prevents diseases, due to its ability to prevent oxidative stress [41].

Oxidative stress induced by exercise depends on the type, intensity, and the length of the exercise. However, interindividual variability is attributed to the level of training, sex, nutrition, and genetic factors [13].

Undesirable effects of exhaustive exercise can be avoided with progress in training. Salminen et al. (1983) [42] showed that training reduces free radical susceptibility to free radicals. On the other hand, Gómez-Cabrera et al. (2008) [43] observed that training increases antioxidant enzymes. These authors previously showed that training protects against glutathione oxidation associated with exhaustive exercise. Regular exercise creates an adaptation against oxidative stress due to a decrease in DNA damage and maintained levels of protein oxidation [44]. There are many studies that confirm that antioxidant supplements can interfere with the free radical metabolism damaged training adaptations. This fact suggests the recommendation of a diet rich in antioxidant compounds (fresh fruits and vegetables)

Antioxidant defenses in the skeletal muscles, heart, and liver are regulated due to the effect of exercise in the body [45, 46] and showed that exhaustive exercise increased the rate of catalase activity in the liver, muscle, and heart. Since then, a great number of works have confirmed the effect of different resistance training in antioxidant defenses [47–50].

Moderate daily exercise and long duration exercise (resistance training) produce an increase in mitochondrial content in the muscle. However, high intensity exercises have demonstrated muscle damage derived from the sensibility increment of oxidant agents, the liberation of proteolytic enzymes in the muscle and liver, and loses in the integrity of membranes.

Ginsburg et al. (1996) [52] described a decrease of 47% ($p<0.001$) in lipid peroxidation in the plasma compared with the result obtained before the test. The same work demonstrated that the lipidic peroxidation values were smaller in basal state in trained subjects than in sedentary subjects. These results indicate that accmmulative effects of training tend to decrease lipidic peroxidation in the plasma.

Criswell et al., (1993) [53] studied the effect of training for 12 weeks and observed favorable changes in the skeletal muscles in rats. The authors demosntrated that 5 min of high intensity exercise, was better for antioxidant defense regulation than continuous exercise with moderate intensity.

Daily exercise is important to mantain and promote the ability to defend the organism against the toxicity of reactive oxygen. In prokaryotes, some of the dependent mechanisms of ROS in the induction of defense antioxidant proteins are known [26]. In mammals, cells have been identifying transcription factors responsible for the activation of protein-1 and NF-kB sensitive to redox balance [27]. The redox-tiol state in the different compartments of these cells seems to be implicated in the regulation of these transcription factors. For example, a high cystosolic concentration of GSSG promotes the deactivation of NF-kB, but low cystosolic concentration of GSSG inhibits the fixation of the activate dimmer to the diana oligonucleotids.

Exercise that promotes changes in the redox-tiol state of the tissues can influence the intracellular signal of the translate process, causing the expression of defense antioxidant proteins [43].

Large amount of works support that chronic exercise increases the antioxiant defenses [47–50]. Erythrocyte catalase activity and glutathione reductase show a significant increase after 10 weeks of training [54].

In other studies, the antioxidant state of highly trained runners (128–230 km/week), moderate and low trained runner (26–70 km/week), and sedentary subjects were studied. The results demosntrated a direct relation between the weekly distance and the erythrocyte activity of the antioxidant enzymes. It was found that trained marathon runners have higher levels of MDA and conjugated dienes (CD) in basal state than sedentary subjects. At the end of the half marathon, trained subjects showed a significant increase in the MDA and CD values, however test values decreased in the recuperation period (24–48 hours) to lower values, even lower than when they were determined in basal state.

These results suggest that aerobic training improves the enzymatic antioxidant activity in erythrocytes in basal state and in the recovery period after exercise. This improvement, along with the increase of muscle blood flow and the activity of mitochondrial deshydrogenase-aldehydo activity in the muscle, could be reponsible for the significant decrease of lipidic peroxidation index after exercise in trained subjects [55].

Lipidic peroxidation in blow decrease in response to the increment of training time in 60-year-old women, indicating an adaptation effect [56].

Another study in rats demonstrated after control their training for 5 days that muscle damage induced because of a race could be eliminated. The experiments conclude significant reductions in the pain sensation and proteolysis after training. The authors suggest that training can induce a protective effect against muscle damage when the intensity and the duration of the exercise was moderate [57].

Child et al. [2, 58] studied trained runners subjected to exhaustive exercise. The study suggested a considerable increase of ROS and observed that variations in oxygen consumed can underestimate the real increase in free radical formation during intensive exercise as a consecuence of the reduction of mitochondrial control repiratory and the increase of the formation of free radicals derived from non-mitochondrial sources [59].

Brites et al. (1999) [60] observed an increase in plasmatic levels of low molecular weight antioxidants (ascorbic acid, uric acid, and $\alpha$-tocopherol) in a footballer trained group with respect to sedentary subjects. This increase can be atributted to a mobilization of the antioxidants from the tissues to the plasma, explaining the improvement of the total plasma antioxidant state with the training [61].

Various authors suggest that physical training promotes parallel adaptation of the mitochondrial antioxidant enzymes and the antioxidant capacity of mitochondrial enzymes. However, Laughlin et al., (1990) [62] studied the relation between the oxidative ability and the antioxidant muscle enzymes and a relation between antioxidant ability and the activity of SOD and catalase was found. Altough training promotes an increase in the muscle's antioxidant ability, there was no effect in the SOD activity, promoting a significant decrease in catalase activity. This coincides with the result found by Ji et al. (1992) [25].

## 4. Antioxidant defenses

The demonstrated contribution of ROS to muscle damage and muscle fatigue as a consequence of intensive or prolongued exercise induces the defense mechanisms in skeletal muscle cells to reduce the risk of oxidative damage [63, 21]. There are two protective mechanisms: enzymatic and nonenzymatic. They act as a unique antioxidant system to reduce the ROS damage in the cells. Antioxidants (enzymatic and nonenzymatic) exist in extracellular and intracellular space [64]. Antioxidants can be both synthesized in vivo and absorbed through diet.

### 4.1. Enzymatic antioxidants

The main antioxidant cellular enzymes are superoxide-dismutase (SOD), catalase (CAT), and glutation-peroxidase (GPx). Each of these enzymes is responsible for the reduction of a different ROS, and they are located in different cellular compartments. (1) SOD has three isoforms, two of them are present within cells, whereas the other one is located in the extracellular space. Specifically in skeletal muscle cells, the highest percentage of SOD (65–85%) is found in the cytosol, and the remaining (15–35%) is present in the mitochondria of the muscles. SOD catalyses the reaction of superoxide radicals into oxygen and hydrogen peroxides ($H_2O_2$).

(2) GPX is located in both the cytosol and the mitochondria of cells. It is responsible for the removal of a wide range of hydroperoxides—from complex organic hydroperoxides to $H_2O_2$ —thus, it may protect membrane lipids, proteins, and nucleic acids from oxidation. GPX is also present in muscle cells, but its activity varies depending on the muscle fiber type, with the greatest activity present in slow twitch muscle fibers (type I) that have higher oxidative capacity. (3) CAT is extensively distributed within the cells and its main function is to degrade $H_2O_2$ into $H_2O$ and $O_2$. Nevertheless, it has a lower affinity for $H_2O_2$ compared with GPX. Similar to the latter, CAT can be found in higher concentrations in type I muscle fibers [22].

The SOD activity shows a significant increase with training, and there is evidence that SOD-Mn is mainly responsible for this increase. The increase in SOD-Mn with training is relatively small compared with the increase in the activity of other mitochondrial enzymes. Furthermore, this rise is not related to a significative improvement in antioxidant protection [65].

Exercise increases SOD activity only in type I muscle fibers, and the SOD activity increase is higher in length than in intensity. An intensive exercise test causes an increase in SOD activity in tissues such as the heart, liver, lungs, and skeletal muscles [25].

GPx activity increases with training only in type II fibers, and this adaptation to training depends on the duration more than the intesity of the exercise. After intensive exercise, GR activity appears to increase in skeletal muscles. GR activity also increase in humans after prolonged exercise [66]. A study on sedentary subjects, marathon athletes, and sprinter trained subjects resulted in a significant increase in GPx compared with sedentary subjects [67].

SOD activity was higher (52%, $p<0.01$) in the trained footballer with respect to sedentary subjects [68].

The effect of training in the catalase activity is controversial. Several studies showed an increase, decrease, and absence of the variation in the catalase activity with chronic exercise. Calderera et al. (1973) [69] observed an important increased CAT activity in the heart, liver, and skeletal muscles after an intensive test exercise.

The activation of the antioxidant enzymatic defenses after intensive exercise can reflect an increase in ROS production. However, due to the differences in oxygen consumption and intrinsic differences in the enzymatic activities, skeletal muscles are subjected to a higher oxidative stress than the liver and heart during exercise [25].

Although evidence has revealed that training controls and regulates antioxidant enzymes in active tissues used in exercise, there is still controversy. In general, antioxidant enzymes of skeletal muscles show the best adaptation response to the training.

In humans, there exists a correlation between the high activity of antioxidant enzymes and the maximum oxygen consumed. Training athletes have a higher SOD and CAT activity in skeletal muscles. Professional and amateur cyclists have higher SOD activity in erythrocytes than sedentary subjects [25]. Due to this, resistance training reduces oxidative damage due to the increase of mitochondrial antioxidant enzymes and a reduction of the oxygen flow in the respiratory chain.

### 4.2. Nonenzymatic antioxidants

The nonenzymatic antioxidant group includes glutathione, vitamin C, vitamin E, carotenoids, uric acid, polyphenols, and others. Similar to enzymatic antioxidants, these are present in different cellular compartments and elicit distinct antioxidant properties that maximize their effectiveness [70].

GSH exerts various essential functions in the body. Amongst these functions is its major antioxidant role. It efficiently scavenges ROS and free radicals, preventing an increase in the oxidative stress process. In these reactions, the reduced GSH is oxidized, via the enzyme glutathione peroxidase, to form glutathione disulfide (GSSG). Note that GSSG is formed by two GSH molecules linked via a disulfide bond due to the oxidation of the thiol (SH) groups. Once oxidized, GSSG can be reduced back to its original GSH form by the enzyme GSSG reductase and nicotinamide adenine dinucleotide phosphate (NADPH). Nevertheless, when there is a high level of oxidative stress, NADPH becomes depleted and there is an intracellular accumulation of GSSG. This excess GSSG can either be exported out of the cell or it can form a mixed disulfide. Measuring the plasma level of GSH or its oxidized form (GSSG) is a widely accepted method of detecting oxidative stress and can be reported as redox potential, GSH or GSSG concentration, or GSH/GSSG ratio. It is not only a good indicator of systemic oxidative status but also a useful indicator to indicate the free radical production during exercise [71, 72–74].

GSH is the major source of tiol groups in the cells. GSH has several defense antioxidant functions. The practise of 90 min of exercise decreases GSH and increases blood levels of GSSG [30].

Sen et al., (1994) [26] made a maximal test for 14 min. After the test, the subject showed no variation in the blood level of (GSH+GSSG) or GSH, but there was an increase of GSSG and also in the relation GSSG/(GSH+GSSG). When the subjects were subjected to a 30 min test (77% $VO_2$ max) of anaerobic exercise, the level of GSH was mantained. However, there was an increase of GSSG, the total glutathione, and also the relation GSSG/(GSH+GSSG). After 24 h, all the results recovered to the levels found before the test.

The human levels of GHS are almost undetectable (<0.01 μM), the oxidation of GSH could be in the erythrocytes, in a specific system dependent of the energy to export the excess intracellular GSSG. Due to fact that the test does not reflect the increase of GSH in the blood, the increasing total glutathione could be due to the GSSG exportation of the tissues and the blood GSSG [26].

Plasma levels of GSH is approximately three times lower than blood levels. Moreover, the changes in GSH plasma due to the accelerated flow of liver GSH, produced during exercise, are not detected in the blood in GSH form or total glutathione. The oxidative stress due to intense physical activity produce a rapid oxidation in intracellular GSH in muscle cells and a GSSG production liberated in blood circulation. Thus, a decrease in intracellular glutathione level is observed. This suggests that the GSSG flow of muscle cells to the blood is due to a mechanism dependent of energy [26].

http://dx.doi.org/10.5772/61547

The effect of the training in GSH content seem to vary in different types of muscle fibers and different tissues.The content of GSG in erythrocytes increased at the same time with glutathione reductase after 20 weeks of training in humans who were previously sedentary [25]. In trained subjects, after 2.5 race, a diminution of plasma GSH and the GSH/GSSG. However, the level of GSSG showed a rise at the end of the test compared with the basal state [75].

Short-term training does not improve the adaptation of antioxidant system. A study made on both sex subjects subjected to 8 weeks of aerobic training (3 times/week) showed no variations in the SOD, CAT, and GPx enzymes activity. Even though were changes in the vitamin E concentration in the muscles and in glutathione levels (GSH, GSSG, total glutation, and GSH/GSSG) [76].

However, it has been demonstrated that training protects against glutathione oxidation associated to exhaustive exercise [32].

Similar to vitamin C, vitamin E has important antioxidant properties. Due to its capacity for scavenging ROS and free radicals, particularly peroxyl radical (R O O ·), it exerts the important function of protecting cellular membranes and plasma lipoproteins against lipid peroxidation. This is possible because vitamin E has a great affinity for reducing peroxyl radicals, preventing their interaction with the membrane phospholipids or lipoproteins [77].

Vitamin E has been found to protect cellular membranes from lipid peroxidation. Hence, it is logical to assume that this vitamin could protect muscle cells against exercise-induced damage. Early studies analyzing the effects of vitamin E supplementation and exercise investigated its effect on performance. Most of the studies, however, report no benefit of vitamin E neither for muscle strength nor for endurance performance [78]. Furthermore, it has been hypothesized that vitamin E supplementation could have a protective effect against the contraction-induced muscle damage oxidative stress that may occur after an intense exercise bout. This rationale is based on the knowledge that this vitamin can stabilize muscle membranes by interacting with its phospholipids that would, this way, provide some protection against the increase in oxidative stress or muscle damage observed after certain types of exercises [78]. Altough vitamin E is an effective capture of free radical, the reaction of vitamin E with radicals produces a functional decrease of vitamin E and the formation of free radicals-vitamin E.

The oxidative stress produces a significative decrease of vitamin E levels in tissues. However, the radical vitamin E can be synthetized with the cooperation of other antioxidants. As a result, the investigations conclude that vitamin E's ability to act as an antioxidant is related with the ability of other antioxidants to recycle vitamin E during stress oxidative periods [79].

The exercise could induce an alteration of plasma levels of vitamin E. During human exercise, an increase in vitamin E concentration in plasma and erytrhocytes was observed, suggesting that exercise could promote vitamin E mobilization from tissues to plasma, and the skeletal muscle could use the circulating vitamin E to protect against oxidative damage [25].

Other authors did not find variations in vitamin E levels in humans after a half marathon race [79]. In trained footballer, the levels of vitamin E were higher (10%) than in sedentary subjects [80].

Vitamin E changes are better appreciated when the results are expressed by unit of mitochondrial ubiquinone. The reduction of vitamin E in the inner mithochondrial membrane can justify the susceptibility of the mitochondria to free radicals damage. The content of vitamin E in the heart can decrease. Vitamin E heart content suffered a light decrease after a training program in treadmill, compared with the disminution in skeletal muscles. Different responses to the training of vitamin E and skeletal muscle can be explained in part due to the high vitamin E content in the heart (≈70nmol/g) [25]. In an ultraresistance race (thriatlon), no variation of vitamin E concentration were found before or after the race [81].

Ascorbic acid is the main form of the vitamin found in vivo. This vitamin, also referred to as ascorbate, is found in relatively high levels in different tissues throughout the body. Ascorbate has clearly been shown to play an essential role in connective tissue biosynthesis. During oxidation reactions, only small amounts of ascorbate are lost because, once it is oxidized, it can be reduced back to ascorbic acid by reductants such as glutathione, nicotinamide adenine dinucleotide (NADH), and NADPH. Similarly, vitamin C is also known to regenerate other antioxidants, such as vitamin E and glutathione, back to their reducing state; thus, maintaining a balanced network of antioxidants. The increase of vitamin C levels can protect against oxidative damage of free radicals. However, high concetration of vitamin C (1 mM) acts as a prooxidant in the presence of metals such as $Fe^{2+}$ or $Cu^{2+}$. Duthie et al. (1990) [79] studied the response of vitamin C after a half marathon race and they observed an increase in the levels of the vitamin. However, Ginsburg et al. (1996) [81] realized a study with triathletes (ultraresistance exercise) and did not observe a variation in the plasma concentration of vitamin C before and after the race.With respect to the trained effect, the vitamin C level of the trained footballer level was higher than in sedentary subjects [81].

As vitamin C, β-carotene can acts as an antioxidant and as a pro-oxidant. A physiological oxygen partial pressure (<100 mm Hg) β-carotene shows a capture activity of radicals. However, the partial pressure >150 mm Hg promotes prooxidant activity of β-carotene [82].

Polyphenolic antioxidants have demonstrated their efficacy against oxidative stress induced by exercise. It has demonstrated the decrease of oxidized proteins in a study subjected to intensive exercise [83, 84].

## 5. Conclusions

During exercise, an important free radical production is predictable and as a consequence a major requirement of defense mechanisms. Some of the antioxidant defenses can be adequated with training and in the presence of an appropriate diet. Defenses can be insufficient when the exercise exceeds the level by which they were adapted.

The knowledge of how antioxidants interact provides rational bases to develop nutritional strategies to put forward the progress in exercise activities and in mantaining the health of amateur and profesional subjects.

http://dx.doi.org/10.5772/61547

## Author details

Juana M. Morillas-Ruiz* and Pilar Hernández-Sánchez

*Address all correspondence to: jmmorillas@ucam.edu

Department of Food Technology and Nutrition, Catholic University of San Antonio, Guadalupe, Murcia, Spain

## References

[1] Niess AM, Simon P. Response and adaptation of skeletal muscle to exercise—the role of reactive oxygen species. Front Biosci. 2007 Sep 1;12:4826-38.

[2] Child RB, Wilkinson DM, Fallowfield JL, Donnelly AE. Elevated serum antioxidant capacity and plasma malondialdehyde concentration in response to a simulated half-marathon run. Med Sci Sports Exerc. 1998;30(11):1603-1607.

[3] Sakellariou GK, Jackson MJ, Vasilaki A. Redefining the major contributors to superoxide production in contracting skeletal muscle. The role of NAD(P)H oxidases.Free Radic Res. 2014 Jan;48(1):12-29.

[4] Kuwahara H, Horie T, Ishikawa S, Tsuda C, Kawakami S, Noda Y, Kaneko T, Tahara S, Tachibana T, Okabe M, Melki J, Takano R, Toda T, Morikawa D, Nojiri H, Kurosawa H, Shirasawa T, Shimizu T. Oxidative stress in skeletal muscle causes severe disturbance of exercise activity without muscle atrophy. Free Radic Biol Med. 2010 May 1;48(9):1252-62.

[5] Yamada M, Suzuki K, Kudo S, Totsuka M, Simoyama T, Nakaji S, Sugawara K. Effect of exhaustive exercise on human neutrophils in athletes. Luminescence. 2000 Jan-Feb; 15(1):15-20.

[6] Neiman D. Immune response to heavy exertion. J Appl Physiol. 1997; 82:1385-1394.

[7] Viña J, Gimeno A, Sastre J, Desco C, Asensi M, Pallardó FV, Cuesta A, Ferrero JA, Terada LS, Repine JE. Mechanism of free radical production in exhaustive exercise in humans and rats; role of xanthine oxidase and protection by allopurinol. IUBMB Life. 2000 Jun;49(6):539-44.

[8] Jiang J, Borisenko GG, Osipov A, Martin I, Chen R, Shvedova AA, Sorokin A, Tyurina YY, Potapovich A, Tyurin VA, Graham SH, Kagan VE. Arachidonic acid-induced carbon-centered radicals and phospholipid peroxidation in cyclo-oxygenase-2-transfected PC12 cells.J Neurochem. 2004 Sep;90(5):1036-49.

[9] Smith ChV. Free radical mechanisms of tissue injury. In: Molen HT, Smith ChV, editors. Free radical mechanisms of tissue injury. Boca Raton; 1992. p. 1-22.

[10] Evans T, Carpenter A, Silue A, Cohen J. Inhibitions of nitric oxide syntethase inespersimetal gram negative sepsis. J Infect Dis. 1994;169:343-349.

[11] Sjódin B, Hellsten Westing Y, Apple FS. Biochemical mechanism for oxygen free radical formation during exercise. J Sports Med. 1990;10:236-254.

[12] Kanter MM, Nolte NA, Holloszy JO. Effect of an antioxidant vitamin micture on lipid peroxidation et rest and post-exercise. J Appl Physiol. 1993;74:965-969.

[13] König D, Berg A. Exercise and oxidative stress: Is there a need for additional antioxidants. Öster J Sport Med. 2002;3(1):6-15.

[14] Wilmore JH, Costill DL. (2005) Physiology of Sport and Exercise: 3rd Edition. Champaign, IL: Human Kinetics.

[15] Keul J, Doll E. Oxydative energy supplementation. In: E.Jokl, editor. Energy Metabolism of Human Muscle. Basel:Karger: 1972;214-237.

[16] Dillard CJ, Litov RE, Savin WM, Dumelin EE, Ttappel AL. Effects of exercise, vitamin E and ozone on pulmonary function and lipid peroxidation. J Appl Physiol. 1978; 45:927-932.

[17] Kumar CT, Reddy VK, Prasad M, Thyagaraju K, Reddanna P. Dietary suppplementation of vitamin E protects heart tissue from exercice-induced oxidant stress. Mol Cell Biochem. 1992;111:109-115.

[18] Diem K, Lentner C. Scientific Tables. Documenta Geigy. 7ª ed. Switzerland: Basel; 1970.

[19] Jenkins RR, Friedland R, Howald H. The relation-ship of oxygen uptake to superoxide dismutase and catalase activity in human skeletal muscle. Int J Sports Med. 1984;5:11-14.

[20] Davies KJA, Quintanilha AT, Brooks GA, Packer L. Free radicals and tissue damage produced by exercice. Biochem Biophys Res Commun. 1982;107(4):1198-1205.

[21] Barclay JK, Hansel M. Free radicals may contribute to oxidative skeletal muscle fatigue. Can J Physiol Pharmacol. 1991;69:279-284.

[22] Powers SK, Jackson MJ.Exercise-induced oxidative stress: Cellular mechanisms and impact on muscle force production. Physiol Rev. 2008 Oct;88(4):1243-76.

[23] Sjogaard G. Exercice-induced muscle fatigue: The significance of potassium. Acta Physiol Scand. 1990;140(593S):1-63.

[24] Strobel NA, Matsumoto A, Peake JM, Marsh SA, Peternelj TT, Briskey D, Fassett RG, Coombes JS, Wadley GD.Altering the redox state of skeletal muscle by glutathione depletion increases the exercise-activation of PGC-1$\alpha$.Physiol Rep. 2014 Dec 23;2(12). pii: e12224.

[25] Ji LL. Exercice-induced oxidative stress in the heart. In: Sen CK, Packer L, Hanninem O, editors. Exercice and Oxygen Toxicity. Amsterdam: Elsevier Science; 1994. p. 249-267.

[26] Sen ChK, Hänninen O. Physiological antioxidants. In: C.K.Sen LPaOH, editor. Exercice and Oxygen Toxicity. Amsterdam: Elsevier Science: 1994c. p. 89-126.

[27] Sen ChK, Marin E, Kretzschmar M, Hanninen O. Skeletal muscle and liver glutathione homeostasis in response to training, exercice and inmovilization. J Appl Physiol. 1992;73:1265-1272.

[28] Viguie CA, Frei B, Shigenaga MK, Ames BN, Packer L, Brooks GA. Antioxidant status and indexes of oxidative stress during consecutive days of exercice. J Appl Physiol. 1993;75:566-572.

[29] Hillman AR, Vince RV, Taylor L, McNaughton L, Mitchell N, Siegler J. Exercise-induced dehydration with and without environmental heat stress results in increased oxidative stress. Appl Physiol Nutr Metab. 2011 Oct;36(5):698-706.

[30] Gohil K, Viguie CA, Stanley WC, Brooks GA, Packer L. Blood glutathione oxidation during human exercice. J Appl Physiol. 1998;64:115-119.

[31] Kretzschmar M, Pfeifer U, Machnik G, Klinger W. Glutathione homeostasis and turnover in the totally hepatectomized rat: Evidence for a high glutathione export capacity of extrahepatic tissues. Exp Toxicol Pathol 1992;44:273-281.

[32] Gómez-Cabrera MC, Gimeno A, Lloret A, Miñana JB, Marquez R, Viña J. Deporte de alta competición y daño oxidativo: Papel de los nutrientes antioxidantes. Antioxidantes y calidad de vida [en línea] [fecha de acceso 21 de mayo de 2001]. URL disponible en: www.antioxidantes.com.ar /12/Art118.htm.

[33] Smith JA, Teldorfd IB, Mason IB, Weidemann MJ. Exercice. Int J Sports Med. 1990;11:179-187.

[34] Nunes-Silva A, Bernardes PT, Rezende BM, Lopes F, Gomes EC, Marques PE, Lima PM, Coimbra CC, Menezes GB, Teixeira MM, Pinho V. Treadmill exercise induces neutrophil recruitment into muscle tissue in a reactive oxygen species-dependent manner. An intravital microscopy study. PLoS One. 2014 May 5;9(5):e96464.

[35] Ames BN, Shigenaga MK, Hagen TM. Oxidants, antioxidants and the degenerative diseases of aging. Proc Natl Acad Sci U S A. 1993;90:7915-7922.

[36] Alessio HM, Cutler RG. Evidence that DNA damage and repair cycle activity increases following a marathon race. Medicine and Science in Sports and Exercise. 1990;22(S126):751-756.

[37] Viguie CA, Frei B, Shigenaga MK, Ames BN, Packer L, Brooks GA. Antioxidant status and indexes of oxidative stress during consecutive days of exercice. J Appl Physiol. 1993;75:566-572.

[38] Aoi W, Naito Y, Yoshikawa T. Potential role of oxidative protein modification in energy metabolism in exercise. Subcell Biochem. 2014;77:175-87.

[39] Reznick AZ, Cross CE, Hu ML, Suzuki YJ, Khwaja S, Safadi A, et al. Modification of plasma proteins by cigarette smoke as measured by protein carbonyl formation. Biochem J. 1992c;286:607-611.

[40] Rajguru SU, Yeargans GS, Seidler NW. Exercice causes oxidative damage to rat skeletal muscle microsomes while increasing cellular sulfhidryls. Life Sci. 1994;54:149-157.

[41] Farney TM, McCarthy CG, Canale RE, Schilling BK, Whitehead PN, Bloomer RJ. Absence of blood oxidative stress in trained men after strenuous exercise. Med Sci Sports Exerc. 2012 Oct;44(10):1855-63.

[42] Salminen A, Vihko V. Lipid peroxidation in exercise myopathy. Exp Mol Pathol. 1983;38:380-388.

[43] Gomez-Cabrera MC1, Domenech E, Viña J. Moderate exercise is an antioxidant: Upregulation of antioxidant genes by training. Free Radic Biol Med. 2008 Jan 15;44(2): 126-31.

[44] Radak Z, Pucsuk J, Boros S. Exercise predonditioning against hydrogen peroxide-induced oxidative damage in proteins of rat myocardium. Arch Biochem Biophys. 2000;376:248-251.

[45] Leeuwenburg C, Fiebig R, Chandwaney R, Ji LL. Aging and exercise training in skeletal muscle: Responses of glutathione and antioxidant enzyme systems. Am J Physiol. 1994;267(R):439-445.

[46] Caldarera CM, Guarnieri C, Lazzari F. Catalase and peroxidase activity of cardiac muscle. Boll Soc Ital Biol Sper. 1973;49:72-77.

[47] Cardoso AM, Bagatini MD, Roth MA, Martins CC, Rezer JF, Mello FF, Lopes LF, Morsch VM, Schetinger MR. Acute effects of resistance exercise and intermittent intense aerobic exercise on blood cell count and oxidative stress in trained middle-aged women.Braz J Med Biol Res. 2012 Dec;45(12):1172-82.

[48] Cunha TF, Bacurau AV, Moreira JB, Paixão NA, Campos JC, Ferreira JC, Leal ML, Negrão CE, Moriscot AS, Wisløff U, Brum PC. Exercise training prevents oxidative stress and ubiquitin-proteasome system overactivity and reverse skeletal muscle atrophy in heart failure. PLoS One. 2012;7(8):e41701.

[49] Hübner-Woźniak E, Morgulec-Adamowicz N, Malara M, Lewandowski P, Okęcka-Szymańska J. Effect of rugby training on blood antioxidant defenses in able-bodied and spinal cord injured players. Spinal Cord. 2012 Mar;50(3):253-6.

[50] Vincent HK, Powers SK, Stewart DJ, Demirel HA, Shanely RA, Naito H. Short-term exercise training improves diaphragm antioxidant capacity and endurance. Eur J Appl Physiol. 2000 Jan;81(1-2):67-74.

[51] Heavens KR, Szivak TK, Hooper DR, Dunn-Lewis C, Comstock BA, Flanagan SD, Looney DP, Kupchak BR, Maresh CM, Volek JS, Kraemer WJ. The effects of high intensity short rest resistance exercise on muscle damage markers in men and women. J Strength Cond Res. 2014 Apr;28(4):1041-9.

[52] Gingsburg GS, Agil A, O'Toole M, Rimm E, Douglas PS. Effects of a single bout of ultraendurance exercise on lipid levels and susceptibility of lipids to peroxidation in Triathletes. J Am Med Assoc. 1996;276(3):221-225.

[53] Criswell D, Powers S, Dodd S, Lawler J, Edwards W, Renshler K, et al. High intensity training-induced changes in skeletal muscle antioxidant enzyme activity. Med Sci Sports Exerc. 1993;25:1135-1140.

[54] Ohno H, Yahata T, Sato Y, Yamamura K, Taniguchi N. Physical training and fasting erythrocyte activities of free radical scavenging enzyme sistems in sedentary men. Eur J Appl Physiol. 1988;57:173-176.

[55] Robertson JD, Maughan RJ, Duthie GG, Morrice PC. Increased blood antioxidant systems of runners in response to training load. Clin Sci. 1991;80:611-618.

[56] Yagi K. Lipid peroxides and exercise. Med Sci Sports Exerc. 1992;30:40-42.

[57] Komulainen J, Vihko V. Training-induced protection and effect of teminated training on exercise-induced damage and water content in mouse skeletal muscles. Int J Sports Med. 1995;16:293-297.

[58] Child RB, Browm S, Day S, Donelly AE, Ropper H, Saxton JM. Changes in indices of antioxidants status, lipid peroxidation and inflamation in human skeletal muscle after eccentric muscle actions. Clin Sci. 1999;96(1):105-115.

[59] Sjódin B, Hellsten Westing Y, Apple FS. Biochemical mechanism for oxygen free radical formation during exercise. J Sports Med. 1990;10:236-254.

[60] Brites FD, Evelson PA, Christiansen MG, Nicol MF, Basílico MJ, Wikinski RW, et al. Soccer Players under regular training Show Oxidative Stress but an Improved Plasma Antioxidant Status. Clin Sci. 1999;96:381-385.

[61] Rosety-Rodríguez M, Díaz-Ordoñez A, Rosety I, Fornieles G, Camacho-Molina A, García N, Rosety MA, Ordoñez FJ. Aerobic training improves antioxidant defense system in women with metabolic syndrome. Medicina (B Aires). 2012;72(1):15-8.

[62] Laughlin MH, Simpson T, Sexton WL, brown OR, Smith JK, Korthuis RJ. Skeletal muscle oxidative capacity, antioxidant enzymes and exercise training. J Appl Physiol. 1990;68:2337-2343.

[63] Spriet LL, Whitfield J. Taurine and skeletal muscle function. Curr Opin Clin Nutr Metab Care. 2015 Jan;18(1):96-101.

[64] Powers SK, Hamilton K. Antioxidants and Exercice. Clin Sports Med. 1999; 18(3): 525-536.

[65] Marsh SA, Laursen PB, Coombes JS. Effects of antioxidant supplementation and exercise training on erythrocyte antioxidant enzymes. Int J Vitam Nutr Res. 2006 Sep; 76(5):324-31.

[66] Marzatico F, Pansarasa O, Bertorelli L, Somenzini L, Della Valle G. Blod free radical antioxidant enzymes and lipid peroxides following long-distance and lactacidemic performances in highly trained aerobic and sprint athletes. J Sports Med Phys Fitness. 1997;37(4):235-239.

[67] Brites FD, Evelson PA, Christiansen MG, Nicol MF, Basílico MJ, Wikinski RW, et al. Soccer Players under regular training Show Oxidative Stress but an Improved Plasma Antioxidant Status. Clin Sci. 1999;96:381-385.

[68] Caldarera CM, Guarnieri C, Lazzari F. Catalase and peroxidase activity of cardiac muscle. Boll Soc Ital Biol Sper. 1973;49:72-77.

[69] Dufaux B, Heine O, Kothe A, Prinz U, Rost R. Blood Glutathione Status Following Distance Running. Int J Sports Med. 1997;18(2):89-93.

[70] Powers SK, DeRuisseau KC, Quindry J, Hamilton KL. Dietary antioxidants and exercise. Journal of Sports Sciences. 2004; 22(1): 81-94.

[71] Gomes EC, Stone V, Florida-James G. Impact of heat and pollution on oxidative stress and CC16 secretion after 8km run. European Journal of Applied Physiology. 2011;111(9):2089-2097.

[72] Sastre J, Asensi M, Gasco E, et al. Exhaustive physical exercise causes oxidation of glutathione status in blood: prevention by antioxidant administration. American Journal of Physiology. 1992;263(5): R992-R995.

[73] Sen CV, Rankinen T, Vaisanen S, Rauramaa R.Oxidative stress after human exercise: Effect of N-acetylcysteine supplementation. Journal of Applied Physiology. 1994;76(6): 2570-2577.

[74] Medved I, Brown MJ, Bjorksten AR, Leppik JA, Sostaric S, McKenna MJ. N-acetylcysteine infusion alters blood redox status but not time to fatigue during intense exercise in humans. Journal of Applied Physiology. 2003;94(4):1572-1582.

[75] Tiidus PM, Pushkarenko J, Houston ME. Lack of antioxidant adaptation to short-term aerobic training in human muscle. Am J Physiol. 1996;271(R):832-836.

[76] Packer L. Protective role of vitamin E in biological systems. Am J Clin Nutr. 1991;53(4 S):1050-1055.

[77] Institute of Medicine, Dietary Intakes for Vitamin C, Vitamin E, Selenium, Carotenoids, National Academy Press, Washigton, DC, USA, 2000.

[78] Urso ML and Clarkson PM. Oxidative stress, exercise, and antioxidant supplementation. Toxicology. 2003;189(1-2):41-54.

[79] Duthie GG, Robertson JD, Maughan RJ, Morrice PC. Blood antioxidant status and erythrocyte lipid peroxidation following distance running. Arch Biochem Biophys. 1990;282(1):78-83.

[80] Brites FD, Evelson PA, Christiansen MG, Nicol MF, Basílico MJ, Wikinski RW, et al. Soccer Players under regular training Show Oxidative Stress but an Improved Plasma Antioxidant Status. Clin Sci. 1998;96:381-385.

[81] Gingsburg GS, Agil A, O'Toole M, Rimm E, Douglas PS. Effects of a single bout of ultraendurance exercise on lipid levels and susceptibility of lipids to peroxidation in Triathletes. J Am Med Assoc. 1996;276(3):221-225.

[82] Palozza P, Luberto C, Calviello G. Antioxidant and prooxidant role of β-carotene in murine normal and tumor thymocytes: Effects of oxygen partial pressure. Free Radic Biol Med. 1997;22:1065-1071.

[83] Morillas-Ruiz JM, Villegas García JA, López FJ, Vidal-Guevara ML, Zafrilla P. Effects of polyphenolic antioxidants on exercise-induced oxidative stress. Clin Nutr. 2006 Jun;25(3):444-53.

[84] Morillas-Ruiz J, Zafrilla P, Almar M, Cuevas MJ, López FJ, Abellán P, Villegas JA, González-Gallego J. The effects of an antioxidant-supplemented beverage on exercise-induced oxidative stress: results from a placebo-controlled double-blind study in cyclists. Eur J Appl Physiol. 2005 Dec;95(5-6):543-9.

Chapter 9

# Antioxidant Status and Sex Hormones in Women with Simple Endometrial Hyperplasia

Snežana Pejić, Ana Todorović, Vesna Stojiljković,
Ivan Pavlović, Ljubica Gavrilović, Nataša Popović and
Snežana B. Pajović

Additional information is available at the end of the chapter

http://dx.doi.org/10.5772/60853

**Abstract**

Cancer of the reproductive tract is an important source of morbidity and mortality among women worldwide. Factors affecting endometrial cancer and endometrial hyperplasia are known to be similar. Endometrial hyperplasia is abnormal proliferation of the glands and the stroma resulting in architectural and cytological modifications. Due to hormonal changes, this condition is most common among women who are nearing the menopause or have reached the menopause. Antioxidant system has a role in preventing cancer initiation and promotion. Since the carcinogenesis occurs in several stages, it is likely that the antioxidant defense depends on the type of cell and tissue. The objective of this study was to investigate whether antioxidant enzymes activities and lipid hydroperoxides concentration in patients with endometrial hyperplasia are influenced by the changes in sex hormones level (estradiol, progesterone, FSH, and LH) during the menstrual cycle and in postmenopause. The material we used consisted of blood and endometrial tissue specimens of women diagnosed with endometrial hyperplasia simplex. Patients were divided in groups depending on the phase of the menstrual cycle: follicular phase, luteal phase and postmenopause. The activities of antioxidant enzymes and the lipid hydroperoxides level were compared among the phases to test the differences and a linear regression model was used to evaluate the associations between hormone levels and antioxidant/oxidant variables. In the blood of examined patients, we observed a phase-related changes of LOOH concentrations. Significant negative correlation between FSH concentration and GR activity ($r= -0.42$, $p<0.05$) and significant positive correlation between LH and

LOOH concentrations ($r = 0.038$, $p < 0.05$) was found. In hyperplasia simplex tissue we recorded significant phase-related changes of LOOH level as well as of AO enzyme activities. SOD and CAT had similar activity pattern, which was higher in luteal phase and in postmenopause, compared to follicular phase ($p < 0.05$). GPx and GR activities did not show any statistical difference. Also, negative correlation between progesterone and GR activity ($r = -0.036$, $p < 0.05$) was observed. Hormonal influence on AO system is of importance in gynecological diseases etiology since they may promote cell proliferation but are also used in conservative therapy, especially for hyperplasia simplex. However, the role of ROS production as a risk factor for endometrial hyperplasia still needs to be clarified as well as the role of AO status in response to gonadotropins and sex steroids.

**Keywords:** antioxidant enzymes, lipid hydroperoxides, gonadotropins, estradiol, progesterone, endometrial hyperplasia

## 1. Introduction

### 1.1. Endometrium

The uterus (womb) is a pelvic organ with reproductive function, i.e., maintenance of pregnancy. The lower, narrow part, which builds on top of the vaginal opening, was marked as *cervix*, and the broader, upper part, as *corpus*. The *corpus* consists of two types of tissue. The smooth-muscular outer layer (*myometrium*) has the function of expanding during pregnancy, and it follows the development of the fetus. The inner layer (*endometrium*) is subjected to a series of so-called cyclic monthly changes known as the menstrual cycle. The endometrium consists of an outer layer, glandular epithelium, below which is an internal part, stroma. This tissue is hormonally regulated by the steroid hormones estrogens (Es) and progestogens (Ps).

## 2. Gonadotropins and steroid hormones in the reproductive period

The reproductive axis consists of the hypothalamus, pituitary, and ovaries. The gonadotropin-releasing hormone (GnRH) acts on the anterior pituitary by regulating the synthesis and storage of gonadotropins, follicle-stimulating hormone (FSH), and luteinizing hormone (LH). GnRH also regulates the movement of gonadotropins from the reserve pool to a readily released point and their secretion. This action requires pulsatile GnRH release [1]. The secretion of FSH and LH takes place in a coordinated manner so as to regulate the growth of ovarian follicles, ovulation, and the maintenance of the corpus luteum and requires constant pulsatile release of LHRH from the hypothalamus [2]. Both estrogens and progestins help regulate the release of gonadotropins, acting through both the hypothalamus and anterior pituitary. High/low levels of either progestins or a combination of progestins and estrogens, as well as the length of exposure to these hormones, inhibit/stimulate the release of GnRH, FSH, and LH from the anterior pituitary – a negative/positive feedback control, respectively [3].

In the reproductive age of women, 17β-estradiol (E2) is a major circulating estrogen that is produced by the granulosa cells of the ovary prior to ovulation and by corpus luteum following ovulation. Almost 95 % of circulating Es in premenopausal women consists of 17β-estradiol and the remaining 5 % originates from the peripheral conversion of the estrone (E1) to estradiol [4]. Although a small amount of estrone, the second most important estrogen, is secreted directly from the ovaries and adrenal glands, its main quantity derives from conversion of androstenedione in adipose tissue [5]. The estrogenic potency of estrone is lesser than estradiol, and both, E2 and E1, are biologically equivalent with subtle structural differences and metabolized by the same pathways. Once a woman has reached menopause and ovaries lose their function, estrone becomes a predominant form of estrogen [6]. Studies have shown the trend to higher mortality rate from coronary heart disease in women with lower estrone level, while patients with higher estrone level had lower body weight, less frequent hypertension and diabetes mellitus, and also a lower triglyceride level [7].

During normal ovulatory cycle, the level of E2 varies individually within the range defined for the follicular phase, mid-cycle, and luteal phase. Most of E2 in the circulation is bound to sex hormone-binding globulin (SHBG) and to a lesser extent, to other serum proteins, such as albumin. Only a very small fraction of this hormone is free and is located in the conjugated form [8,9].

During a normal menstrual cycle, E2 secretion is biphasic, and the highest concentration is recorded just prior to ovulation. This growth affects the pituitary gland secretion of FSH and LH by a positive feedback. After ovulation, E2 level rapidly decreases and luteal cells, by their activities, cause mild, subsequent rise and a plateau of E2 during the luteal phase [2]. During pregnancy, the level of E2 in serum increases to much higher values than recorded in the preovulatory peak, and it is maintained during pregnancy [10].

Progesterone (P) belongs to a group of steroid hormones called progestogens, and it is secreted by the *corpus luteum* in the ovary during the second half of the menstrual cycle. During pregnancy, the high levels of progesterone are provided by the secretion of placenta. Contrary to stimulating, proliferative effect of estradiol, progesterone induces secretory activity of the endometrium and has the role of accepting a fertilized egg and beginning pregnancy. In the circulation, progesterone is bound to the corticosteroid-binding globulin and albumin [2,11]. During normal ovulatory cycle, the increase of serum levels of P induces an increase of LH concentration and together with E2, regulates in this way the preovulatory peak of gonadotropins [12]. In vivo studies in humans suggest that P stimulates its own production during the periovulatory and middle luteal period through self-priming [13]. Besides steroids, the ovary secretes peptide hormones (the inhibins) under gonadotropin control. The inhibins belong to transforming growth factors and have the ability to inhibit gonadotropin (FSH) secretion and may also play an important role in ovarian carcinogenesis [14]. The concentrations of inhibin A and inhibin B in circulation fluctuate during normal menstrual cycle. During the follicular phase, inhibin B is dominant, and during the luteal phase, inhibin A dominates [15].

## 3. Gonadotropins and steroid hormones in menopause

The transition from the reproductive period to the menopause is a gradual process that takes place over many years and is referred to as perimenopause. It starts with the first symptoms of changes in the cyclic occurrence of menstruation and/or bleeding, which may be accompanied by physical and psychological symptoms and ends with the last menstruation. In terms of morphology, this phase is characterized by a sudden drop in the number of primordial follicles in the ovaries, as well as extreme fluctuations in hormone levels [16], so that the frequency of normal ovulatory cycles decreases [2].

It has been shown that some women experience an increase in serum FSH concentrations before the age of 40, especially during the mid-follicular and early luteal phase [17]. Similar increase of FSH was also detected through regular cycles, although there were no clinical manifestations of approaching menopause [18].

Generally, a significant increase in the concentration of FSH is observed approximately 5 years before the onset of menopause and it is positively correlated with age [19,20]. With the onset of menopause, there is an additional increase in FSH levels in serum for about six months, and the peak concentration is detected 3–4 years after menopause. After this period, a slight decline in serum FSH was detected. However, compared with fertile women, levels of gonadotropins remained at elevated levels even 10 years after menopause [19]. Besides FSH, the LH concentration also changes during this period. It has been shown that serum LH increases slightly during 4–5 years of perimenopause in women who still regularly cycled [17]. During the first six months from the onset of menopause, there is an increase in serum concentrations of LH, and the highest level is recorded during the first year of menopause. Over the next 8 years, there is a continuous fall, but as in the case of FSH, the LH level remained elevated compared with fertile period [19]. These data represent the results which should not be generalized and considered as absolute parameters that apply to the period of perimenopause and menopause, since clear markers still have to be identified. In addition, they cannot be reliably interpreted since ovulatory (potentially fertile) cycles can normally take place immediately after the detection of postmenopausal levels of FSH. Both estradiol and inhibin are important regulators of the negative feedback loop of circulating FSH [21-23].

As a consequence of declined follicular function during menopause, the concentration of Es in circulation also decreases. The level of estradiol in the serum of postmenopausal women is less than 15 pg/ml, and the level of estrone is about 30 pg/ml, so that the ratio E1/E2 is 2:1 [11, 24]. The main source of E1, which is the principal form of the postmenopausal estrogen, derives from androstenedione in peripheral adipose tissue and liver [2]. In this period, 95 % of the total synthesis of androstenedione occurs in the adrenal glands and only 5 % in the ovaries [25,26]. Increased conversion of androstenedione to estrone is proportional to the increase of body weight, and it consequently increases the amount of estrogen in the bloodstream [2,26]. The main source of E2 in postmenopausal women originates from the peripheral conversion of E1. During and after menopause, the concentration of E1 decreases as well as the concentration of E2, so that both forms of estrogen are strongly correlated [27,28]. The concentration of estrone sulfate, which is a metabolite of those estrogens, shows a similar trend of decline in menopausal women. Although it does not belong to the active Es, it can be activated by hydrolysis of the sulfate group [27]. Since premenopause leads to inadequate luteal function or anovulation, progesterone is also lowered in the serum. The level of P is further reduced during the aging process, so it is very low in postmenopausal women [19]. Statistically, approximately 2 % to 3 % of women will develop uterine cancer during lifetime. About 97 % of all uterine cancers originate from endometrial glands and represent endometrial carcinomas [29]. Endometrial carcinoma is the fourth common cancer after breast, bowel, and lung carcinoma [30].

## 4. Endometrial hyperplasia

Endometrial proliferation is a normal part of the menstrual cycle that occurs during the follicular/estrogen phase of the cycle [31]. If the endometrium is exposed to continuous endogenous or exogenous estrogen in the absence of progesterone, simple proliferation can advance to endometrial hyperplasia, which is the most common precursor of endometrioid adenocarcinoma. Generally, endometrial hyperplasia is the abnormal proliferation of the glands and the stroma characterized by the presence of architectural and cytological changes [32].

## 5. Classification and histology

As an attempt to correlate morphological features with clinical outcome, the World Health Organization (WHO) classified endometrial hyperplasia:

| |
|---|
| Nonatypical hyperplasias (typical) |
| Simple hyperplasia without atypia |
| Complex hyperplasia without atypia (syn. adenomatous hyperplasia without atypia) |
| Atypical hyperplasias |
| Simple atypical hyperplasia |
| Complex atypical hyperplasia (syn. atypical adenomatous hyperplasia) |

**Table 1.** WHO classification of endometrial hyperplasia [33]

The normal proliferative endometrium is characterized by no crowding of glands within the stroma. Morphological features of all endometrial hyperplasia forms include an increase in the gland-stroma ratio, irregularities in gland shape, and variation in gland size. Regardless of the presence of atypia, simple and complex forms of hyperplasia are distinguished by architectural alterations characterized by glandular complexity and the amount of stroma separating the glands [34]. Hyperplasia generally involves much of the whole endometrium, but sometimes it may be present as a localized lesion and might be associated within an endometrial polyp. Most endometrial hyperplasias are estrogen driven and related to type 1 endometrial carcinoma, the endometrioid endometrial adenocarcinoma [35].

*Simple hyperplasia,* formerly cystic or mild hyperplasia, is a proliferative lesion with minimal glandular complexity and crowding. Histologically, glands are of irregular size from small to those with cystic appearance and shape, separated by abundant stroma. The glandular architectural changes are characterized by varying degrees of irregular branching. Cytologically, the glandular epithelium resembles to proliferative endometrium. It is considered as the least significant form which is not commonly associated with progression to endometrial carcinoma [36].

*Complex hyperplasia,* previously adenomatous hyperplasia or moderate hyperplasia, represents a proliferative lesion with severe glandular complexity and more densely crowded glands. The glands can vary in size and may demonstrate increased structural complexity. Usually, the glands are closely packed, frequently appearing almost back to back and with gland-stroma ratio of more than 2:1 [37,38]. As the severity of hyperplasia increases, the glands become more crowded and more structurally transformed. The complex hyperplasia is considered as the true intraepithelial neoplastic process. Occasionally, this form of hyperplasia may be found coexisting with areas of endometrial carcinoma [39]. Endometrial hyperplasia is further classified based on the presence of cytologic atypia and disordered maturation. Cytologic atypia refers to enlarged epithelial cells that are hyperchromatic with prominent nucleoli, an increased nuclear-to-cytoplasmic ratio, and loss of cellular polarity. Cytologic atypia is the most important prognostic factor for progression to endometrial carcinoma [40].

Thus, the WHO classification also includes lesions termed *simple atypical hyperplasia and complex atypical hyperplasia.* Simple atypical hyperplasia is rare, so the term atypical hyperplasia is widely used to refer to all women with simple or complex atypical hyperplasia. The glands in atypical hyperplasia are very closely packed, and endometrial stroma might be seen, separating them [41]. Less than 2 % of hyperplasias without atypia progress to carcinoma, and the mean duration of progression takes almost 10 years. Atypical hyperplasia progresses to carcinoma in 23 % of cases over a mean duration of 4 years [42].

## 6. Endometrial intraepithelial neoplasia system

There is a discussion to replace the WHO classification of type 1 endometrial carcinoma precursors with the endometrial intraepithelial neoplasia classification system. This system was proposed in 2000 by an international group of gynecologic pathologists, and it defines two classes of endometrial changes, endometrial hyperplasia (EH) and endometrial intraepithelial neoplasia (EIN) [43]. In this classification, endometrial hyperplasia refers to changes observed with anovulation or other etiologies of prolonged estrogen exposure. Morphologi-

http://dx.doi.org/10.5772/60853

cally, EH varies from proliferative endometrium with a few cysts to endometria with many dilated glands. This type is also known as cystic glandular hyperplasia, mild hyperplasia, or simple hyperplasia [44]. The term EIN represents monoclonal endometrial preinvasive glandular proliferation as the immediate precursor of endometrial type 1 adenocarcinoma. In EIN, the proliferation of endometrial glands exceeds the stroma (gland/stroma >1) [45]. EIN categories do not correspond directly to the WHO system of classification. Most simple and some complex hyperplasias fall into EH category and many complex hyperplasias with or without atypia are in the EIN category.

## 7. Epidemiology and risk factors

A well-documented study regarding the epidemiology of endometrial hyperplasia included women aged 18 to 90 over the 18-year period. The diagnosis was mostly made in women aged 50–54 years and rarely was found in women under the age of 30. The incidence of simple and complex hyperplasia was 142 and 213 per 100,000 women-years, respectively. The rate of atypical hyperplasia was highest in older women aged 60–64 years, and it was 56 per 100,000 women-years. This rate seems to correlate with age of peak incidence for endometrial cancer [46,47]. Age-specific cancer incidence was demonstrated for the pancreas, bladder, stomach, lung, prostate, ovary, colorectal, and uterine endometrium. One explanation for increased cancer incidence with age is the latency period required for damage to occur and cancer to develop, including the time necessary for accumulation of carcinogen-induced genetic mutations like in oncogenes and tumor suppressor genes but also as a maladaptive response to replicative senescence due to telomere shortening. Also, a deterioration of the innate and the adaptive immune response with aging, referred to as immunosenescence, must be considered [48].

Symptoms of endometrial hyperplasia include heavy or prolonged menstrual periods, intermenstrual bleeding, and prolonged amenorrhea. Postmenopausal women with hyperplasia may experience vaginal bleeding or spotting. However, only minority of women with abnormal uterine bleeding (AUB) are subsequently diagnosed with endometrial hyperplasia [49].

The risk factors for endometrial hyperplasia are the same as for endometrial carcinoma. Most of them include exposure of endometrium to continuous estrogen unopposed by progestin. Unopposed estrogen may be of various sources like early menarche (beginning menstruation before age 12), hormone replacement therapy (HRT) with exogenous estrogen, late menopause (after 52 years of age), estrogen-secreting tumor (some breast cancer types), and nulliparity or low parity. Medical conditions such as diabetes mellitus, polycystic ovary syndrome, or thyroid diseases also increase the risk for hyperplasia and cancer of the uterus. Endometrial hyperplasia is also more likely to occur in women with personal history of breast, colorectal, or ovarian cancer and in women of white race. Endometrial cancer and hyperplasia are more common in Caucasian women, while uterine sarcoma is more common in African American women [50,51].

## 8. Molecular pathogenesis of endometrial hyperplasia and cancer

Although the findings suggest that there are certain molecular characteristics which distinguish types and degrees of endometrial cancer, the molecular mechanisms that underlie the endometrial carcinogenesis are still unclear. Cell changes can begin with genetic aberrations and continue with uncontrolled growth stimulated by tumor promoters.

## 9. Hormone receptors and growth factors

Endometrial tissue is the target tissue for steroid hormones produced by ovaries. Both epithelium and stroma contain receptors for Es and Ps, and ovarian steroids have a fundamental role in the regulation of growth and differentiation of endometrial cells [2]. It seems this influence is partly preserved in well-differentiated tumors of the lower grade, as suggested by data which showed that these tumors are frequently receptor positive than the advanced tumors [52]. Growth factors are, among other influences, regulated by steroid hormones, and they are involved in a paracrine and autocrine regulation of endometrial proliferation. The most often mentioned are the epidermal growth factor (EGF) and transforming growth factor-$\alpha$ (TGF-$\alpha$). Both factors are single-chain peptides that exert their effect through the EGF receptor. They were shown to be expressed in normal endometrial tissue [53] and to stimulate growth of cultured endometrial cancer cells [54]. In addition to these two factors, it is considered that the transforming growth factor-β (TGF-β) is also involved in the carcinogenesis. This factor is expressed in normal human endometrium and certain endometrial cancer cell lines. In some of these cell lines, like RL95-2, SPEC-2, and KLE, the TGF-β inhibits their growth [55]. Among the other growth factors which affect endometrial carcinogenesis, the basic fibroblast growth factor (bFGF) and insulin-like growth factor I (IGF-I) should also be mentioned [56].

## 10. Activation of oncogenes

The most frequently altered oncogenes in endometrial cancer are the point-mutational activation of K-ras. Point mutations of K-ras were found in approximately 10–30 % of endometrial cancers [57]. Also, K-ras mutations have been identified in endometrial hyperplasia and more frequently in complex atypical hyperplasia, suggesting that K-ras mutations may be an early event in endometrial carcinogenesis [58].

In addition to this oncogene, the amplification and overexpressed HER-2/neu (c-erb B-2) was found in about 10–20 % of sporadic endometrial carcinoma cases [59-61]. HER-2/neu gene encodes a membrane receptor protein which is structurally similar to the receptor for epidermal growth factor (EGF-R). In some endometrial carcinomas, the overexpression of oncogenes Myb, Fos, Myc, and fms, as well as their correlation with advanced stages of carcinogenesis and poor prognosis of the outcome of survival, was recorded [57, 62]. Results of some endometrial carcinoma studies detected the overexpression of oncogenes Myb, Fos, Myc, and fms, as well as their correlation with advanced stages of carcinogenesis and poor prognosis of the outcome of survival [57, 62].

## 11. Inactivation of tumor suppressor genes

Until now, it is observed that mutations in PTEN (phosphatase and tenzin homologue deletion on chromosome 10) tumor suppressor gene, also known as MMAC1 and TEP1, are detected in approximately 50 % of endometrial cancers [63], as well as in 20 % of endometrial hyperplasias, suggesting that these mutations occur relatively early in pathogenesis of this cancer type [64, 65]. PTEN is a dual-specificity protein phosphatase which dephosphorylates tyrosine-, serine- and threonine-phosphorylated proteins. Acting as lipid phosphatase, which is critical for its tumor suppressor function, it removes the phosphate in the D3 position of the inositol ring from phosphatidylinositol 3,4,5-trisphosphate; phosphatidylinositol 3,4-diphosphate; phosphatidylinositol 3-phosphate; and inositol 1,3,4,5-tetrakisphosphate. PTEN is crucial in the control of PI3K-AKT/PKB signaling pathway by dephosphorylating phosphoinositides and thereby modulating cell cycle progression and cell survival [66, 67]. There is a wide spectrum of PTEN mutations in endometrial cancer, which occur in exons 3, 4, 5, 7, and 8 and targeting the phosphatase domain and regions that control the stability and localization of proteins. The consequence of these mutations is reduced or completely absent expression of PTEN [68]. It was shown that progesterone treatment of cultured endometrial stromal cells induces an increase in PTEN levels, while estradiol induces the PTEN phosphorylation. This indicates an outstanding role of PTEN in the development and/or progression of endometrial cancer [69]. Although loss of PTEN function was implicated in the pathogenesis of many different tumors [70], it is believed that the altered expression of PTEN can be a diagnostic marker for the early precancerous conditions of the endometrium [43].

Mutations of the p53 tumor suppressor gene have been found in approximately 10–20 % of all endometrial cancers, with the greatest frequency in the high-grade tumors. Approximately 50 % of grade III tumors type 1 and the rare tumors of type 2 contain mutations in p53, but they have not been reliably detected within the tumor of grade I or hyperplasia [68, 71, 72], so it is considered that they occur in the late stages of endometrial carcinogenesis [56, 68]. The partial role of the p53 in the cell cycle regulation is mediated through the transcriptional activation of other genes, such as p21, followed by inhibition of the cyclin-dependent kinases [73]. Thus, inactivation of p21 could potentially lead to tumor progression. Studies have shown that in approximately 15–40 % of endometrial cancer cases, a loss of p21 gene expression can be detected [74-76]. In addition to p53 and p21, the alterations of p16INK4a (CDKN2A) tumor suppressor gene were also observed. This gene encodes the p16 protein that specifically binds to CDK4 cyclin-dependent kinases, thereby inhibiting the catalytic activity of the CDK4-cyclin D complexes. Until now, it is observed that methylation, mutations, and deletions of p16INK4a gene are rare, and they were detected in approximately 2–6 % of endometrial cancer cases [56, 77], while the loss of expression was found in 20–70 % of cases [78-80].

Endothelins (ETs), ET-1, ET-2, and ET-3, are potent vasoconstricting peptides involved in the pathophysiology of many human malignancies by activating G protein-coupled receptor

(GPCR) subtypes, $ET_A$ and $ET_B$ [81]. Expression of ET-1 was detected in normal human endometrium and in endometrial adenocarcinoma. Also, $ET_AR$ and $ET_BR$ expression was decreased in endometrial cancer tissue compared with that of normal endometrium [82].The ET-1-ETRA axis is frequently dysfunctional in numerous types of carcinomas and contributes to the promotion of cell growth and migration [83].

## 12. Microsatellite instability

In addition to mutations of the PTEN gene, microsatellite instability (MSI) is often detected in type 1 of endometrial cancer. MSI was first demonstrated in patients with hereditary nonpolyposis colorectal carcinoma (HNPCC), in which endometrial cancer is often an associated phenomenon. Additional studies have shown that MSI is detected in approximately 25 % of sporadic cases of endometrial cancer [84] or by other studies in 9–45 % of cases [56]. Unlike hereditary forms of nonpolyposis colorectal carcinoma, where subjects with this type of cancer carry mutations of one of the DNA mismatch repair genes, hMLH1, hMSH2, and hMSH6 [85, 86], promoter hypermethylation of the gene hMLH1 represents the predominant cause of MSI only in sporadic cases [87]. There are also data on the hypermethylation of this gene promoter in hyperplasia and in the absence of cancer, which suggests that inactivation of mismatch repair genes precedes the formation of MSI [88].

## 13. Reactive oxygen species

Oxygen may be a source of reactive oxygen species (ROS) due to its incomplete reduction mostly by the oxidoreductase complex I and III of the mitochondrial respiratory chain [89], forming the superoxide anion radical ($O_2^{\bullet-}$). ROS molecules are characterized by a higher reactivity than oxygen in its ground state. The ROS include free radicals (a term that refers to molecules with one unpaired electron in the outer orbital), like superoxide anion radical ($O_2^{\bullet-}$), hydroxyl radical (•OH), peroxyl radical (ROO• ), as well as reactive nonradical molecules such as singlet oxygen ($^1O_2$), peroxynitrite (ONOO-), or hydrogen peroxide ($H_2O_2$). Their half-life varies from a few nanoseconds for the most reactive molecules up to a few seconds or hours for stable radicals [90].

There are a few major sources of $O_2^{\bullet-}$ in the cell: the respiratory chain in mitochondria, endoplasmic reticulum cytochromes (cytochrome P-450-dependent oxygenase, NADPH-cytochrome P-450 reductase), as well as the oxidase contained in the cell cytoplasm and membranes (NADPH oxidase of polymorphonuclear leucocytes, macrophages, and endothelial cells) [91, 92]. The resulting $O_2^{\bullet-}$ may be converted to $H_2O_2$ by spontaneous dismutation, as well as by the enzyme superoxide dismutase (SOD). In addition, the $H_2O_2$ may originate from the monoamine oxidase activity [93] or from the beta-oxidation of fatty acids in peroxisomes [94]. Its reduction is carried out by the enzyme catalase (CAT) and glutathione peroxidase (GPx), which can be considered as the main way of detoxification. $H_2O_2$ may also be

http://dx.doi.org/10.5772/60853

reduced by the neutrophil myeloperoxidase which catalyzes the conversion of $H_2O_2$ and $Cl^-$ to hypochlorous acid (HOCl) and in the presence of transition metals ($Fe^{2+}$ or $Cu^+$), producing •OH [95]. The hydroxyl radical is a highly reactive oxidant that reacts almost instantaneously with the surrounding molecules abstracting the hydrogen atom (RH). The resulting free radical (R•) is more stable and therefore has usually longer half-life compared to the •OH [96]. Peroxyl radicals have a relatively long half-life, and they are formed in the process of lipid peroxidation, which begins with removal of the hydrogen atom of polyunsaturated fatty acids [97]. Lipid peroxidation in cell membranes can significantly damage their function due to the formation of irreversible disturbance of fluidity and elasticity, which can lead to impairment of cellular homeostasis.

ROS are constantly produced in the body as a result of normal metabolic processes, but there is also a significant influence of external factors. Many chemical and biological agents which are prooxidants under certain conditions can lead to increased production of free radicals. If their production exceeds the capacity of the antioxidant defense, the oxidative stress occurs [96]. ROS can react with any molecules in the cell, thus causing considerable damage which results in cellular dysfunction. These processes are increasingly studied today in the framework of the mechanisms of etiopathogenesis of various diseases. Also, their role in cell signaling, proliferation, differentiation, and programmed cell death – apoptosis – is intensively examined.

## 14. Antioxidant System (AOS)

The term antioxidant refers to a substance that, when present in small amounts compared with the substrate to be oxidized, inhibits or prevents its oxidation. The antioxidant system can be divided into two categories: nonenzymatic antioxidants, which include various compounds of low molecular weight (vitamin E, vitamin C, carotenoids, polyphenols, ubiquinone, and glutathione), and the AO enzyme system [98].

## 15. Nonenzymatic antioxidants

Vitamin E (tocopherol-OH, vitamin E) is a generic name for a group of compounds known as the tocopherols and tocotrienols, and it includes all forms which exhibit biological activity of natural vitamin E (d-alpha-tocopherol) [98]. Vitamin C (ascorbic acid) is the most important hydrophilic antioxidant. Their main function is to prevent peroxidation of lipids in the membrane and, consequently, cell damage. The carotenoids are the vitamin A, which also possess antioxidant properties. Beta-carotene is one of the most studied forms, and its antioxidant function is based on its attribute to quench the singlet oxygen and remove free radicals, thus protecting the cell membrane lipids from oxidative degradation. Polyphenols are a group of compounds with antioxidant capacity to prevent formation of ROS production through inhibition of the enzyme, as well as trace elements, involved in their formation [99].

Ubiquinone prevents lipid peroxidation in liposomes, lipid emulsions, phospholipids, and LDL particles [100]. Glutathione (GSH) is a tripeptide consisting of L-glutamine, L-cysteine, and L-glycine. In addition to its role as a substrate of GSH redox cycles, it also removes the hydroxyl radicals and singlet oxygen and maintains the enzymes and other cellular components in a reduced state [98].

## 16. Antioxidant enzymes

In mammals, three types of SODs have been identified, depending on the cellular localization and prosthetic groups. In the cytoplasm, the predominant form is copper-zinc-superoxide dismutase (CuZnSOD, SOD1), which represents a stable dimeric protein with molecular mass of 32 kDa. It contains copper and zinc in its active site. Copper is considered necessary for the catalytic activity of this enzyme, whereas zinc contributes to its stability [101]. CuZnSOD is also located in the extracellular matrix, and this form is known as the extracellular superoxide dismutase (EC-SOD, SOD3). This form of CuZnSOD is a tetrameric protein with molecular mass of 135 kDa, and it possesses a heparin-binding domain that affects its extracellular distribution [102]. Manganese superoxide dismutase (MnSOD SOD2) is a tetramer enzyme with molecular weight of 88 kDa, containing manganese atom in the active sites and it is located in the mitochondria.

CAT is homo-tetramer enzyme with molecular weight of 240 kDa, with each subunit containing the heme prosthetic group and also the attached NADPH that protects the enzyme from oxidative damage. CAT has a function to decompose $H_2O_2$ to $O_2$ and $H_2O$ [103].

GPx family can be divided into two groups: selenium-independent peroxidase presented glutathione S-transferase (GST) and selenium-dependent peroxidases (GPx).

Glutathione S-transferase belongs to the so-called phase II detoxifying enzymes that are involved in conjugation reactions of a wide range of electrophilic xenobiotics (including carcinogens and mutagens). Several selenoprotein glutathione peroxidases are present in human tissues, cell GPx (GPx-1, CGP-x), gastrointestinal GPx (GPx-2, giGPx), plasma (extracellular) GPx (GPx-3, eGPx), and phospholipid hydroperoxide GPx (GPx-4, PHGPx) and GPx-6, which is only expressed in the epithelium of the olfactory system [104]. With the exception of PHGPx which is a monomer (19 kDa), other forms of GPx are composed of four identical subunits of a molecular weight of 19–25 kDa. Each subunit in its active site contains a selenocysteine ($Cys^{Se}$). The enzyme uses a reduced GSH as a source of reducing equivalents (electrons) to regenerate $Cys^{Se}$ to the reduced state [105]. Glutathione reductase (GR) is an enzyme that catalyzes the reduction of oxidized glutathione GSSG to GSH and it is essential for the GSH redox cycle [106].

## 17. Oxidative stress and cell signaling

Because of their high reactivity, elevated ROS concentrations represent a great danger for biomolecules. At physiological concentrations, these molecules are often necessary for normal functioning of cells as second messengers in the transduction of the cell signaling [107]. They can be activated in such a way as to prevent or potentiate the cell death. Many signaling pathways in the cell can be activated in both directions (cell survival or apoptosis), which depends on the type and duration of oxidative stress or cell types. Also, some of these pathways can affect the activation or suppression of other signaling pathways in the cell.

It is difficult to determine which type of ROS activates signaling pathways, because of their extremely rapid conversion to other forms or due to the conversion of acid conjugates or complexes with transition metals [108]. It is believed that $H_2O_2$ is highly suitable as a secondary messenger because it does not randomly react with all of the molecules like other forms of ROS, but tends to oxidize the -SH group of cysteine (Cys), which is then reduced by GSH [109].

In this way, by redox cycling of Cys, many transcription factors are regulated, such as activating protein 1 (AP-1) [110], nuclear factor NF-IL6 [111], and proteins important in cell signaling and cancerogenesis: protein kinase C (PKC), $Ca^{2+}$-ATPase, collagenases, and SRC tyrosine kinase [108]. It is known that ROS are critical molecules in regulation not only of the AP-1 but also AP-2 [112] and of nuclear factor NF-kappaB [113] transcription families, which have a decisive role in cell proliferation, differentiation, and morphogenesis.

Other processes induced by hydrogen peroxide included activation of the stress-activated protein kinase/c-Jun N-terminal kinases (SAPK/JNK), the increased c-Jun phosphorylation, activation of caspase 3 (CPP32), and decomposition of poly(ADP-ribose) polymerase (PARP), which are associated with the apoptosis process [114]. Besides regulating the activity of cell proteins, $H_2O_2$ also induces the expression of many genes [115]. In addition, these molecules are responsible for the disruption of cell signaling and regular patterns of gene expression [116], which can lead to a number of pathological processes including carcinogenesis. The process of carcinogenesis is complex and consists of a series of changes at the cellular and molecular levels and in at least three stages: initiation, promotion, and malignant conversion, i.e., progression [117].

## 18. AO antioxidant status and carcinogenesis

In relation to carcinogenesis, it is known that the AO system has a role in preventing its occurrence and promotion. The studies AO status in tumor tissues have not yet yielded results that could lead to general conclusions about AO defense in tumor tissues. Since the carcinogenesis occurs in several stages, it is likely that the antioxidant defense depends on the type of cell and tissue [118]. Mammalian cells and tissues differ significantly in the generation of ROS. They also vary in antioxidant activity, induction capability, and cell repair capacities which altogether results in a different susceptibility of mammalian tissues for tumor induction [119-121].

Our earlier studies indicated a significant role of oxidative-induced injury in the breast carcinogenesis, particularly during the later stages of aging [122]. It was also observed that chemotherapy and radiotherapy promote further oxidative shift, which potentiates already existing chronic oxidative stress linked to breast cancer [123]. It is believed that the high antioxidant capacity protects DNA from oxidative damage and mutagenesis but also can protect the cells in the stage of initiation of increased oxidative toxicity, thus favoring their clonal expansion and tumor progression [124]. It has long been known that oxidizing agents may be cytotoxic, although under certain circumstances, can promote cell growth and facilitate the clonal expansion of the initiated cells in carcinogenesis [125].

## 19. AO enzymes in gynecologic disorders

Some previous studies have shown that compared to healthy people, women with benign and malignant changes in the genital tract have increased level of lipid peroxidation and altered activity of AO enzymes in peripheral blood and tissue. Chiou and Hu [126] have detected that the activity of SOD in plasma and erythrocytes of patients with cervicitis and uterine myoma was lower compared to that of healthy women. At the same time, patients with cervicitis had an increased level of CAT and GPx activity, while their activity in patients with uterine fibroids (leiomyoma) was reduced. Similar results regarding the activities of SOD, CAT, and GPx in erythrocytes of patients with cervicitis were obtained by Manoharan et al. [127]. These authors also found that the activity of these enzymes was lower in patients with cervical cancer. Research of Kolanjiappan et al. [128] and Manoharan et al. [127] showed that the level of lipid peroxidation increased and the concentration of the antioxidant GSH, vitamin E, and CAT decreased in erythrocytes of patients with cervical cancer. These patients had altered activity of $Na^+K^+$-ATPase in erythrocytes compared to healthy persons. Our previous results showed that AO status in blood of gynecological patients varies with diagnosis and the enzyme type. Generally, both reduction in antioxidants and elevation of lipid peroxidation were observed. Lipid hydroperoxide level was negatively correlated to SOD and GPx activities and concurrently positively correlated with CAT activity. In addition, the lipid hydroperoxides/glutathione peroxidase ratio increased, according to the type of uterine disorder [129-131]. The perturbation of antioxidant status was more pronounced in blood of patients with hyperplastic and adenocarcinoma lesions compared to those with benign uterine changes such as polypus and myoma. Our results of AO status in endometrial tissue showed significant decrease of SOD activity in women with hyperplasia and adenocarcinoma. In both types of hyperplasia, activities of GPx and GR were increased to 60 % and 100 % on average, while in adenocarcinoma patients, only GR activity was elevated to 100 %. CAT activity was significantly decreased in adenocarcinoma patients (47 %). Lipid hydroperoxides level was negatively correlated to SOD and CAT activities and positively correlated to GPx and GR activities [132]. Since association of different clinical risk factors and various types of gynecologic pathologies is still not fully known as well as their influence on AO status, in our latest study, we evaluated the influence of diagnostic categories, age, and reproductive factors on AO status in blood of gynecological patients [133].The obtained results showed that reproductive and other factors may be associated, at least partially, with AO capacity and ability to defend against the oxidative damage in gynecological patients.

The AO status and hormone influence were studied during the menstrual cycle and postmenopause in healthy women and those with gynecologic disorders. The SOD was found to have a role in maintaining luteal cell integrity and steroidogenic capacity in fertile women [134]. An increase in the GPx activity was observed during the menstrual cycle, from the late follicular to the early luteal phase. The rise in GPx activity is related to increased ovarian production of estrogen that occurs in that particular period of menstrual cycle [135]. Decrease in GPx activity has been noted in the endometrium and blood in late-menopausal women [136]. Menopause is accompanied by hormone imbalance. A significant fall of the estrogen serum level with rise of follicle-stimulating hormone (FSH) has been recorded in postmenopausal women compared to premenopausal women [137]. Hormone replacement therapy (HRT) shows protective antioxidant role by reduction of lipid peroxide (LOOH) serum levels [138]. It is also found that HRT positively correlates with SOD activity in postmenopausal women [139].

We have shown that AO enzyme activity and lipid hydroperoxide level in patients with endometrial polyps are influenced by the changes in sex hormones during the menstrual cycle and in menopause [140]. In this study, we aimed to examine the AO status in menstrual cycle and postmenopause of women with endometrial hyperplasia simplex as well as the relationship between sex hormones and AO parameters.

## 20. Methods

*Subjects.* The material used in this study consisted of 35 blood and tissue specimens of women admitted to the Department of Gynecology and Obstetrics for gynecological evaluation within routine checkups or for abnormal uterine bleeding (prolonged menstrual bleeding and postmenopausal bleeding). On the basis of diagnosis and histological examination, subjects were diagnosed with *hyperplasia simplex endometrii*, and the specimens were taken after obtaining the informed consent. The study was conducted prospectively and it was approved by the Human Studies Ethics Committee of the Clinical Center. The protocol was consistent with the World Medical Association Declaration of Helsinki (Ethical Principles for Medical Research Involving Human Subjects). None of them had undergone hormone therapy or any other medical treatment in the last six months. Patients were divided as follows: 10 in the proliferative (follicular phase, F) (age, 40–52 years; median 46 years), 15 in the secretory (luteal phase, L) (age, 27–53 years; median 44 years), and 10 in the postmenopause (PM) (age, 47–60 years; median 53 years).

*Samples.* Samples were collected and prepared for enzyme assays according to the procedures described previously [129,131]:

Venous blood samples were collected into heparinized tubes on the same day of uterine biopsy and aliquoted immediately. For SOD assay (OxisResearch™), blood was centrifuged at 2500 g for 5 min. Plasma was discarded and pellet was resuspended in 4 packed-cell volume of ice-cold demineralized ultrapure water (MilliQ reagent grade water system, Millipore Corp., Bedford, MA, USA). After addition of ethanol/chloroform extraction reagent (62.5/37.5 vol/vol) to remove hemoglobin interference, samples were centrifuged at 3000 g for 10 min

(Eppendorf centrifuge 5417, Eppendorf AG, Hamburg, Germany). Upper aqueous layer was collected and kept at -70 ºC until assay.

Fresh endometrial tissue samples were washed in saline solution and homogenized in phosphate buffer containing 0.05M $KH_2PO_4$ and 1 mM EDTA, pH 7.8 (1 g tissue per 2 ml buffer) in a Teflon/glass homogenizer (Spindler & Hoyer, Göttingen, Germany) and frozen at -70 ºC for 20 h in order to disrupt cell membranes. For SOD assay (OxisResearch™), thawed homogenates were vortexed 1 min and centrifuged at 8600 g, for 20 min at 4 ºC (Eppendorf centrifuge 5417, Eppendorf AG, Hamburg, Germany). According to manufacturer's recommendation, after addition of ethanol/chloroform extraction reagent (62.5/37.5 vol/vol) to completely remove hemoglobin interference, samples were centrifuged at 6000 g for 20 min, at 4 ºC (Beckman centrifuge J2-21, Beckman Instruments Inc., Palo Alto, CA, USA). Upper aqueous layer was collected and kept at -70 ºC until assay. The enzyme activities and lipid hydroperoxide (LOOH) concentration were monitored spectrophotometrically (Perkin Elmer Spectrophotometer, Lambda 25, Perkin Elmer Instruments, Norwalk, CT,USA).

The specific enzyme activities were expressed as Units (U) or mU per milligram of total cell protein (U or mU/mg protein), and LOOH concentration was expressed as nmol/mg protein. Protein concentration in tissue homogenates was performed by the method of Lowry et al. [141] and expressed as mg/ml. Plasma follicle-stimulating hormone (FSH), luteinizing hormone (LH), estradiol (E), and progesterone (P) levels were analyzed using standard radioimmunoassay (RIA) methods by the hormone analysis laboratory.

*Enzyme Assays.* Enzyme assays were performed as described previously [132]:

*Assay of SOD activity.* Determination of SOD activity was performed using Oxis Bioxytech® SOD-525™ Assay (Oxis International, Inc., Portland, OR, USA). The method is based on SOD-mediated increase of autoxidation of 5,6,6a11b-tetrahydro-3,9,10-tryhydroxybenzo[c]fluorene in aqueous alkaline solution to yield a chromophore with maximum absorbance at 525 nm. The SOD activity is determined from the ratio of the autoxidation rates in the presence (Vs) and in the absence (Vc) of SOD. One SOD-525 activity unit is defined as the activity that doubles the autoxidation rate of the control blank.

*Assay of CAT activity.* CAT activity was determined by the method of Beutler [142].The reaction is based on the rate of $H_2O_2$ degradation by catalase contained in the examined samples. The reaction was performed in an incubation mixture containing 1 M Tris-HCl, 5 mM EDTA, pH 8.0, and monitored spectrophotometrically at 230 nm. One unit of CAT activity is defined as 1 μmol of $H_2O_2$ decomposed per minute under the assay conditions.

*Assay of GPX activity.* GPx activity was assessed using the Oxis Bioxytech® GPx-340™ Assay (Oxis International, Inc., Portland, OR, USA), based on the principle that oxidized glutathione (GSSG) produced upon reduction of an organic peroxide by GPx, is immediately recycled to its reduced form (GSH) with concomitant oxidation of NADPH to NADP+. The oxidation of NADPH was monitored spectrophotometrically as a decrease in absorbance at 340 nm. One GPx-340 unit is defined as 1 μmol of NADH oxidized per minute under the assay conditions.

*Assay of GR activity.* Activity of GR was measured using the Oxis Bioxytech® GR-340™ Assay (Oxis International, Inc., Portland, OR, USA). Assay is based on the oxidation of NADPH to

NADP+ during the reduction of oxidized glutathione (GSSG), catalyzed by a limiting concentration of glutathione reductase. The oxidation of NADPH was monitored spectrophotometrically as a decrease in absorbance at 340 nm. One GR-340 unit is defined as 1 μmol of NADH oxidized per minute under the assay conditions.

*Lipid hydroperoxides.* Concentration of LOOH was measured by Oxis Bioxytech® LPO-560™ Assay (Oxis International, Inc., Portland, OR, USA), which is based on the oxidation of ferrous ($Fe^{2+}$) ions to ferric ($Fe^{3+}$) ions by hydroperoxides under acidic conditions. Ferric ions then bind with the indicator dye, xylenol orange, and form a colored complex. The absorbance of the complex was measured at 560 nm. Since hydrogen peroxide content in many biological samples is much higher than that of other hydroperoxides, samples were pretreated with catalase to decompose the existing $H_2O_2$ and eliminate the interference.

*Statistics.* Statistical analysis was carried out by the use of the Kruskal-Wallis test followed by the Dunn's *post hoc test*, which considered the unequal and small sample sizes we used in this study. A linear regression model was used to evaluate associations between hormonal and antioxidant variables. Before plotting the data in the regression study, the normality test on the variables was performed, and the values of estradiol and progesterone were log transformed. The 95 % confidence intervals (CIs) for the regression lines were calculated. Two-tailed *p* values are given throughout. All data were analyzed using GraphPad Prism software.

## 21. Results

The phase-related concentrations of gonadotropins and sex hormones are reported in Table 2. Significant changes were observed in FSH (H=12.75, p<0.01, Kruskal-Wallis), LH (H=8.98, p<0.01, Kruskal-Wallis), and estradiol (H=7.93, p<0.05, Kruskal-Wallis) concentrations.

| | Follicular phase | Luteal phase | Postmenopause |
|---|---|---|---|
| FSH (U/L)** | 14.30±3.51 | 13.20±2.21 | 38.88±7.88 |
| Median | 11.50 | 10.80 | 32.15 |
| (Min/max) | (7.50–31.50) | (0.1–31.50) | (14.30–75.00) |
| LH (U/L)** | 3.33±1.70 | 3.93±0.96 | 11.88±2.47 |
| Median | 2.50 | 3.10 | 11.25 |
| (Min/max) | (0.60–9.40) | (0.60–11.20) | (1.30–24.00) |
| Estradiol (pg/ml)* | 39.30±7.59 | 71.41±14.07 | 5.16±1.16 |
| Median | 48.80 | 56.50 | 3.80 |
| (Min/max) | (12.60–57.20) | (10.00–208.10) | (0.70–11.40) |
| Progesterone (nmol/L) | 7.40±1.09 | 8.83±2.68 | 5.16±1.16 |
| Median | 6.30 | 5.30 | 3.80 |
| (Min/max) | (5.20–12.10) | (1.30–41.60) | (0.70–11.40) |

**Table 2.** Changes in hormone levels during follicular phase, luteal phase, and in postmenopause (data are expressed as mean ± SEM; * p<0.05, **p<0.01)

### 21.1. Antioxidant parameters and correlation with sex hormones in blood

Figure 1 shows the phase-related changes of LOOH concentrations and AO enzyme activities in the blood of examined patients. The significant change with respect to the phase was observed in LOOH concentrations (H=5.76, p<0.05, Kruskal-Wallis). In the follicular phase, it was significantly lower than in the postmenopause (p<0.05, Dunn test). There were no significant changes of AO enzymes in the examined phases.

The linear regression analysis of individual hormonal variables against antioxidant parameters in blood (Figure 2) showed a significant negative correlation between FSH concentrations and GR activity (r=-0.42, p<0.05), as well as a significant positive correlation between LH and LOOH concentrations (r=0.38, p<0.05). No significant correlations were found between other hormones and antioxidant variables.

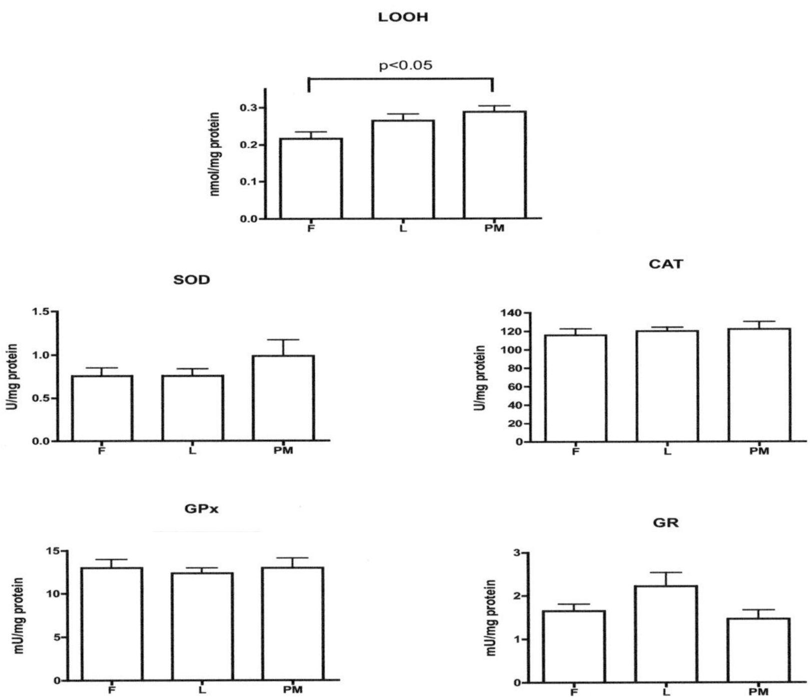


**Figure 1.** Changes in blood LOOH concentrations and AO enzyme activities in follicular phase (F), luteal phase (L), and postmenopause (PM) in blood of patients with hyperplasia simplex. Data are shown as mean ± SEM. *P* values refer to the results of the Dunn test

### 21.2. Antioxidant parameters and correlation with sex hormones in hyperplasia simplex tissue

The phase-related changes of LOOH concentrations and AO enzyme activities in hyperplasia simplex tissue are shown in Figure 3. The LOOH concentration significantly differed with

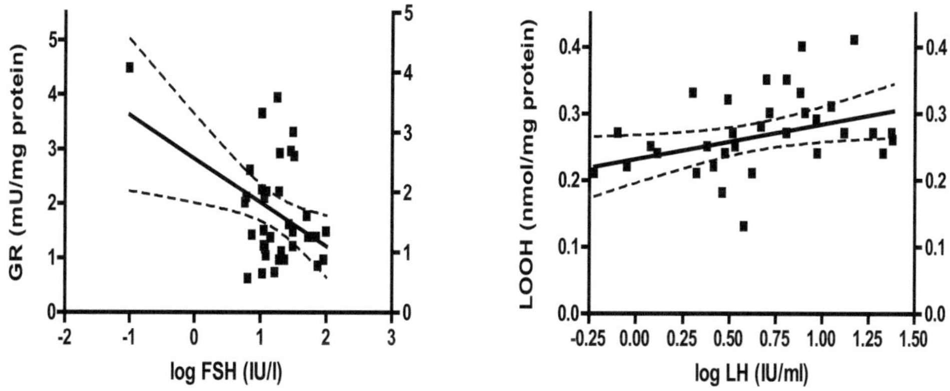


**Figure 2.** Linear regression line and 95 % CI to study the relationship between log FSH and GR activity; log LH and LOOH concentration in the blood of patients with hyperplasia simplex

respect to the phase (H=7.74, $p<0.05$, Kruskal-Wallis), and it was significantly elevated in luteal phase and in postmenopause, in comparison to the follicular phase ($p<0.05$, Dunn test).

Unlike blood, where no changes in AO enzyme activities were recorded, we found significant phase-related changes of SOD (H=9.11, p=0.01, Kruskal-Wallis) and CAT activity H=7.60, $p<0.05$, Kruskal-Wallis). Both enzymes had similar activity pattern, which was higher in luteal phase and in postmenopause, compared to follicular phase ($p<0.05$, Dunn test). The phase-related activity of GPx and GR did not show any statistical difference.

The linear regression analysis of hormone levels on the examined AO parameters in hyperplasia simplex tissue showed a negative correlation between progesterone and GR activity (Figure 4) (r=-0.36, $p<0.05$).

## 22. Discussion

Studies have shown a different AO status and sex hormone influence during menstrual cycle and postmenopause in healthy women and those with ovarian disorders [143-146], but we found no data regarding that relation in patients with endometrial hyperplasia simplex.

In the blood of these patients, we detected a lower level of LOOH in the F phase in comparison to the postmenopause. In hyperplastic tissue, LOOH level was lower in the F phase than in L phase and postmenopause. The activities of SOD and CAT were also lower in F phase when compared to the L phase and postmenopause. There was a negative correlation between FSH/P concentrations and GR activity in the blood and hyperplastic tissue, respectively. Positive correlation between LH and LOOH concentrations was recorded in the blood.

Similar pattern of LOOH concentration and SOD activity in endometrium of healthy women throughout the menstrual cycle was also observed in [134]. They found that LOOH concentration increased from early proliferative phase to mid-late proliferative phase and further

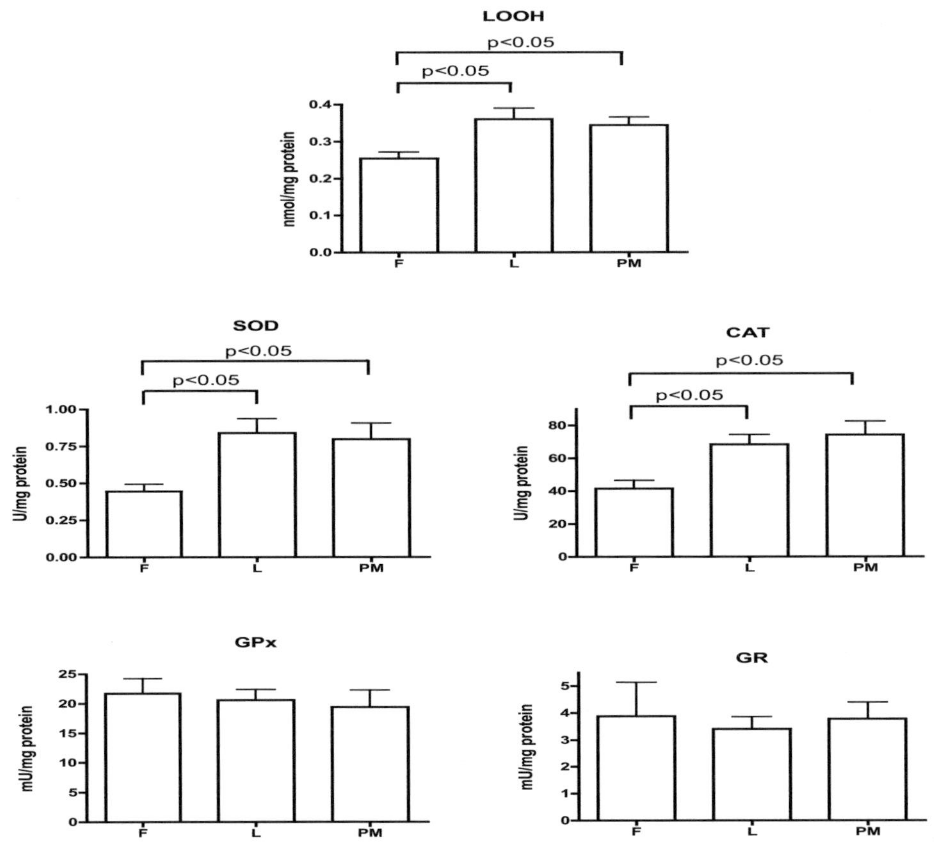


**Figure 3.** Changes in endometrial LOOH concentrations and AO enzyme activities in follicular phase (F), luteal phase (L), and postmenopause (PM) in hyperplasia simplex tissue. Data are shown as mean ± SEM. *P* values refer to the results of the Dunn test

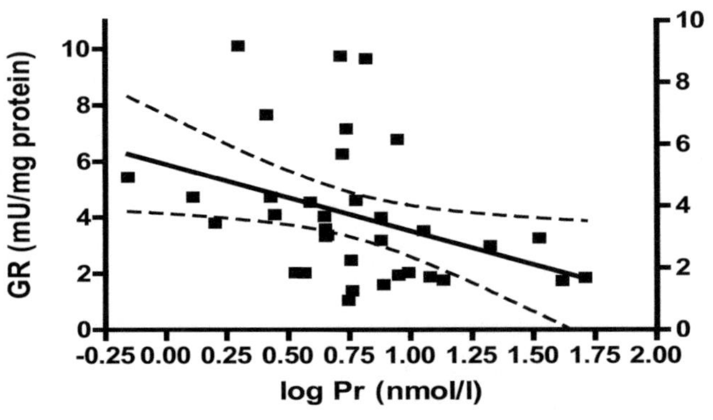


**Figure 4.** Linear regression line and 95 % CI to study the relationship between log Pr and GR activity in hyperplasia simplex tissue.

increased in the late secretory phase. The SOD activity increased from early proliferative phase to mid-late proliferative phase, further increased in the mid-secretory phase, and then decreased in the late secretory phase. Previous investigations of immunohistochemical distribution of SOD in human endometrium during menstrual cycle also showed that surface and glandular epithelia contain SOD during proliferative and secretory phases except just prior to the menstruation [147].

The study of Ota et al. [148] regarding SOD expression in endometrium during the menstrual cycle of healthy fertile women and women with diagnosed endometriosis and adenomyosis have shown the phase-dependent changes of SODs in glandular and surface epithelia in healthy women. Specifically, the expression of copper, zinc SOD was lowest during the early and mid-proliferative phases and then gradually increased and was most pronounced in the early and mid-secretory phases. The expression of manganese SOD reached a peak in the late secretory phase. In women with endometriosis and adenomyosis, the expression of both SODs was constantly elevated compared to healthy women throughout the menstrual cycle, which suggested a key role of superoxide in infertility caused by endometriosis and adenomyosis [148]. Our recent findings in women with endometrial polyp showed the opposite pattern of LOOH concentration and SOD activity in blood and polyp tissue than in women with hyperplasia simplex. Both parameters were higher in the proliferative phase compared to the secretory phase or postmenopause in blood and endometrium of the examined women [140].

Regarding CAT, in [149], it was found that CAT expression in healthy women fluctuated greatly during the menstrual cycle and the surface epithelium showed a similar pattern to that in the glandular epithelium. The expression was the lowest during the early proliferative phase, increased during the mid-proliferative phase, and peaked in the late secretory phase. In patients with endometriosis, the CAT expression did not fluctuate during the cycle, but it was consistently elevated throughout the menstrual cycle when compared to healthy women. Likewise, in women with adenomyosis, the CAT expression did not vary during the cycle in comparison to healthy ones, and it was significantly higher than in patients with endometriosis [149]. In women with endometrial polyp, we found no significant change of CAT activity in different phases [140]. In this study, however, the CAT activity in endometrium of patients with simple hyperplasia was also similar to the healthy women.

Studies in women with gynecologic disorders indicate a different AO status, as one of the possible factors contributing to the development of oxidative stress [150]. There are also studies which investigated the role of oxidative stress and hormones in development of gynecologic pathologies. For example, in [151], it was found that FSH, LH, and estrogen could induce ROS production at different levels in ovarian epithelial carcinoma and may therefore participate in cancer development process. FSH was found to increase cell proliferation in ovarian epithelial carcinoma (OEC) [152], and LH may also be involved in OEC development under pathological conditions [151].

Simple hyperplasia is the most common type of endometrial hyperplasia and the type most likely to spontaneously regress, and it rarely progresses to endometrial cancer [42, 51, 153]. The LOOH concentrations and AO enzyme activities in this study which were similar to the healthy women point to the preserved cellular AO status in these patients. Endometrial hyperplasias are generally considered as precancerous lesions and are treated either conservatively or surgically. The regression of hyperplastic to normal endometrium is the main purpose of any conservative treatment. It is based on the administration of agents, like progestogens [154], which have an indirect antiestrogenic action and also a direct antiproliferative effect on the endometrium [155]. Also, therapeutic application of gonadotropin-releasing hormone analogue (GnRHa) in women with hyperplasia was associated with high regression rates. The regression to normal endometrium is considered to be due to decreased gonadotropin levels as a result of pituitary downregulation or to the decreased ovarian steroidogenesis following low gonadotropin levels [156, 157]. The results of this study also showed that gonadotropins and progesterone influenced oxidant/antioxidant parameters in hyperplastic patients. Although we found no significant changes of GR activity among the menstrual cycle phases, FSH/P was negatively correlated with GR activity in the blood and hyperplastic tissue, while positive correlation between LH and LOOH concentrations was recorded in the blood. Our previous study in women with endometrial polyp also showed the influence of gonadotropins on AO status. In these patients, we observed a negative correlation between FSH/LH and GPx activity and also between LH and SOD activity [140].

The role of gonadotropins in gynecological diseases in not fully clarified. In ovarian epithelial cancer (OEC), gonadotropin theory proposes that elevated serum FSH and LH levels contribute significantly to its development [158]. FSH generally acts through its membrane-bound receptor which activates the intracellular signaling cascade, starting with cyclic AMP/protein kinase A (cAMP/PKA) that is followed by phosphorylation of specific transcriptional factors, like cAMP-response element-binding protein (CRE), or p38 MAPK, which controls other kinase cascades. The FSH receptor can also activate extracellular signal-regulated protein kinases (ERK-s) [159].

It was shown that synthesis of antioxidants, such as glutathione in the ovary, is regulated by gonadotropins, but exact mechanisms are still unknown [160]. One of the mechanisms behind FSH and antioxidants interaction is through activation of transcriptional factors, like Nrf2. The induced Nrf2 binds to the antioxidant-response element (ARE), thus coordinately regulates the expression of AO genes [161].

The pathogenesis of endometrial hyperplasia is still not fully understood. Prolonged estrogen stimulation is considered as one of the factors related to the etiology. This study showed that patients with endometrial hyperplasia simplex have similar AO status like healthy women, and it also demonstrated the relation of hormones and prooxidant/antioxidant parameters in this gynecologic disorder. Since simple hyperplasia may spontaneously regress, these results point to the preserved AO capacity as a potentially important factor in the regression mechanisms. However, the role of ROS production as a risk factor for endometrial hyperplasia still needs to be clarified as well as the role of AO status in response to gonadotropins and sex steroids.

http://dx.doi.org/10.5772/60853

## Acknowledgements

This work was financially supported by the Ministry of Education, Science and Technological Development, Republic of Serbia (Grants 41027, 41022, 173041).

## Author details

Snežana Pejić*, Ana Todorović, Vesna Stojiljković, Ivan Pavlović, Ljubica Gavrilović, Nataša Popović and Snežana B. Pajović

*Address all correspondence to: snezana@vinca.rs

Laboratory of Molecular Biology and Endocrinology, "Vinča" Institute of Nuclear Sciences, University of Belgrade, Belgrade, Serbia

## References

[1] Beshay VE, Carr BR. Hypothalamic-Pituitary-Ovarian Axis and Control of the Menstrual Cycle. In: Falcone T, Hurd WW (eds.) Clinical Reproductive Medicine and Surgery. A Practical Guide. New York: Springer; 2013. pp. 31–42.

[2] Carr BR. Disorders of the ovary and female reproductive tract. In: Wilson JD, Foster DW, Kronenberg HM, Larsen PR (eds.) Williams Textbook of Endocrinology. 9th ed. Philadelphia: Saunders; 1998. pp. 733–98.

[3] Fink G. Gonadotropin secretion and its control. In: Knobil E, Neil J. (eds.) The Physiology of Reproduction. New York: Raven Press;1988. pp.1349–77.

[4] Baird DT, Fraser IS. Blood production and ovarian secretion rates of estradiol-17 beta and estrone in women throughout the menstrual cycle. J Clin Endocrinol Metab. 1974 Jun;38(6):1009–17. PubMed PMID: 4598662. Epub 1974/06/01. eng.

[5] Forney JP, Milewich L, Chen GT, Garlock JL, Schwarz BE, Edman CD, et al. Aromatization of androstenedione to estrone by human adipose tissue in vitro. Correlation with adipose tissue mass, age, and endometrial neoplasia. J Clin Endocrinol Metab. 1981 Jul;53(1):192–9. PubMed PMID: 7240376. Epub 1981/07/01. eng.

[6] Brucker MC, Likis FE. Steroid hormones. In: King TL, Brucker MC (eds.) Pharmacology for Women's Health. Sudbury: Jones and Bartlett Publishers, LLC; 2011. pp. 366–71.

[7] Isayeva GS, Martynenko AV, Beloded OA, Struk TyA, Vovchenko MV, Maloy L. Effect of age, sex hormones and aldosterone on SCORE in perimenopausal women. Life Sci J. 2015;12(1s):44–49.

[8] Martin B, Rotten D, Jolivet A, Gautray JP. Binding of steroids by proteins in follicular fluid of the human ovary. J Clin Endocrinol Metab. 1981 Aug;53(2):443–7. PubMed PMID: 7195910. Epub 1981/08/01. eng.

[9] Siiteri PK, Murai JT, Hammond GL, Nisker JA, Raymoure WJ, Kuhn RW. The serum transport of steroid hormones. Recent Prog Horm Res. 1982;38:457–510. PubMed PMID: 6750727. Epub 1982/01/01. eng.

[10] Simpson E, McDonald P. Endocrinology of pregnancy. Textbook of Endocrinology. Philadelphia: Saunders Company; 1981. pp. 412–22.

[11] Kuhl H. Pharmacokinetics of oestrogens and progestogens. Maturitas. 1990 Sep;12(3): 171–97. PubMed PMID: 2170822. Epub 1990/09/01. eng.

[12] Mahesh VB, Brann DW. Regulation of the preovulatory gonadotropin surge by endogenous steroids. Steroids. 1998 Dec;63(12):616–29. PubMed PMID: 9870258. Epub 1998/12/31. eng.

[13] Natraj U, Richards JS. Hormonal regulation, localization, and functional activity of the progesterone receptor in granulosa cells of rat preovulatory follicles. Endocrinology. 1993 Aug;133(2):761–9. PubMed PMID: 8344215. Epub 1993/08/01. eng.

[14] Walentowicz P, Krintus M, Sadlecki P, Grabiec M, Mankowska-Cyl A, Sokup A, et al. Serum inhibin A and inhibin B levels in epithelial ovarian cancer patients. PloS One. 2014;9(3):e90575. PubMed PMID: 24599287. Pubmed Central PMCID: PMC3944095. Epub 2014/03/07. eng.

[15] Luisi S, Florio P, Reis FM, Petraglia F. Inhibins in female and male reproductive physiology: role in gametogenesis, conception, implantation and early pregnancy. Hum Reprod Update. 2005 Mar–Apr;11(2):123–35. PubMed PMID: 15618291. Epub 2004/12/25. eng.

[16] Richardson SJ, Nelson JF. Follicular depletion during the menopausal transition. Ann N Y Acad Sci. 1990;592:13–20; discussion 44–51. PubMed PMID: 2197939. Epub 1990/01/01. eng.

[17] Lee SJ, Lenton EA, Sexton L, Cooke ID. The effect of age on the cyclical patterns of plasma LH, FSH, oestradiol and progesterone in women with regular menstrual cycles. Hum Reprod. 1988 Oct;3(7):851–5. PubMed PMID: 3141454. Epub 1988/10/01. eng.

[18] Sherman BM, West JH, Korenman SG. The menopausal transition: analysis of LH, FSH, estradiol, and progesterone concentrations during menstrual cycles of older women. J Clin Endocrinol Metab. 1976 Apr;42(4):629–36. PubMed PMID: 1262439. Epub 1976/04/01. eng.

[19] Rannevik G, Jeppsson S, Johnell O, Bjerre B, Laurell-Borulf Y, Svanberg L. A longitudinal study of the perimenopausal transition: altered profiles of steroid and pituitary hormones, SHBG and bone mineral density. Maturitas. 1995 Feb;21(2):103–13. PubMed PMID: 7752947. Epub 1995/02/01. eng.

[20] Burger HG, Dudley EC, Hopper JL, Groome N, Guthrie JR, Green A, et al. Prospectively measured levels of serum follicle-stimulating hormone, estradiol, and the dimeric inhibins during the menopausal transition in a population-based cohort of women. J Clin Endocrinol Metab. 1999 Nov;84(11):4025–30. PubMed PMID: 10566644. Epub 1999/11/24. eng.

[21] Metcalf MG, Donald RA, Livesey JH. Pituitary-ovarian function in normal women during the menopausal transition. Clin Endocrinol (Oxf). 1981 Mar;14(3):245–55. PubMed PMID: 6790204. Epub 1981/03/01. eng.

[22] Hee J, MacNaughton J, Bangah M, Burger HG. Perimenopausal patterns of gonadotrophins, immunoreactive inhibin, oestradiol and progesterone. Maturitas. 1993 Dec; 18(1):9–20. PubMed PMID: 8107620. Epub 1993/12/01. eng.

[23] Burger HG. Diagnostic role of follicle-stimulating hormone (FSH) measurements during the menopausal transition – an analysis of FSH, oestradiol and inhibin. Eur J Endocrinol. 1994 Jan;130(1):38–42. PubMed PMID: 8124478. Epub 1994/01/01. eng.

[24] Cauley JA, Gutai JP, Kuller LH, LeDonne D, Powell JG. The epidemiology of serum sex hormones in postmenopausal women. Am J Epidemiol. 1989 Jun;129(6):1120–31. PubMed PMID: 2729251. Epub 1989/06/01. eng.

[25] Meldrum DR, Davidson BJ, Tataryn IV, Judd HL. Changes in circulating steroids with aging in postmenopausal women. Obstet Gynecol. 1981 May;57(5):624–8. PubMed PMID: 7219911. Epub 1981/05/01. eng.

[26] Siiteri PK. Adipose tissue as a source of hormones. Am J Clin Nutr. 1987 Jan;45(1 Suppl):277–82. PubMed PMID: 3541569. Epub 1987/01/01. eng.

[27] Longcope C, Franz C, Morello C, Baker R, Johnston CC, Jr. Steroid and gonadotropin levels in women during the peri-menopausal years. Maturitas. 1986 Oct;8(3):189–96. PubMed PMID: 3097458. Epub 1986/10/01. eng.

[28] Longcope C. Hormone dynamics at the menopause. Ann N Y Acad Sci. 1990;592:21–30; discussion 44–51. PubMed PMID: 2375582. Epub 1990/01/01. eng.

[29] Beckmann CRB, Ling FW, Herbert WNP, Laube DW, Smith RP, Casanova R, Chuang A, Goepfert AR, Hueppchen NA, Weiss PM. Cancer of the uterine corpus. In: Obstetrics and Gynecology. Philadelphia: Wolters Kluwer Health/Lippincott Williams & Wilkins; 2014. pp. 427–34.

[30] Jemal A, Siegel R, Xu J, Ward E. Cancer statistics, 2010. CA Cancer J Clin. 2010 Sep–Oct;60(5):277–300. PubMed PMID: 20610543. Epub 2010/07/09. eng.

[31] Nair A, Taylor H. The mechanism of menstruation. In: Santoro NF, Neal-Perry G (eds.) Amenorrhea. Contemporary Endocrinology. Totowa: Humana Press; 2010. pp. 21–34.

[32] Kurman R, Norris H. Endometrial hyperplasia and related cellular changes. In: Kurman R (ed.) Blaustein's Pathology of the Female Genital Tract. New York: Springer; 1994. pp. 411–37.

[33] Silverberg SG, Kurman RJ, Nogales F, Mutter GL, Kubik-Huch RA, Tavassoli FA. Tumors of the uterine corpus: epithelial tumors and related lesions. In: Tavassoli FA, Devilee P, (eds.) Pathology and Genetics of Tumours of the Breast and Female Genital Organs (WHO Classification of Tumours), Lyon, France: IARC Press; 2003. pp. 222–32.

[34] Mazur MT. Endometrial hyperplasia/adenocarcinoma. a conventional approach. Ann Diagn Pathol. 2005 Jun; 9(3):174–81. PubMed PMID: 15944963. Epub 2005/06/10. eng.

[35] Horn LC, Meinel A, Handzel R, Einenkel J. Histopathology of endometrial hyperplasia and endometrial carcinoma: an update. Ann Diagn Pathol. 2007 Aug;11(4):297–311. PubMed PMID: 17630117. Epub 2007/07/17. eng.

[36] Ellenson LH, Ronnett BM, Kurman RJ. Precursor lesions of endometrial carcinoma. Blaustein's Pathology of the Female Genital Tract. New York: Springer; 2011. pp. 359–91.

[37] Dietel M. The histological diagnosis of endometrial hyperplasia. Virchows Arch. 2001;439(5):604–8.

[38] Lacey JV, Jr., Chia VM. Endometrial hyperplasia and the risk of progression to carcinoma. Maturitas. 2009 May 20;63(1):39–44. PubMed PMID: 19285814. Epub 2009/03/17. eng.

[39] Trimble CL, Kauderer J, Zaino R, Silverberg S, Lim PC, Burke JJ, et al. Concurrent endometrial carcinoma in women with a biopsy diagnosis of atypical endometrial hyperplasia. Cancer. 2006;106(4):812–9.

[40] Horn LC, Schnurrbusch U, Bilek K, Hentschel B, Einenkel J. Risk of progression in complex and atypical endometrial hyperplasia: clinicopathologic analysis in cases with and without progestogen treatment. Int J Gynecol Cancer. 2004 Mar–Apr;14(2): 348–53. PubMed PMID: 15086736. Epub 2004/04/17. eng.

[41] Wheeler DT, Bristow RE, Kurman RJ. Histologic alterations in endometrial hyperplasia and well-differentiated carcinoma treated with progestins. Am J Surg Pathol. 2007 Jul;31(7):988–98. PubMed PMID: 17592264. Epub 2007/06/27. eng.

[42] Kurman RJ, Kaminski PF, Norris HJ. The behavior of endometrial hyperplasia. A long-term study of "untreated" hyperplasia in 170 patients. Cancer. 1985 Jul 15;56(2): 403–12. PubMed PMID: 4005805. Epub 1985/07/15. eng.

[43] Mutter GL. Endometrial intraepithelial neoplasia (EIN): will it bring order to chaos? The Endometrial Collaborative Group. Gynecol Oncol. 2000 Mar;76(3):287–90. PubMed PMID: 10684697. Epub 2000/02/24. eng.

[44] Baak JP, Orbo A, van Diest PJ, Jiwa M, de Bruin P, Broeckaert M, et al. Prospective multicenter evaluation of the morphometric D-score for prediction of the outcome of endometrial hyperplasias. Am J Surg Pathol. 2001 Jul;25(7):930–5. PubMed PMID: 11420465. Epub 2001/06/23. eng.

[45] Mutter GL, Baak JP, Crum CP, Richart RM, Ferenczy A, Faquin WC. Endometrial precancer diagnosis by histopathology, clonal analysis, and computerized morphometry. J Pathol. 2000 Mar;190(4):462–9. PubMed PMID: 10699996. Epub 2000/03/04. eng.

[46] Reed SD, Newton KM, Clinton WL, Epplein M, Garcia R, Allison K, et al. Incidence of endometrial hyperplasia. Am J Obstet Gynecol. 2009 Jun;200(6):678 e1–6. PubMed PMID: 19393600. Pubmed Central PMCID: PMC2692753. Epub 2009/04/28. eng.

[47] Moore E, Shafi M. Endometrial hyperplasia. Obstet Gynaecol Reprod Med. 2013;23(3):88–93.

[48] Cramer DW, Finn OJ. Epidemiologic perspective on immune-surveillance in cancer. Curr Opin Immunol. 2011 Apr;23(2):265–71. PubMed PMID: 21277761. Pubmed Central PMCID: PMC3073666. Epub 2011/02/01. eng.

[49] Palmer JE, Perunovic B, Tidy JA. Endometrial hyperplasia. The Obstetrician & Gynaecologist. 2008;10(4):211–6.

[50] Farquhar CM, Lethaby A, Sowter M, Verry J, Baranyai J. An evaluation of risk factors for endometrial hyperplasia in premenopausal women with abnormal menstrual bleeding. Am J Obstet Gynecol. 1999 Sep;181(3):525–9. PubMed PMID: 10486458. Epub 1999/09/16. eng.

[51] Epplein M, Reed SD, Voigt LF, Newton KM, Holt VL, Weiss NS. Risk of complex and atypical endometrial hyperplasia in relation to anthropometric measures and reproductive history. Am J Epidemiol. 2008 Sep 15;168(6):563–70; discussion 71–6. PubMed PMID: 18682485. Pubmed Central PMCID: PMC2727194. Epub 2008/08/07. eng.

[52] Ehrlich CE, Young PC, Stehman FB, Sutton GP, Alford WM. Steroid receptors and clinical outcome in patients with adenocarcinoma of the endometrium. Am J Obstet Gynecol. 1988 Apr;158(4):796–807. PubMed PMID: 2966586. Epub 1988/04/01. eng.

[53] Murphy LJ, Gong Y, Murphy LC, Bhavnani B. Growth factors in normal and malignant uterine tissue. Ann N Y Acad Sci. 1991;622:383–91. PubMed PMID: 2064196. Epub 1991/01/01. eng.

[54] Anzai Y, Gong Y, Holinka CF, Murphy LJ, Murphy LC, Kuramoto H, et al. Effects of transforming growth factors and regulation of their mRNA levels in two human en-

dometrial adenocarcinoma cell lines. J Steroid Biochem Mol Biol. 1992 Jun;42(5):449–55. PubMed PMID: 1616874. Epub 1992/06/01. eng.

[55] Boyd JA, Kaufman DG. Expression of transforming growth factor beta 1 by human endometrial carcinoma cell lines: inverse correlation with effects on growth rate and morphology. Cancer Res. 1990 Jun 1;50(11):3394–9. PubMed PMID: 2334934. Epub 1990/06/01. eng.

[56] Salvesen HB, Akslen LA. Molecular pathogenesis and prognostic factors in endometrial carcinoma. APMIS. 2002 Oct;110(10):673–89. PubMed PMID: 12583434. Epub 2003/02/14. eng.

[57] Inoue M. Current molecular aspects of the carcinogenesis of the uterine endometrium. Int J Gynecol Cancer. 2001 Sep–Oct;11(5):339–48. PubMed PMID: 11737463. Epub 2001/12/12. eng.

[58] Dobrzycka B, Terlikowski SJ, Mazurek A, Kowalczuk O, Niklinska W, Chyczewski L, et al. Mutations of the KRAS oncogene in endometrial hyperplasia and carcinoma. Folia Histochem Cytobiol. 2009;47(1):65–8. PubMed PMID: 19419940. Epub 2009/05/08. eng.

[59] Berchuck A, Boyd J. Molecular basis of endometrial cancer. Cancer. 1995 Nov 15;76(10 Suppl):2034–40. PubMed PMID: 8634996. Epub 1995/11/15. eng.

[60] Saffari B, Jones LA, el-Naggar A, Felix JC, George J, Press MF. Amplification and overexpression of HER-2/neu (c-erbB2) in endometrial cancers: correlation with overall survival. Cancer Res. 1995 Dec 1;55(23):5693–8. PubMed PMID: 7585656. Epub 1995/12/01. eng.

[61] Niederacher D, An HX, Cho YJ, Hantschmann P, Bender HG, Beckmann MW. Mutations and amplification of oncogenes in endometrial cancer. Oncology. 1999;56(1):59–65. PubMed PMID: 9885379. Epub 1999/01/14. eng.

[62] Leiserowitz GS, Harris SA, Subramaniam M, Keeney GL, Podratz KC, Spelsberg TC. The proto-oncogene c-fms is overexpressed in endometrial cancer. Gynecol Oncol. 1993 May;49(2):190–6. PubMed PMID: 8504987. Epub 1993/05/01. eng.

[63] Tashiro H, Blazes MS, Wu R, Cho KR, Bose S, Wang SI, et al. Mutations in PTEN are frequent in endometrial carcinoma but rare in other common gynecological malignancies. Cancer Res. 1997 Sep 15;57(18):3935–40. PubMed PMID: 9307275. Epub 1997/10/27. eng.

[64] Levine RL, Cargile CB, Blazes MS, van Rees B, Kurman RJ, Ellenson LH. PTEN mutations and microsatellite instability in complex atypical hyperplasia, a precursor lesion to uterine endometrioid carcinoma. Cancer Res. 1998 Aug 1;58(15):3254–8. PubMed PMID: 9699651. Epub 1998/08/12. eng.

[65] Maxwell GL, Risinger JI, Gumbs C, Shaw H, Bentley RC, Barrett JC, et al. Mutation of the PTEN tumor suppressor gene in endometrial hyperplasias. Cancer Res. 1998 Jun 15;58(12):2500–3. PubMed PMID: 9635567. Epub 1998/07/04. eng.

[66] Di Cristofano A, Pandolfi PP. The multiple roles of PTEN in tumor suppression. Cell. 2000 Feb 18;100(4):387–90. PubMed PMID: 10693755. Epub 2000/02/29. eng.

[67] Georgescu MM. PTEN tumor suppressor network in PI3K-Akt pathway control. Genes Cancer. 2010 Dec;1(12):1170–7. PubMed PMID: 21779440. Pubmed Central PMCID: PMC3092286. Epub 2011/07/23. eng.

[68] Ellenson LH, Wu TC. Focus on endometrial and cervical cancer. Cancer Cell. 2004 Jun;5(6):533–8. PubMed PMID: 15193256. Epub 2004/06/15. eng.

[69] Terakawa N, Kanamori Y, Yoshida S. Loss of PTEN expression followed by Akt phosphorylation is a poor prognostic factor for patients with endometrial cancer. Endocr Relat Cancer. 2003 Jun;10(2):203–8. PubMed PMID: 12790783. Epub 2003/06/07. eng.

[70] Djordjevic B, Hennessy BT, Li J, Barkoh BA, Luthra R, Mills GB, et al. Clinical assessment of PTEN loss in endometrial carcinoma: immunohistochemistry outperforms gene sequencing. Mod Pathol. 2012 May;25(5):699–708. PubMed PMID: 22301702. Pubmed Central PMCID: PMC3341518. Epub 2012/02/04. eng.

[71] Lax SF, Kendall B, Tashiro H, Slebos RJ, Hedrick L. The frequency of p53, K-ras mutations, and microsatellite instability differs in uterine endometrioid and serous carcinoma: evidence of distinct molecular genetic pathways. Cancer. 2000 Feb 15;88(4): 814–24. PubMed PMID: 10679651. Epub 2000/02/19. eng.

[72] Kaaks R, Lukanova A, Kurzer MS. Obesity, endogenous hormones, and endometrial cancer risk: a synthetic review. Cancer Epidemiol Biomarkers Prev. 2002 Dec;11(12): 1531–43. PubMed PMID: 12496040. Epub 2002/12/24. eng.

[73] Jacks T, Weinberg RA. Cell-cycle control and its watchman. Nature. 1996 Jun 20;381(6584):643–4. PubMed PMID: 8649505. Epub 1996/06/20. eng.

[74] Ito K, Sasano H, Matsunaga G, Sato S, Yajima A, Nasim S, et al. Correlations between p21 expression and clinicopathological findings, p53 gene and protein alterations, and survival in patients with endometrial carcinoma. J Pathol. 1997 Nov;183(3):318–24. PubMed PMID: 9422988. Epub 1998/01/10. eng.

[75] Palazzo JP, Mercer WE, Kovatich AJ, McHugh M. Immunohistochemical localization of p21(WAF1/CIP1) in normal, hyperplastic, and neoplastic uterine tissues. Hum Pathol. 1997 Jan;28(1):60–6. PubMed PMID: 9013833. Epub 1997/01/01. eng.

[76] Salvesen HB, Iversen OE, Akslen LA. Prognostic significance of angiogenesis and Ki-67, p53, and p21 expression: a population-based endometrial carcinoma study. J Clin Oncol. 1999 May;17(5):1382–90. PubMed PMID: 10334522. Epub 1999/05/20. eng.

[77] Peiffer SL, Bartsch D, Whelan AJ, Mutch DG, Herzog TJ, Goodfellow PJ. Low frequency of CDKN2 mutation in endometrial carcinomas. Mol Carcinog. 1995 Aug; 13(4):210–2. PubMed PMID: 7646759. Epub 1995/08/01. eng.

[78] Shiozawa T, Nikaido T, Shimizu M, Zhai Y, Fujii S. Immunohistochemical analysis of the expression of cdk4 and p16INK4 in human endometrioid-type endometrial carcinoma. Cancer. 1997 Dec 15;80(12):2250–6. PubMed PMID: 9404701. Epub 1997/12/24. eng.

[79] Nakashima R, Fujita M, Enomoto T, Haba T, Yoshino K, Wada H, et al. Alteration of p16 and p15 genes in human uterine tumours. Br J Cancer. 1999 May;80(3–4):458–67. PubMed PMID: 10408854. Pubmed Central PMCID: PMC2362344. Epub 1999/07/17. eng.

[80] Salvesen HB, Das S, Akslen LA. Loss of nuclear p16 protein expression is not associated with promoter methylation but defines a subgroup of aggressive endometrial carcinomas with poor prognosis. Clin Cancer Res. 2000 Jan;6(1):153–9. PubMed PMID: 10656444. Epub 2000/02/03. eng.

[81] Masaki T. The endothelin family: an overview. J Cardiovasc Pharmacol. 2000;35(4 Suppl 2):S3–5. PubMed PMID: 10976772. Epub 2000/09/08. eng.

[82] Bagnato A, Spinella F, Rosano L. The endothelin axis in cancer: the promise and the challenges of molecularly targeted therapy. Can J Physiol Pharmacol. 2008 Aug;86(8): 473–84. PubMed PMID: 18758494. Epub 2008/09/02. eng.

[83] Tsai KW, Hu LY, Chen TW, Li SC, Ho MR, Yu SY, et al. Emerging role of microRNAs in modulating endothelin-1 expression in gastric cancer. Oncol Rep. 2015 Jan;33(1): 485–93. PubMed PMID: 25394359. Epub 2014/11/15. eng.

[84] Gurin CC, Federici MG, Kang L, Boyd J. Causes and consequences of microsatellite instability in endometrial carcinoma. Cancer Res. 1999 Jan 15;59(2):462–6. PubMed PMID: 9927063. Epub 1999/02/02. eng.

[85] Peltomaki P, Vasen HF. Mutations predisposing to hereditary nonpolyposis colorectal cancer: database and results of a collaborative study. The International Collaborative Group on Hereditary Nonpolyposis Colorectal Cancer. Gastroenterology. 1997 Oct;113(4):1146–58. PubMed PMID: 9322509. Epub 1997/10/10. eng.

[86] Wijnen J, de Leeuw W, Vasen H, van der Klift H, Moller P, Stormorken A, et al. Familial endometrial cancer in female carriers of MSH6 germline mutations. Nat Genet. 1999 Oct;23(2):142–4. PubMed PMID: 10508506. Epub 1999/10/03. eng.

[87] Sobczuk A, Romanowicz-Makowska H, Smolarz B, Pertynski T. Microsatellite instability (MSI) and MLH1 and MSH2 protein expression analysis in postmenopausal women with sporadic endometrial cancer. J Exp Clin Cancer Res. 2007 Sep;26(3):369–74. PubMed PMID: 17987798. Epub 2007/11/09. eng.

[88] Esteller M, Levine R, Baylin SB, Ellenson LH, Herman JG. MLH1 promoter hypermethylation is associated with the microsatellite instability phenotype in sporadic endo-

metrial carcinomas. Oncogene. 1998 Nov 5;17(18):2413–7. PubMed PMID: 9811473. Epub 1998/11/12. eng.

[89] Nishikawa T, Edelstein D, Du XL, Yamagishi S, Matsumura T, Kaneda Y, et al. Normalizing mitochondrial superoxide production blocks three pathways of hyperglycaemic damage. Nature. 2000 Apr 13;404(6779):787–90. PubMed PMID: 10783895. Epub 2000/04/28. eng.

[90] Kuppusamy P. EPR spectroscopy in biology and medicine. Antioxid Redox Signal. 2004 Jun;6(3):583–5. PubMed PMID: 15130284. Epub 2004/05/08. eng.

[91] Babior BM. The NADPH oxidase of endothelial cells. IUBMB Life. 2000 Oct–Nov; 50(4–5):267–9. PubMed PMID: 11327320. Epub 2001/05/01. eng.

[92] Vignais PV. The superoxide-generating NADPH oxidase: structural aspects and activation mechanism. Cell Mol Life Sci. 2002 Sep;59(9):1428–59. PubMed PMID: 12440767. Epub 2002/11/21. eng.

[93] Cadenas E, Davies KJ. Mitochondrial free radical generation, oxidative stress, and aging. Free Radic Biol Med. 2000 Aug;29(3–4):222–30. PubMed PMID: 11035250. Epub 2000/10/18. eng.

[94] Reddy JK, Hashimoto T. Peroxisomal beta-oxidation and peroxisome proliferator-activated receptor alpha: an adaptive metabolic system. Annu Rev Nutr. 2001;21:193–230. PubMed PMID: 11375435. Epub 2001/05/26. eng.

[95] Kettle AJ, Winterbourn CC. Superoxide modulates the activity of myeloperoxidase and optimizes the production of hypochlorous acid. Biochem J. 1988 Jun 1;252(2): 529–36. PubMed PMID: 2843172. Pubmed Central PMCID: PMC1149176. Epub 1988/06/01. eng.

[96] Sies H. Oxidative stress: introduction. In: Sies H (ed.) Oxidants and Antioxidants. London: Academic Press; 1991. pp. 1–8.

[97] Porter NA. Chemistry of lipid peroxidation. Methods Enzymol. 1984;105:273–82. PubMed PMID: 6727666. Epub 1984/01/01. eng.

[98] Chakraborty P, Kumar S, Dutta D, Gupta V. Role of antioxidants in common health diseases. Res J Pharm Tech. 2009;1:239–44.

[99] Aherne SA, O'Brien NM. Dietary flavonols: chemistry, food content, and metabolism. Nutrition. 2002 Jan;18(1):75–81. PubMed PMID: 11827770. Epub 2002/02/06. eng.

[100] Stocker R, Bowry VW, Frei B. Ubiquinol-10 protects human low density lipoprotein more efficiently against lipid peroxidation than does alpha-tocopherol. Proc Natl Acad Sci U S A. 1991 Mar 1;88(5):1646–50. PubMed PMID: 2000375. Pubmed Central PMCID: PMC51081. Epub 1991/03/01. eng.

[101] Hasnain SS, Strange RW. Marriage of XAFS and crystallography for structure-function studies of metalloproteins. J Synchrotron Radiat. 2003 Jan 1;10(Pt 1):9–15. PubMed PMID: 12511785. Epub 2003/01/04. eng.

[102] Marklund SL. Extracellular superoxide dismutase in human tissues and human cell lines. J Clin Invest. 1984 Oct;74(4):1398–403. PubMed PMID: 6541229. Pubmed Central PMCID: PMC425307. Epub 1984/10/01. eng.

[103] Kirkman HN, Rolfo M, Ferraris AM, Gaetani GF. Mechanisms of protection of catalase by NADPH. Kinetics and stoichiometry. J Biol Chem. 1999 May 14;274(20): 13908–14. PubMed PMID: 10318800. Epub 1999/05/13. eng.

[104] Brigelius-Flohe R. Glutathione peroxidases and redox-regulated transcription factors. Biol Chem. 2006 Oct–Nov;387(10–11):1329–35. PubMed PMID: 17081103. Epub 2006/11/04. eng.

[105] Ursini F, Maiorino M, Brigelius-Flohe R, Aumann KD, Roveri A, Schomburg D, et al. Diversity of glutathione peroxidases. Methods Enzymol. 1995;252:38–53. PubMed PMID: 7476373. Epub 1995/01/01. eng.

[106] Go Y-M, Jones DP. Redox control systems in the nucleus: mechanisms and functions. Antioxid Redox Signal. 2010;13(4):489–509.

[107] Hoidal JR. Reactive oxygen species and cell signaling. Am J Respir Cell Mol Biol. 2001 Dec;25(6):661–3. PubMed PMID: 11726388. Epub 2001/12/01. eng.

[108] Dalton TP, Shertzer HG, Puga A. Regulation of gene expression by reactive oxygen. Annu Rev Pharmacol Toxicol. 1999;39:67–101. PubMed PMID: 10331077. Epub 1999/05/20. eng.

[109] Reth M. Hydrogen peroxide as second messenger in lymphocyte activation. Nat Immunol. 2002 Dec;3(12):1129–34. PubMed PMID: 12447370. Epub 2002/11/26. eng.

[110] Devary Y, Gottlieb RA, Lau LF, Karin M. Rapid and preferential activation of the c-jun gene during the mammalian UV response. Mol Cell Biol. 1991 May;11(5):2804–11. PubMed PMID: 1901948. Pubmed Central PMCID: PMC360059. Epub 1991/05/01. eng.

[111] Hsu W, Kerppola TK, Chen PL, Curran T, Chen-Kiang S. Fos and Jun repress transcription activation by NF-IL6 through association at the basic zipper region. Mol Cell Biol. 1994 Jan;14(1):268–76. PubMed PMID: 8264594. Pubmed Central PMCID: PMC358376. Epub 1994/01/01. eng.

[112] Grether-Beck S, Olaizola-Horn S, Schmitt H, Grewe M, Jahnke A, Johnson JP, et al. Activation of transcription factor AP-2 mediates UVA radiation- and singlet oxygen-induced expression of the human intercellular adhesion molecule 1 gene. Proc Natl Acad Sci U S A. 1996 Dec 10;93(25):14586–91. PubMed PMID: 8962096. Pubmed Central PMCID: PMC26177. Epub 1996/12/10. eng.

[113] Muller JM, Ziegler-Heitbrock HW, Baeuerle PA. Nuclear factor kappa B, a mediator of lipopolysaccharide effects. Immunobiology. 1993 Apr;187(3–5):233–56. PubMed PMID: 8330898. Epub 1993/04/01. eng.

[114] Yang Y, Cheng JZ, Singhal SS, Saini M, Pandya U, Awasthi S, et al. Role of glutathione S-transferases in protection against lipid peroxidation. Overexpression of hGSTA2-2 in K562 cells protects against hydrogen peroxide-induced apoptosis and inhibits JNK and caspase 3 activation. J Biol Chem. 2001 Jun 1;276(22):19220–30. PubMed PMID: 11279091. Epub 2001/03/30. eng.

[115] Pahl HL, Baeuerle PA. Oxygen and the control of gene expression. Bioessays. 1994 Jul;16(7):497–502. PubMed PMID: 7945278. Epub 1994/07/01. eng.

[116] Suzuki YJ, Forman HJ, Sevanian A. Oxidants as stimulators of signal transduction. Free Radic Biol Medic. 1997;22(1–2):269–85. PubMed PMID: 8958153. Epub 1997/01/01. eng.

[117] Pitot HC, Goldsworthy T, Moran S. The natural history of carcinogenesis: implications of experimental carcinogenesis in the genesis of human cancer. J Supramol Struct Cell Biochem. 1981;17(2):133–46. PubMed PMID: 7033553. Epub 1981/01/01. eng.

[118] Ray G, Husain SA. Oxidants, antioxidants and carcinogenesis. Indian J Exp Biol. 2002 Nov;40(11):1213–32. PubMed PMID: 13677623. Epub 2003/09/19. eng.

[119] Lenzen S, Drinkgern J, Tiedge M. Low antioxidant enzyme gene expression in pancreatic islets compared with various other mouse tissues. Free Radic Biol Medic. 1996;20(3):463–6. PubMed PMID: 8720919. Epub 1996/01/01. eng.

[120] Surai P. Tissue-specific changes in the activities of antioxidant enzymes during the development of the chicken embryo. Br Poult Sci. 1999;40(3):397–405.

[121] Mailloux RJ, Jin X, Willmore WG. Redox regulation of mitochondrial function with emphasis on cysteine oxidation reactions. Redox Biol. 2014;2:123–39. PubMed PMID: 24455476. Pubmed Central PMCID: PMC3895620. Epub 2014/01/24. eng.

[122] Kasapovic J, Pejic S, Todorovic A, Stojiljkovic V, Pajovic SB. Antioxidant status and lipid peroxidation in the blood of breast cancer patients of different ages. Cell Biochem Funct. 2008 Aug;26(6):723–30. PubMed PMID: 18636415. Epub 2008/07/19. eng.

[123] Kasapovic J, Pejic S, Stojiljkovic V, Todorovic A, Radosevic-Jelic L, Saicic ZS, et al. Antioxidant status and lipid peroxidation in the blood of breast cancer patients of different ages after chemotherapy with 5-fluorouracil, doxorubicin and cyclophosphamide. Clin Biochem. 2010 Nov;43(16–17):1287–93. PubMed PMID: 20713039. Epub 2010/08/18. eng.

[124] Hileman EO, Liu J, Albitar M, Keating MJ, Huang P. Intrinsic oxidative stress in cancer cells: a biochemical basis for therapeutic selectivity. Cancer Chemother Pharmacol. 2004 Mar;53(3):209–19. PubMed PMID: 14610616. Epub 2003/11/12. eng.

[125] Muehlematter D, Larsson R, Cerutti P. Active oxygen induced DNA strand breakage and poly ADP-ribosylation in promotable and non-promotable JB6 mouse epidermal cells. Carcinogenesis. 1988 Feb;9(2):239–45. PubMed PMID: 3338107. Epub 1988/02/01. eng.

[126] Chiou JF, Hu ML. Elevated lipid peroxidation and disturbed antioxidant enzyme activities in plasma and erythrocytes of patients with uterine cervicitis and myoma. Clin Biochem. 1999 Apr;32(3):189–92. PubMed PMID: 10383079. Epub 1999/06/26. eng.

[127] Manoharan S, Kolanjiappan K, Kayalvizhi M. Enhanced lipid peroxidation and impaired enzymic antioxidant activities in the erythrocytes of patients with cervical carcinoma. Cell Mol Biol Lett. 2004;9(4A):699–707. PubMed PMID: 15647792. Epub 2005/01/14. eng.

[128] Kolanjiappan K, Manoharan S, Kayalvizhi M. Measurement of erythrocyte lipids, lipid peroxidation, antioxidants and osmotic fragility in cervical cancer patients. Clin Chim Acta. 2002 Dec;326(1–2):143–9. PubMed PMID: 12417105. Epub 2002/11/06. eng.

[129] Pejic S, Kasapovic J, Todorovic A, Stojiljkovic V, Pajovic SB. Lipid peroxidation and antioxidant status in blood of patients with uterine myoma, endometrial polypus, hyperplastic and malignant endometrium. Biol Res. 2006;39(4):619–29.

[130] Pajovic S, Saicic Z, Pejic S, Kasapovic J, Stojiljkovic V, Kanazir D. Antioxidative biomarkers and carcinogenesis. Jugoslov Med Biochem. 2006;25:397–402.

[131] Pejic S, Todorovic A, Stojiljkovic V, Cvetkovic D, Lucic N, Radojicic RM, et al. Superoxide dismutase and lipid hydroperoxides in blood and endometrial tissue of patients with benign, hyperplastic and malignant endometrium. An Acad Bras Cienc. 2008 Sep;80(3):515–22. PubMed PMID: 18797802. Epub 2008/09/18. eng.

[132] [132] Pejic S, Todorovic A, Stojiljkovic V, Kasapovic J, Pajovic SB. Antioxidant enzymes and lipid peroxidation in endometrium of patients with polyps, myoma, hyperplasia and adenocarcinoma. Reprod Biol Endocrinol. 2009;7:149. PubMed PMID: 20030853. Pubmed Central PMCID: PMC2805669. Epub 2009/12/25. eng.

[133] [133] Pejic S, Stojiljkovic V, Todorovic A, Gavrilovic L, Popovic N, Pavlovic I, Pajovic SB. Antioxidant status in blood of gynecological patients: influence of diagnosis and reproductive factors. Folia Biol (Praha) 2015; 61(1): 26–32.

[134] Sugino N, Shimamura K, Takiguchi S, Tamura H, Ono M, Nakata M, et al. Changes in activity of superoxide dismutase in the human endometrium throughout the menstrual cycle and in early pregnancy. Hum Reprod. 1996 May;11(5):1073–8. PubMed PMID: 8671393. Epub 1996/05/01. eng.

[135] Massafra C, Gioia D, De Felice C, Picciolini E, De Leo V, Bonifazi M, et al. Effects of estrogens and androgens on erythrocyte antioxidant superoxide dismutase, catalase

and glutathione peroxidase activities during the menstrual cycle. J Endocrinol. 2000 Dec;167(3):447–52. PubMed PMID: 11115771. Epub 2000/12/15. eng.

[136] Gurdol F, Oner-Yyidothan Y, Yalcyn O, Genc S, Buyru F. Changes in enzymatic antioxidant defense system in blood and endometrial tissues of women after menopause. Res Commun Mol Pathol Pharmacol. 1997 Jul;97(1):38–46. PubMed PMID: 9507566. Epub 1997/07/01. eng.

[137] Bednarek-Tupikowska G, Tworowska U, Jedrychowska I, Radomska B, Tupikowski K, Bidzinska-Speichert B, et al. Effects of oestradiol and oestroprogestin on erythrocyte antioxidative enzyme system activity in postmenopausal women. Clin Endocrinol (Oxf). 2006 Apr;64(4):463–8. PubMed PMID: 16584521. Epub 2006/04/06. eng.

[138] Clemente C, Caruso MG, Berloco P, Notarnicola M, D'Attoma B, Osella AR, et al. Antioxidant effect of short-term hormonal treatment in postmenopausal women. Maturitas. 1999 Jan 4;31(2):137–42. PubMed PMID: 10227007. Epub 1999/05/05. eng.

[139] Unfer TC, Conterato GM, da Silva JC, Duarte MM, Emanuelli T. Influence of hormone replacement therapy on blood antioxidant enzymes in menopausal women. Clin Chim Acta. 2006 Jul 15;369(1):73–7. PubMed PMID: 16472795. Epub 2006/02/14. eng.

[140] Pejic SA, Kasapovic JD, Todorovic AU, Stojiljkovic VR, Gavrilovic LV, Popovic NM, et al. Antioxidant enzymes in women with endometrial polyps: relation with sex hormones. Eur J Obstet Gynecol Reprod Biol. 2013 Sep;170(1):241–6. PubMed PMID: 23871381. Epub 2013/07/23. eng.

[141] Lowry OH, Rosebrough NJ, Farr AL, Randall RJ. Protein measurement with the Folin phenol reagent. J Biol Chem. 1951 Nov;193(1):265–75. PubMed PMID: 14907713. Epub 1951/11/01. eng.

[142] Beutler E. Red Cell Metabolism: A Manual of Biochemical Methods. Orlando: Grune & Stratton; 1984.

[143] Bednarek-Tupikowska G, Bohdanowicz-Pawlak A, Bidzinska B, Milewicz A, Antonowicz-Juchniewicz J, Andrzejak R. Serum lipid peroxide levels and erythrocyte glutathione peroxidase and superoxide dismutase activity in premenopausal and postmenopausal women. Gynecol Endocrinol. 2001 Aug;15(4):298–303. PubMed PMID: 11560104. Epub 2001/09/19. eng.

[144] Sabuncu T, Vural H, Harma M, Harma M. Oxidative stress in polycystic ovary syndrome and its contribution to the risk of cardiovascular disease. Clin Biochem. 2001 Jul;34(5):407–13. PubMed PMID: 11522279. Epub 2001/08/28. eng.

[145] Lutoslawska G, Tkaczyk J, Panczenko-Kresowska B, Hubner-Wozniak E, Skierska E, Gajewski AK. Plasma TBARS, blood GSH concentrations, and erythrocyte antioxidant enzyme activities in regularly menstruating women with ovulatory and anovu-

latory menstrual cycles. Clin Chim Acta. 2003 May;331(1–2):159–63. PubMed PMID: 12691877. Epub 2003/04/15. eng.

[146] Bellanti F, Matteo M, Rollo T, De Rosario F, Greco P, Vendemiale G, et al. Sex hormones modulate circulating antioxidant enzymes: impact of estrogen therapy. Redox Biol. 2013;1:340–6. PubMed PMID: 24024169. Pubmed Central PMCID: PMC3757703. Epub 2013/09/12. eng.

[147] Narimoto K, Noda Y, Shiotani M, Tokura T, Goto Y, Takakura K, et al. Immunohistochemical assessment of superoxide dismutase expression in the human endometrium throughout the menstrual cycle. Acta Histochem Cytochem. 1990;23(4):487–98.

[148] Ota H, Igarashi S, Hatazawa J, Tanaka T. Immunohistochemical assessment of superoxide dismutase expression in the endometrium in endometriosis and adenomyosis. Fertil Steril. 1999;72(1):129–34.

[149] Ota H, Igarashi S, Sato N, Tanaka H, Tanaka T. Involvement of catalase in the endometrium of patients with endometriosis and adenomyosis. Fertil Steril. 2002;78(4): 804–9.

[150] Agarwal A, Aponte-Mellado A, Premkumar BJ, Shaman A, Gupta S. The effects of oxidative stress on female reproduction: a review. Reprod Biol Endocrinol. 2012;10:49. PubMed PMID: 22748101. Pubmed Central PMCID: PMC3527168. Epub 2012/07/04. eng.

[151] Liao H, Zhou Q, Zhang Z, Wang Q, Sun Y, Yi X, et al. NRF2 is overexpressed in ovarian epithelial carcinoma and is regulated by gonadotrophin and sex-steroid hormones. Oncol Rep. 2012 Jun;27(6):1918–24. PubMed PMID: 22378150. Epub 2012/03/02. eng.

[152] King ER, Wong K-K. Steroid Hormones and Ovarian Cancer. In: Abduljabbar H (ed.) Steroids-Clinical Aspect (vol ISBN: 978-953-307-705-5). Rijeka: InTech; 2011. pp 111-40.

[153] Ferenczy A, Gelfand M. The biologic significance of cytologic atypia in progestogen-treated endometrial hyperplasia. Am J Obstet Gynecol. 1989 Jan;160(1):126–31. PubMed PMID: 2912075. Epub 1989/01/01. eng.

[154] Gallos ID, Shehmar M, Thangaratinam S, Papapostolou TK, Coomarasamy A, Gupta JK. Oral progestogens vs levonorgestrel-releasing intrauterine system for endometrial hyperplasia: a systematic review and metaanalysis. Am J Obstet Gynecol. 2010 Dec;203(6):547 e1–10. PubMed PMID: 20934679. Epub 2010/10/12. eng.

[155] Yang S, Thiel KW, De Geest K, Leslie KK. Endometrial cancer: reviving progesterone therapy in the molecular age. Discov Med. 2011 Sep;12(64):205–12. PubMed PMID: 21955848. Epub 2011/10/01. eng.

[156] Conn PM, Crowley WF, Jr. Gonadotropin-releasing hormone and its analogs. Annu Rev Med. 1994;45:391–405. PubMed PMID: 8198390. Epub 1994/01/01. eng.

http://dx.doi.org/10.5772/60853

[157] Grimbizis G, Tsalikis T, Tzioufa V, Kasapis M, Mantalenakis S. Regression of endometrial hyperplasia after treatment with the gonadotrophin-releasing hormone analogue triptorelin: a prospective study. Hum Reprod. 1999 Feb;14(2):479–84. PubMed PMID: 10099998. Epub 1999/04/01. eng.

[158] Wang J, Luo F, Lu JJ, Chen PK, Liu P, Zheng W. VEGF expression and enhanced production by gonadotropins in ovarian epithelial tumors. Int J Cancer. 2002;97(2):163–7.

[159] Telikicherla D, Ambekar A, Palapetta SM, Dwivedi SB, Raju R, Sharma J, et al. A comprehensive curated resource for follicle stimulating hormone signaling. BMC Res Notes. 2011;4:408. PubMed PMID: 21996254. Pubmed Central PMCID: PMC3204250. Epub 2011/10/15. eng.

[160] Hoang YD, Nakamura BN, Luderer U. Follicle-stimulating hormone and estradiol interact to stimulate glutathione synthesis in rat ovarian follicles and granulosa cells. Biol Reprod. 2009 Oct;81(4):636–46. PubMed PMID: 19516019. Pubmed Central PMCID: PMC2754881. Epub 2009/06/12. eng.

[161] Zhang Z, Wang Q, Ma J, Yi X, Zhu Y, Xi X, et al. Reactive oxygen species regulate FSH-induced expression of vascular endothelial growth factor via Nrf2 and HIF1 alpha signaling in human epithelial ovarian cancer. Oncol Rep. 2013 Apr;29(4):1429–34. PubMed PMID: 23404377. Epub 2013/02/14. eng.

Section 4

# Oxidative Stress and Therapy

Chapter 10

# Oxidative Stress-Based Photodynamic Therapy with Synthetic Sensitizers and/or Natural Antioxidants

Rodica-Mariana Ion and
Ioana-Raluca Şuică-Bunghez

Additional information is available at the end of the chapter

http://dx.doi.org/10.5772/60872

**Abstract**

Photodynamic therapy (also called PDT, photoirradiation therapy, phototherapy, or photochemotherapy) as a new modality for cancer treatment involves the combination of light with a photosensitizing drug in an oxygen-rich environment. PDT consists in the systemic administration of a photosensitizer (PS), which is selectively retained by tumor cells and the subsequent irradiation with visible light of the affected area. Basically two types of reactions can occur after photoactivation of the photosensitizer: the first involves the generation of free radicals (type-I photochemical reaction) and the other the production of singlet molecular oxygen, (type II) as the main species responsible for cell inactivation. The mechanism of PDT effects may involve a direct tumor cell injury and also an indirect cell killing via microcirculatory changes resulting in reduced blood flow in the tumor. Also, PDT may be considered an oxidative stress that induces cellular death in different types of cancerous cells both in vitro and in vivo. Oxidation or oxidative stress leads to the production of free radicals. Generated reactive oxygen species (ROS) cause oxidative stress in the cells targeting mainly at cellular macromolecules, such as lipids, nucleic acids and proteins. These oxidizing agents can damage cells by starting chain reactions such as lipid peroxidation, or oxidation of proteins or DNA, causing mutations and even major diseases, while protein oxidation can lead to distortion and degradation. To evaluate the presence of oxidative stress in PDT, some methods could be used: detection of malondialdehyde (MDA), the carbonylated proteins (CP), the hydrogen donating capacity (HDA), detection pf the -SH groups. By reducing basal levels of ROS in cells may facilitate the therapeutic effects of oxidative stress-based therapies. Natural antioxidants may also protect healthy tissues and lower the incidence of treatment-related side effects, and the chapter shows some results about hesperidin, hesperitine, diosmin and also, some tea rxtracts based on tea leaves and fruits. The Green tea polyphenols have been shown to have a protective effect in different forms of cancer in a variety of pre-clinical animal models. Green tea is composed of several catechins, including (-)-Epigallocatechin-3-gallate (EGCG), epicatechin (EC), epicatechin-3-gallate (ECG), and epigallocatechin (EGC). Among them, (-)-Epigallocatechin-3-gallate (EGCG), the major catechin found in green tea, has been recognized as a potential therapeutic agent. In the context where many clinical studies with respect to the application of antioxidants as sensitizers are lacking, this chapter shows a systematic review by putting into evidence the antioxidant action in photochemotherapy and their comparison with synthetic sensitizers (porphyrins and phthalocyanines). Some up-to-date results of photodynamic therapy with synthetic sensitizers and/or coupled with some natural antioxidants are shown and discussed.

**Keywords:** photodynamic therapy, sensitizers, natural compounds, antioxidants

## 1. Introduction

Photodynamic therapy (PDT) (also known as photoirradiation therapy, phototherapy or photochemotherapy) is a new modality for cancer treatment and involves the combination of light with a photosensitizing drug in an oxygen-rich environment [1, 2]. The photodynamic therapy has been successfully used in the last decade for the therapy of solid malignant tumors and non-tumoral diseases such as psoriasis, age-related macular degeneration, actinic keratosis, ageing [3]. The PDT mechanism could be a direct one (by apoptosis and necrosis), or an indirect one (by immune response on tumor vasculature). An ideal sensitizer should have a defined pure chemical structure, a proper absorption in red/near-infrared region, and a good capacity of singlet oxygen generation. After irradiation, the photosensitizer (PS) passes into the singlet oxygen excited state and afterwards it can pass into the triplet state; in this state it can react with molecular oxygen, generating singlet oxygen or it can react with different molecules from the tissue generating radical forms of PS [4]. This final form can react with oxygen, leading to the formation of different reactive oxygen species such as the hydroxyl radical, hydrogen peroxide and the superoxide anion, which in their turn may oxidize the macromolecular cellular components, resulting in cellular death through either apoptosis or necrosis. PDT induces oxidative stress at specific subcellular sites through the light activation of organelle-associated photosensitizers, and is used in the destruction of tumor cells.

This chapter offers the most up-to-date results of photodynamic therapy with synthetic sensitizers and/or coupled with some natural antioxidants, by using our data correlated with literature reports.

http://dx.doi.org/10.5772/60872

## 2. General concept of photodynamic therapy

Photodynamic therapy (PDT) is a relatively new type of treatment for cancer, and makes use of a photosensitizer, visible light and molecular oxygen. Reactive oxygen species (ROS) are generated, causing the death of tumor cells by apoptosis or necrosis. Significant research efforts are nowadays focused on finding new photosensitizers with antineoplastic activity.

Over the last decade, there is an increasing interest in photosensitization mechanisms in biological systems, in relation to the therapeutic aspects of this phenomenon [5, 6]. The most used photosensitizers in photodynamic therapy are porphyrins, phthalocyanines and related compounds (Figure 1). These compounds are capable of generating a long-lived triplet excited state, responsible for facile energy transfer to molecular oxygen, leading to the formation of singlet oxygen. The photosensitization reactions include free radical reactions (type I) and singlet oxygen reactions ($^1O_2$) (type II) (Figure 2). The mechanism of PDT may lead to a direct tumor cell injury and also an indirect cell death via microcirculatory changes, resulting in reduced blood flow to the tumor [7-10].

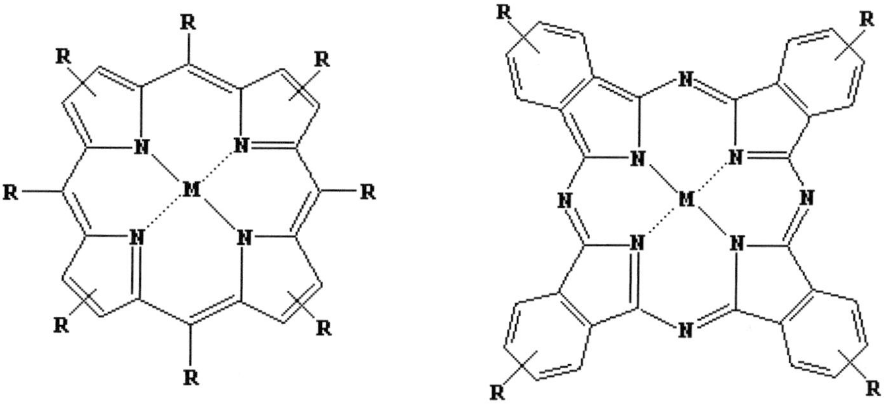


**Figure 1.** The structure of porphyrin (left) and phthalocyanine (right). (R=meso-substituent; M=central metal)

The study of radical-induced damage in living systems is a topic of great interest in biology and medicine. Attention has been increasingly focused on the role of free radicals in normal physiological conditions and in different pathologies, with an essential role in cellular processes resulting in damages [11].

## 3. Mechanisms of photodynamic therapy

### 3.1. Photophysical and photochemical mechanism

Porphyrins are used in medical and biological applications as they can generate new sensitizer structures with extensive and versatile photophysical and photochemical properties. Synthetic

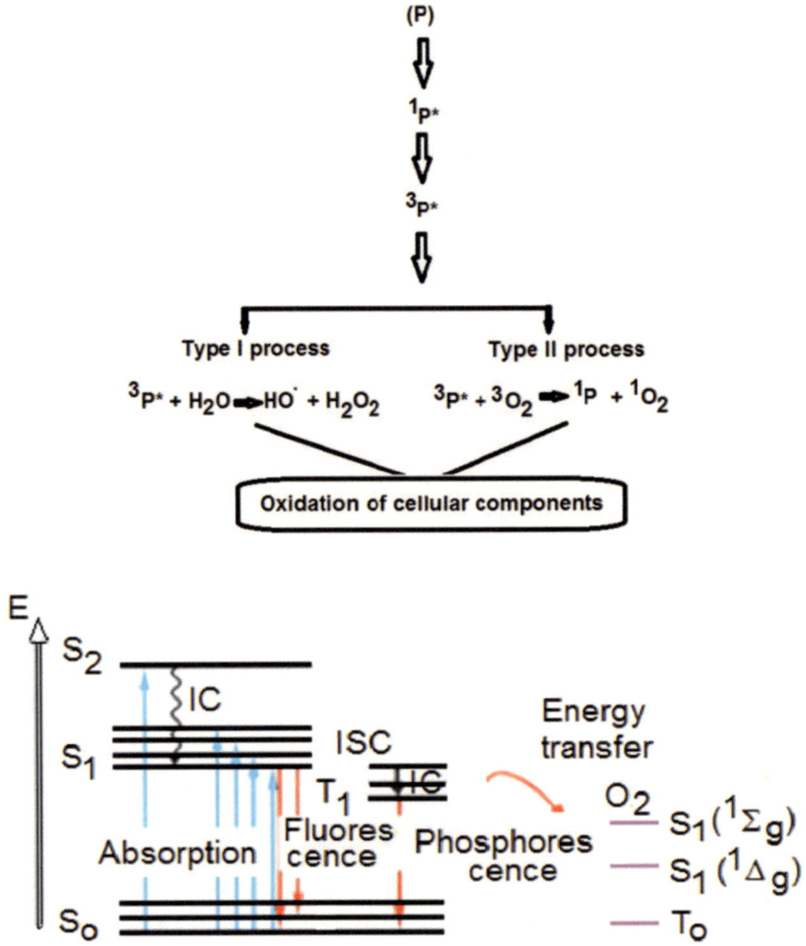


**Figure 2.** The dual (up) and singlet mechanism (down) of PDT

porphyrins and phthalocyanines appear to be good models in order to create new efficient drugs for photodynamic therapy of cancer [12]. The photodynamic therapy of cancer is based on the ability of some sensitizers to be retained in larger amounts and longer time in neoplastic than in normal cells as well as on the possibility to generate singlet oxygen after the activation of the porphyrin with red light [13].

In our previous investigations on PDT sensitizers [14, 15], we have used several spectroscopic methods to evaluate the efficiency of triplet state generation of some sensitizers [16]. The absorption spectra of the dye incorporated into the cells provide information with low accuracy because of small absorption amplitudes and perturbation of the spectra by light scattering. Therefore fluorescence methods are widely used [17]. The fluorescence of photosensitizers observed under a confocal microscope enables to establish the localization of dye in the cells [18–20]. Changes in the fluorescence of cell material due to illumination of samples depend on

http://dx.doi.org/10.5772/60872

the sensitizer localization in the cells, but for a group of similar dye molecules, the cells, stained to a higher degree, usually show higher fluorescence intensity [21].

### 3.2. Photobiological mechanism

Within PDT, a sensitizer, light and oxygen are used to cause photochemically induced cell death via apoptotic pathways. In vitro and in vivo photosensitizing efficacy of certain synthesized photosensitizers are usually investigated by their interaction with some proteins (human serum albumin (HSA) (Site II) for binding affinity, intracellular localization and DNA attack [22, 23].

The photobiological mechanisms by which some sensitizers induce tumor necrosis are very different: they strongly associate with serum albumins and it is assumed that they kill neoplastic cells indirectly by damaging blood vessels and interrupting the supply of oxygen [24]. A good example is water-soluble porphyrins, tetrakis (4-sulfonatophenyl) porphyrin ($H_2$TSPP) as one of the best tumor localizer. It has a very low affinity for human plasma lipoproteins but binds strongly to human serum albumin (HSA), Figure 3. The association between $H_2$TSPP and HSA has a spectacular effect on the singlet-state lifetime of porphyrins: it decreases from 5.2 ns to about 1.6 ns [25]. This marked decrease is due to the enhancement of one of the molecular processes involved in the deactivation of the singlet state. The shortening of the singlet state lifetime causes an important decrease of the quantum yield of triplet state ($\phi_T$). This assumption is in agreement with the result of Davila and Harriman [26] that measured the $\phi_T$ of the complex $H_2$TSPP–HSA and reported a value of 0.70. The shortening of the singlet state lifetime of this complex is compensated by a corresponding increase of the intersystem crossing rate. The photosensitizing efficiency of bound $H_2$TSPP has a reduced mobility, characteristic of molecules bound deeply into the protein structure. Such molecule is expected to be less accessible by oxygen and thus less efficient in producing singlet oxygen. The number of binding sites per protein molecule is greater than unity ($n$=3 and $k$=1.5 × $10^6$ $M^{-1}$ for one binding site and $k$=2x$10^5$ $M^{-1}$ for the other two). The binding of $H_2$TSPP to HSA significantly modifies the photophysics of porphyrin.

On irradiation with light (laser), in the presence of a photosensitizer, DNA undergoes several modifications including chain breaks, DNA–protein cross-links, and basic sites changes coupled with oxidized DNA bases, which have been shown to be able to induce point mutations [27]. The single strand breaks (SSB) are mainly formed through the attack of OH radicals, where the hydrogen bonds between purine bases play an important role. OH radicals react with DNA to remove an H atom, leading to strand rupture. Mitochondria play a central role in the control of apoptosis induced by PDT, which through the mitochondrial permeability transition pore (PTP) lose the integrity of the outer mitochondrial membrane, thus releasing the intermembrane proteins, such as cytochrome *c*, into the cytosol to form the "apoptosome". The apoptosome attacks procaspase-3 and cleaves it to form active caspase-3, which can lead to DNA breakage and nuclear chromatin condensation and cause cell death [28].

The porphyrins are able to induce changes of melting points and viscosity of DNA, which will be fragmented into small chains (Table 1). In this case, two modes of interaction between dye

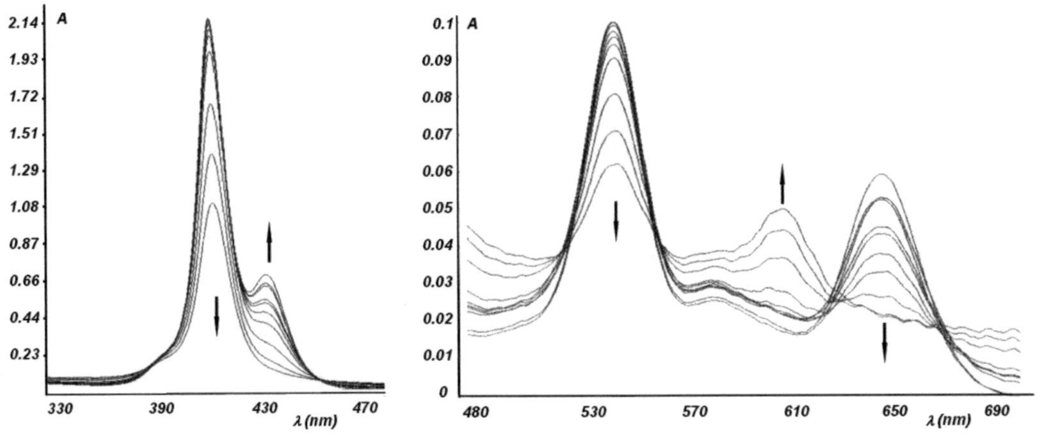


**Figure 3.** The changes of absorption spectra during the interaction between $H_2TSPP$ and HSA

molecules and DNA are distinguished: a strong binding mode involving about 20-23% of the DNA phosphate groups and a weak binding mode involving the remaining phosphate groups.

| DNA (50 μg/ml) | DNA (25 μg/ml) | | | DNA + TSPP (25 μg/ml + 8 μg/50 ml) | | |
|---|---|---|---|---|---|---|
| Sample | 1 | 2 | 3 | 4 | 5 | 6 |
| Irradiation time (min) | 0 | 0 | 0 | 30 | 45 | 60 |
| Flow time (min) | 49,85 | 49,38 | 49,46 | 48,86 | 48,61 | 48,36 |
| $\eta$ rel | 1,009 | 1 | 1,002 | 0.989 | 0,984 | 0,979 |

**Table 1.** The temperature and viscosity changes of DNA during the photodynamic process

An analysis on DNA degraded by sensitized attack of methylene blue (MB) showed that it had 80% of the guanine residues [29], and similar value for hematoporphyrins and furrocoumarins [30]. Apoptosis was evident in the post-PDT cells through the TdT-mediated dUTP nick end labeling (TUNEL) method and DNA fragmentation assay. Apoptosis was determined by cell morphology with light microscopy and transmission electron microscopy [31].

The two known modes of interaction between dye molecules and DNA are a strong binding mode involving about 20–23% of the DNA phosphate groups and a weak binding mode involving the remaining phosphate groups [32].

Based on the interaction sensitizer-DNA, three groups of porphyrins are known:

- Group I porphyrins, which induce changes characteristic of intercalation in DNA samples with greater 40% GC base composition with an increasing of the linear viscosity of CT DNA, strongest intercalative binding in GC regions and the localization of AT regions only outside binding.

- Group II porphyrins are believed to be outside binders, with an additional self-stacking features that induce DNA aggregation and a small viscosimetric changes of DNA with greater than 40% GC content.

- Group III porphyrins give results characteristic of outside binding at both AT and GC sites, identified by the absence of an increasing of the solution viscosity for any of the linear or super helical DNA samples; they being able to bind with a preferential orientation.

Cationic porphyrins can interact with DNA either by intercalation between the G–C base-pairs or by outside binding in the minor groove with A–T sites selectively. An exemplification has been checked by personal data for cytosine (C) and guanine(G) by using as sensitizers methylene blue (MB), $H_2$TSPP and Rhodamine (Rh) (Figure 4).

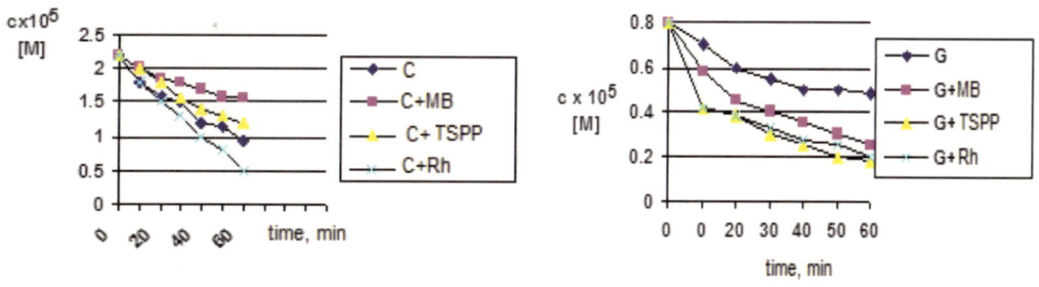


**Figure 4.** Cytosine (C)- and guanine(G)-sensitized photooxidation by using as sensitizers MB, $H_2$TSPP and Rh

The weakly bound dye molecules are identified to be attached to the helix exterior by means of electrostatic interaction with the phosphate groups. A red-shifted absorption is due to porphyrin–porphyrin interactions within long-range assemblies on the polymer in exterior. When the porphyrin is intercalated, a strong hypochromicity and a strong bathochromic shift for the absorption bands are observed. As a possibility of external association a small hypochromicity (or even a hyperchromicity) and a small batochromic shift are observed. If an external association is possible in concert with an aggregation process, only hypochromicity (or hyperchromicity) is important [28]. Some important conclusions had been stated: porphyrins intercalate in DNA with a binding constant of around $10^3$–$10^6$ $M^{-1}$; the porphyrin complex appears to be stabilized by extensive electrostatic interactions, especially in the minor groove; the porphyrin does not form van der Waals stacking contacts with adjacent bases; DNA may require conformational distortion reaching the limit of DNA melting; the resulting severe conformational distortion not resolved upon achievement of the ground-state complex; steric clashing between the DNA backbone and the porphyrin extending the DNA along its helical axis; and de-stacking the interior of the complex.

The aggregation process is very important in this interaction study. If the sensitizer is in a monomer form, the external association with DNA could be favored. The association process increases its rigidity and changes the distance between the bases creating the adequate distances unfavorable to the dimerization. The Soret band is batochromically shifted from 412 nm to 433 nm, and the Q-band decreases its intensity. New bands appear at 644 and 707 nm. Also, a new band appears at 490 nm (Figure 5). The band at 490 nm arises from the J-aggregate (edge-to-edge interaction) of porphyrin molecules [33]. Their structures have been identified by atomic force microscopy (AFM) (Figure 6). The porphyrins are externally bound when their planar structure fit into the major or minor groove and interact with DNA electrostatically. Also, hyperchromicity is observed (the greatest absorption value at 260 nm), which is similar with some mono- and di-catenare destructions. It could be presumed that during the laser irradiation support a photooxidation reaction takes place preferably at guanine site. By irradiation with DNA, an increase of viscosity is observed as an effect of a chain-breaks and helix destruction.

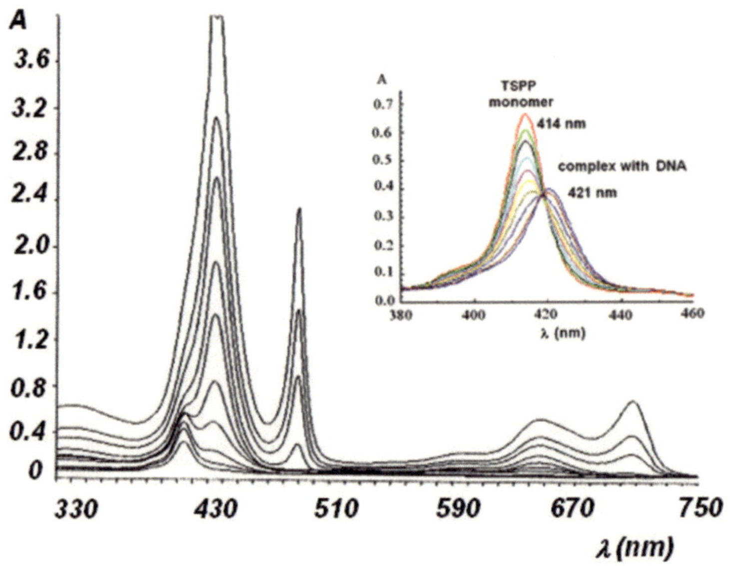


**Figure 5.** The changes of absorption spectra during the interaction between $H_2$TSPP and DNA

The fluorescence emission spectra of $H_2$TSPP, DNA and their complex consist of two bands with maxima little affected by DNA; however, the emission intensity is markedly reduced (Figure 7). For $H_2$TSPP, in the presence of DNA, the maxima from 605 and 640 nm increase simultaneously, as a proof for the external binding between both compounds. From the amplitude ration $I_{604}/I_{640}$, it was concluded that the porphyrin/DNA this ratio increases without irradiation and by irradiation with laser beam, this ratio registered a significant increase at the high time of irradiation. A spatial arrangement of porphyrin and DNA takes place on the DNA strand exterior based on electrostatic interaction between phosphate groups of DNA and positive charges of J-aggregated porphyrin. If $H_2$TSPP has a J-aggregate form, it is possible to cover DNA molecule like two concentric bodies. DNA has a 20 Å diameter, while $H_2$TSPP -J-aggregate has 0.35–0.40 Å interplanar distances and a 40 Å diameter [34].

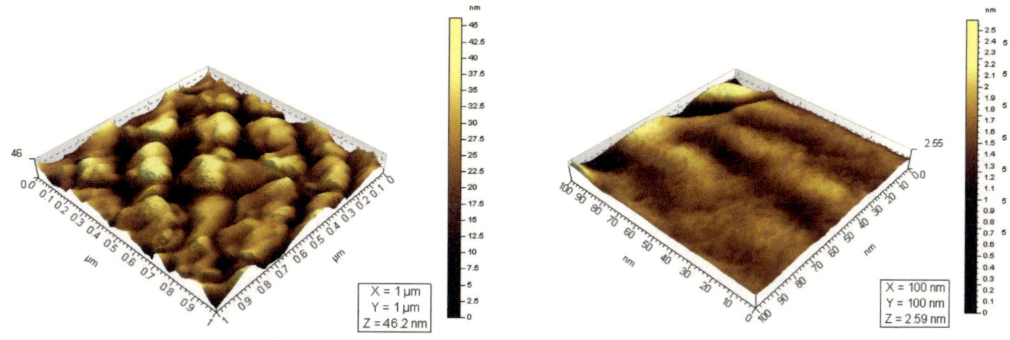


**Figure 6.** The structure of $H_2TSPP$ in monomer (left) and J-aggregate (right) form

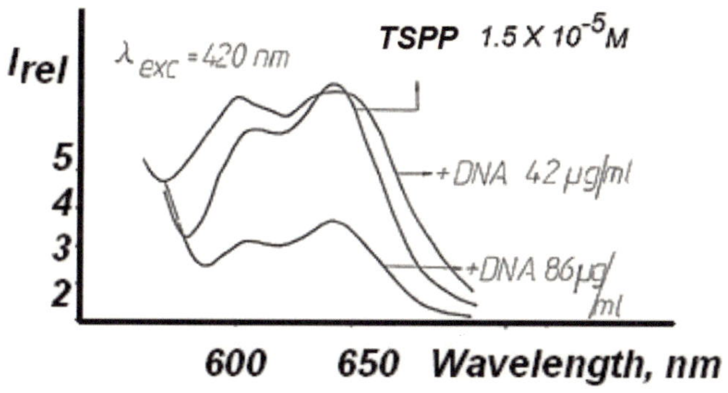


**Figure 7.** The fluorescence of $H_2TSPP$ and its mixture with DNA at different concentrations.

## 4. Photosensitizers

### 4.1. Conventional photosensitizers

The most used and most efficient sensitizers are synthetic compounds, as porphyrins (P) and phthalocyanines (Pc).

Porphyrins constitute a class of the molecules that contain four pyrrole rings linked by the methane carbon bridges. A large group of porphyrins can be chemically modified by introducing the metal at the center of the pyrrole rings or by attaching the peripheral groups to the outer rings of the methane bridges units, respectively. The main characteristic of free-base porphyrin consists in the absorption maxima: a Soret band around 400 nm and four Q-bands in the region of 500–700 nm. Although the porphyrins absorb light poorly in this wavelength region (650 nm, $\varepsilon$= 30000 $M^{-1}\cdot\bullet$ $cm^{-1)}$ [35], as a result of increased transparency of biological tissues at longer wavelength; red light is normally used for PDT. Some exemplification of porphyrins and phthalocyanines are shown in Figure 8.

Phthalocyanines (Pc)/naphthalocyanines (Nc) are molecules composed of four indole units: pyrrole rings linked by nitrogen atoms conjugated with benzene rings characterized by a strong, isolated Q band in the red region of the UV/Vis spectrum, while the less intense B band is found at higher energies (Figure 9). The Q band is characterized by a high molar absorptivity ($\varepsilon = 10^5\ M^{-1} \cdot cm^{-1}$) [36], and it is accompanied by a series of vibrational bands. The B band is broad due to the super-positioning of B1 and B2 [37].

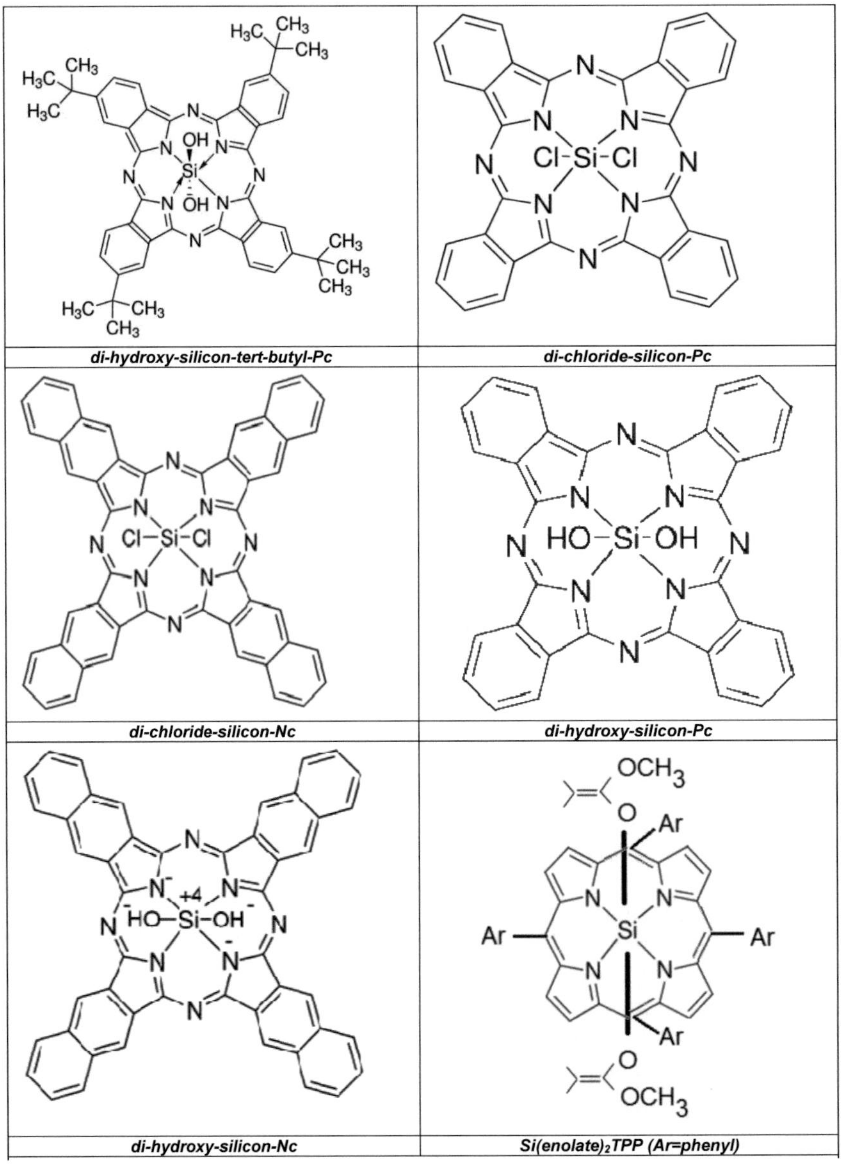


**Figure 8.** The structure of some SiPc, SiNc and SiTPP

Porphyrins act as free bases and chelated with a variety of metals, the diamagnetic ones enhancing the phototoxicity. Paramagnetic metals shorten the lifetime of the triplet state and as result can make the dyes photoinactive [38]. The photosensitizing activity is quenched by the presence of transition metal ions (as central ions) with a d-electron configuration [39]. An exemplification is Si(enolate)2 5, 10, 15, 20-tetra-*p*-phenyl porphyrin (Si(enolate)$_2$TPP) (Figure 8).

The presence of axial ligands to the centrally coordinated metal ion is often advantageous, since it generates some degree of steric hindrance to intermolecular aggregation, without impairing the photophysical properties of the dye. Several photophysical parameters for tetrasulfonated aluminum porphyrins (Table 2) have been evaluated by means of some in vitro experiments on EL-4 cells [40].

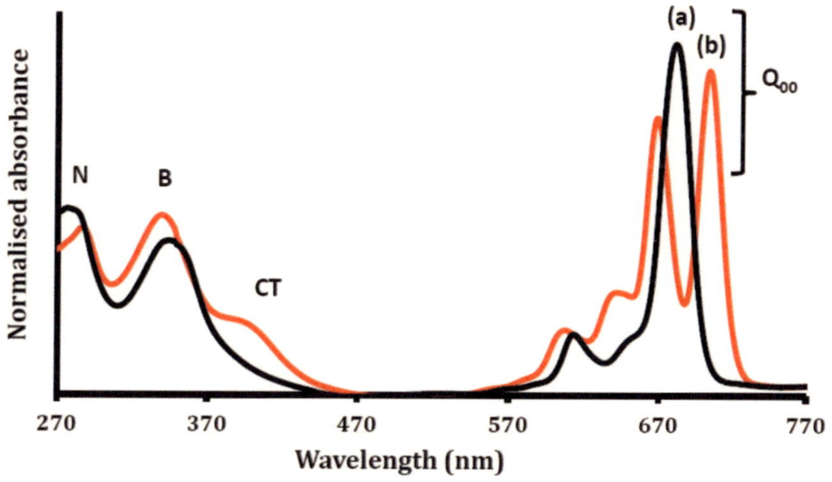


**Figure 9.** The spectra of Pc as free base (a) and metallic complex (b)

| Drug | absorption | excitation | fluorescence ($\lambda_{ex}$ = 427 nm) | phosphorescence ($\lambda_{ex}$= 427 nm) | Partition coefficient |
|---|---|---|---|---|---|
| $C_2$AlTSPP | 215; 230; 263; 266; 432; 570; 603; 645; 827 | 236; 254; 300; 330; 368; 438; 564; 644 | 344; 425;588; 651; 676;708; 854; | 950; 1075 | 0.23 |
| $C_8$AlTSPP | 212; 222; 264; 428; 564; 606; 645 | 250; 300; 422; 564; 608 | 426; 598; 652; 826; 706; 776; | 950 | 0.37 |
| $C_{12}$AlTSPP | 212; 222; 264; 428; 564; 606; 645 | 442; 644 | 426; 598; 652; 856 | 1075 | 0.32 |

**Table 2.** Spectral properties and data for partitioning coefficients of some metallo-porphyrins

The quantum yield for singlet oxygen generation was evaluated in this experiments by using DPBF method [41], and an exemplification is shown in Table 3.

| Porphyrin | $^1O_2$ quantum yield | Lifetime for singlet excited states of porphyrins (ps) |
|---|---|---|
| $C_2$AlTSPP | 0, 96 | 5.03 |
| $C_4$AlTSPP | 0.78 | 40, 8 |
| $C_8$AlTSPP | 0.83 | 32, 9 |
| $C_{12}$AlTSPP | 0.665 | 47, 5 |

**Table 3.** The singlet oxygen quantum yields and the lifetime values for the first excited states of Al porphyrins

The cellular uptake of different drugs seems to be correlated to their hydrophobicity only when the drugs are very closely related to chemical structure. Aluminum ion has a great influence on the sensitizer hydrophobicity without a corresponding effect on the cellular uptake (Table 4).

| Porphyrin | $K_{octanol}$ |
|---|---|
| TSPP | 0.47 |
| $C_2$AlTSPP | 0.23 |
| $C_8$AlTSPP | 0.37 |
| $C_{12}$AlTSPP | 0.32 |

**Table 4.** The data for cellular uptake of the aluminum porphyrins

Nevertheless, $C_2$AlTSPP is not as well uptaken as $H_2$TSPP, even in the studied series, porphyrin is the most reactive.The uptake ratio of porphyrins in vitro increases when increasing the lipophilic property of the drug. The uptake of aluminum porphyrins increases linearly with increasing lipophilicity. The inactivation kinetics for EL-4 cells is shown in Figure 10 and their microscopic aspects are shown in Figure 11.

Apoptosis (type I cell death) is different from necrosis (type III cell death) [42, 45]. If apoptosis is a controlled suicide pathway, involving DNA breaks, membrane blebbing, cell shrinkage and phagocytosis, necrosis involves membrane damage, local inflammation and injury, which lead to oncosis. Except of these types, there is type II cell death, which is characterized by an enormous increase of two-membrane autophagic vacuoles in the cytoplasm, which are finally catalyzed by lysosomal hydrolases [43]. Autophagy is a convertible process, which can provoke both survival and death pathways, in contrast to the apoptotic irreversible process leading only to cell death. Apoptotic cell death is the most preferable effect of various anti-cancer therapies, which leads to destruction and elimination of pathological cells. Inflammation does not occur through apoptosis in cancer cells and surroundings tissue [44].

http://dx.doi.org/10.5772/60872

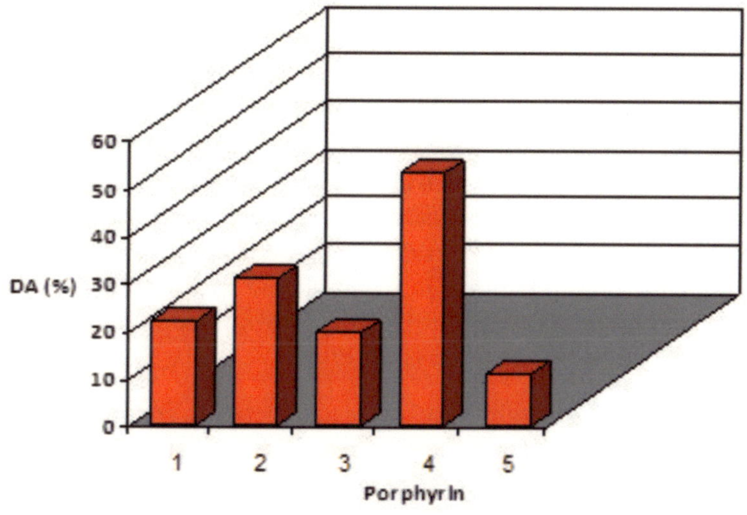


**Figure 10.** The inactivation diagram for EL-4 cells with aluminum porphyrins (1=$AlC_{12}TSPP$; 2=$AlC_8TSPP$; 3=Ali-BuTSPP; 4=$AlC_2TSPP$; 5=Control

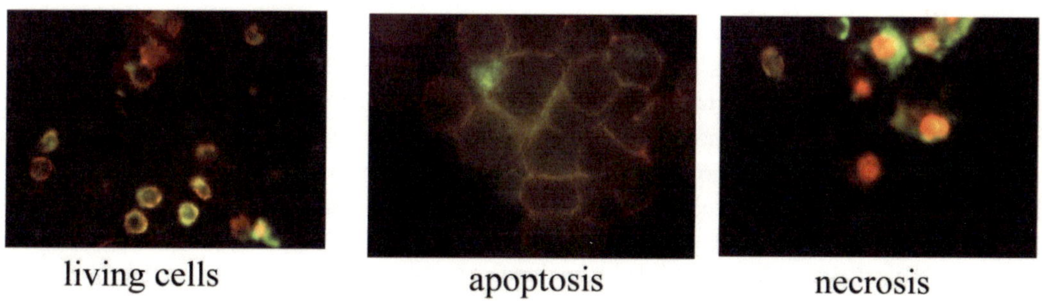


**Figure 11.** The microscopic aspects of cells before (left) and after PDT (middle and right)

## 5. Concept of oxidative stress in photodynamic therapy

### 5.1. Types of Reactive Oxygen Species (ROS)

PDT is a concerted action of a sensitizer, a light source in the presence of oxygen, based on singlet oxygen and ROS production, leading to cell death by different mechanism [46].

Reactive oxygen species (ROS) is a collective term used for a group of oxidants, which are either free radicals or molecular species able to generate free radicals. The most important free radicals occurring in the human body are hydroxyl radical (*OH), singlet oxygen ($^1O_2$) and superoxide ($O_2^{\bullet -}$). Except these species, the intracellular generation of ROS mainly comprises nitric oxide ($NO^\bullet$) radicals, which together with $O_2^{\bullet -}$ radicals, are converted to powerful oxidizing radicals like hydroxyl radical ($^\bullet OH$), alkoxy radicals ($RO^\bullet$), peroxyl radicals ($ROO^\bullet$), singlet oxygen ($^1O_2$) by complex transformation reactions. Some of the radical species are converted into molecular oxidants like hydrogen peroxide ($H_2O_2$), peroxynitrite ($ONOO^-$) and hypochlorous acid (HOCl). The oxidative stress damage is targeted mainly at cellular macromolecules, such as lipids, nucleic acids and proteins. Hydrophobic photosensitizers accumulate mainly in cell membranes and they are primarily attacked by free radicals. These oxidizing agents can damage cells by starting chain reactions such as lipid peroxidation, or oxidation of proteins or DNA. Damage to DNA can cause mutations and even major diseases (cancer), while protein oxidation can lead to distortion and degradation. Oxidation or oxidative stress leads to the production of free radicals, for example the hydrogen and oxygen molecules, which are not related to other factors. In their free form these molecules react with other molecules, and contribute to oxidative stress [47-49].

Oxidative stress, arising as a result of an imbalance between free radical production and antioxidant defenses, is associated with damage to a wide range of molecular species including lipids, proteins, and nucleic acids. Oxidative stress is a factor that initiates cell death after photodynamic reaction [50–54]. The formation of reactive oxygen species (ROS) with cytotoxic effects is the key involved in the death of the tumor cells by PDT [55]. PDT may be considered an oxidative stress that induces apoptosis in different types of cancerous cells both in vitro and in vivo [56].

Cells have a highly developed and regulated antioxidant defense system to maintain appropriate intracellular ROS levels and prevent oxidative damage. This system includes antioxidant enzymes such as superoxide dismutase (SOD), catalase and various peroxidases and non-enzymatic systems (GSH, thioredoxin, uric acid, vitamins, coenzyme Q) that effectively remove ROS. Under normal conditions, antioxidant mechanisms scavenge ROS and protect the organism from the damaging effects of ROS. However, under conditions of excessive oxidative stress, cellular antioxidant mechanisms may be unable to prevent the adverse impact of ROS on critical cellular processes. ROS can interact with cellular macromolecules, including DNA, protein and lipids, and interfere with vital cellular functions [35]. Mutations caused by ROS can result in malignant transformation and the development of cancer [57].

To evaluate the presence of oxidative stress in the plasma, we used indirect methods. These quantify the lesions produced by the reactive oxygen species on the organism's biomolecules.

To evaluate the presence of oxidative stress in PDT, some of the following methods could be used [58–64]:

- Detection of malondialdehyde (MDA), the marker used most frequently for lipid peroxidation, using the fluorimetric method with 2-thiobarbituric acid (TBA);
- The carbonylated proteins (CP) with 2, 4-dinitrophenylhydrazine using the Reznick method. To determine the level of proteins from the homogenate samples Bradford method was used.
- The hydrogen-donating capacity (HDA) using the stable free radical: 1, 1 diphenyl-picrylhydrasyl (DPPH);

http://dx.doi.org/10.5772/60872

- Detection of the –SH groups with 2, 2-dithiobisnitrobenzoic acid (Ellman's reagent). The zymographic method was used to determine the MMPs activity.

Some synthetic PSs such as porphyrins have the capacity to generate reactive oxygen species with cytotoxic effects leading to the necrosis of tumor cells, and induced a significant oxidative stress response, with peak intensity at 24-hour post-exposure. The most significant responses were the increase of CP, MDA, MMP-2 (matrix metalloproteinase-2) activity and the decrease in HDA levels. The experiment also induced less significant, but present, reduction of the thiol groups.

As an exemplification, PDT with $H_2$TSPP increases ROS production in plasma and tumor tissue and determines oxidative alterations of biomolecules (lipids, proteins) (Figure 12).

Oxidative stress, responsible for tissues injured in different pathologies, involves the none-quilibrium between the produced radical species and antioxidant defense agents.

There are some priority parameters which should be determined and evaluated:

1. Cell viability test measured by the lactate dehydrogenase release in the supernatant of cell culture as a marker of cell integrity. Viability results were expressed as % of live cells from the suspension subjected to 24 h incubation in the tested agents and/or irradiation.

2. Cell proliferation test measuring the number of viable cells as live cells reduce the kit reagent into a formazan compound that is colorimetric measured at 490 nm. From this point of view the test measures the quantity of viable cells in culture and thus the proliferative capacity of the tested cells.

3. RNA synthesis achieved by tritium-labeled uridine incorporation method, which involves beta-radioactivity measurement for radiolabeled cell cultures.

4. Total cellular RNA offer the concentration of total RNA, measured with SV total RNA isolation system.

5. TUNEL and immunofluorescence essential for apoptosis detection by observing the DNA fragmentation assay and TdT-mediated dUTP nick-end labeling (TUNEL) assay. TUNEL staining was performed to detect internal and end-strand breaks, which often occur in the early stages of apoptosis. TUNEL staining was carried out according to the manufacturer's instructions (Promega, Madison, USA). The procedure is carried out as follows. Biotinylated nucleotide is incorporated at the 3'-OH DNA ends using the terminal deoxynucleotidyl transferase (rTdT) recombinant enzyme. The apoptotic cells were counted under the microscope and photographed.

6. Analysis of genomic DNA fragmentation visualized by staining with ethidium bromide (0.5 μg/ml) and photographed under UV illumination. DNA was separated using standard 1.5% agarose (Bio-Rad) gel electrophoresis at 10 V/cm.

7. Cellular morphology visible by electron microscopy necessary for identifying the apoptotic cells. After PDT treatment, cells are developing morphological features characteristic of apoptosis: chromatin compaction into uniform electron-dense masses with

nuclear margination, nuclear fragmentation, cellular shrinkage, cell membrane vacuolization and blebbing, and the increase in electron-density of the cytosol [63-65].

**8.** Photodegradation reaction of sensitizers, which are indicators for lifetime of the sensitizers, time efficiency and potential side effects of them (Figure 13).

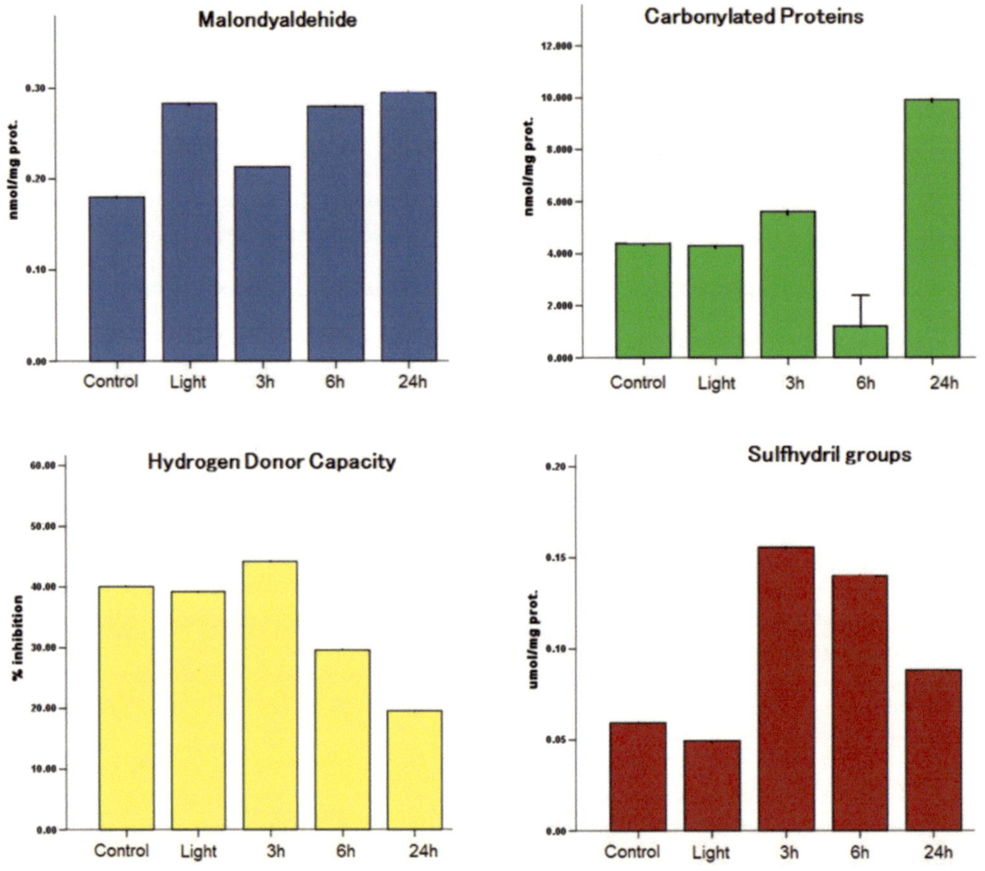


**Figure 12.** The oxidative stress parameters for PDT with $H_2$TSPP. Adapted after [55]

Some of these parameters have been calculated for PDT with different sensitizers and for different cells; B16, K562, EL-4, etc. [64–69]. During irradiation, after the first 30 minutes, a strong cellular degradation, especially for TNP-loaded cells, is visible (Figure 14). The results indicate that a free-base porphyrin as 5, 10, 15, 20-tetra-*p*-naphthyl-porphyrin (TNP) induces apoptosis in K562 cells in a time-dependent manner. There were no changes in tail moment of K562 cells in the absence of light, whereas TNP (10 μg/mL) leads to DNA damage at different time incubation (1–4 h), light dose (0, 10 or 25 J/cm$^2$) and time irradiation (5–50 minutes). Less genotoxicity will appear for higher light dose that for higher sensitizer concentration. Unloaded cells subjected to irradiation have a slightly higher LDH release, 56% proliferation capacity and 13% Urd incorporation compared to control cells. After 48 h of irradiation (Figure 14) cells have a high LDH release compared to control and a reduced proliferative capacity (100 %). After irradiation, the cells were subjected for 24 h, 48 h and 72 h to the LDH, MTS and uridine incorporation tests in order to study the capacity of remaining cells to proliferate after treatment. A significant reduction of RNA has been registered for TNP loaded and non-loaded cells (Figure 15). The viability of the irradiated K562 tumor cells was assessed by the increase of the LDH release and the associated decline of the MTS reduction (Figure 16). We also observed that cells that survive after the PDT procedure are more stable than the irradiated control.

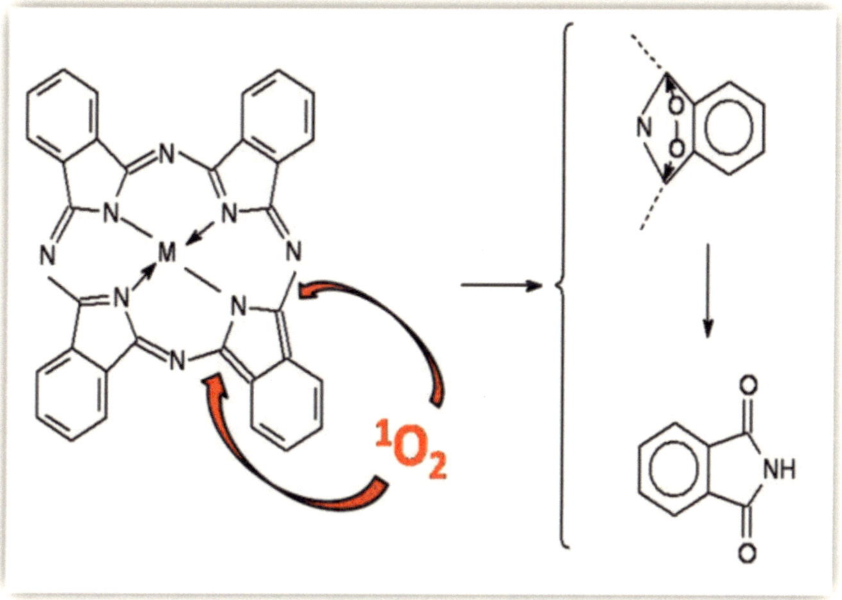


**Figure 13.** The photo degradation scheme of a sensitizer

After irradiation, the K562 cell viability is strongly affected both for TNP-loaded cells and for those unloaded. TNP activated increases the cell mortality by comparison with control cells (Mi) (Figure 15).

At different time post-irradiation, TNP intracellular loaded and activated shows a homogeneous effect on the cells' capacity to deliver LDH (Figure 16).

RNA isolation from K562 cells has been achieved in the first 4 hours after irradiation. A 35% decrease of RNA content has been registered in the case of $H_2TSPP$ (Figure 17) [72], Also, there is a good correlation between uridine incorporation and number of live cells in 24 h post-irradiation in the presence of the same sensitizer, both of them decreading (Figure 18). Also, the amount of total cellular RNA isolated from the K562 cells drastically decreased after irradiation (Figure 19).

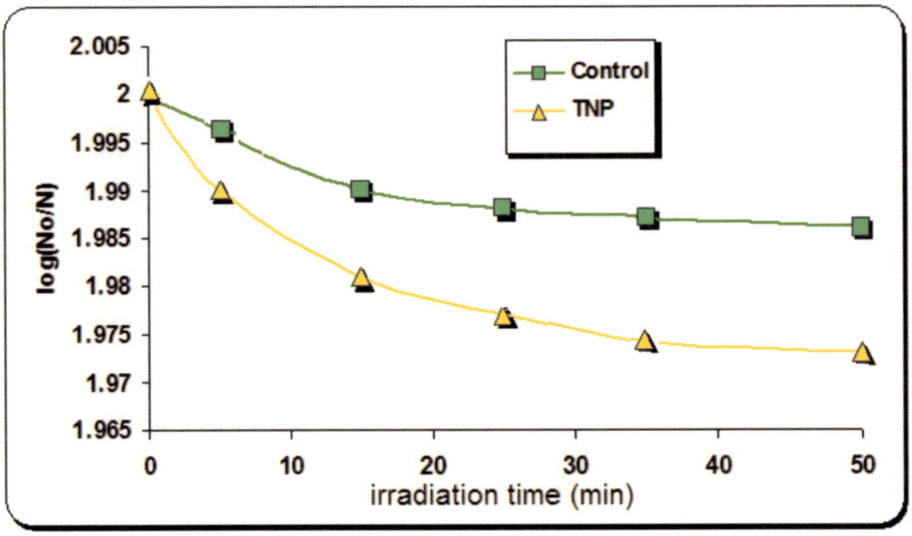


**Figure 14.** Time evolution of the cellular density for TNP-loaded K562 cells (10 μg/ml TNP)

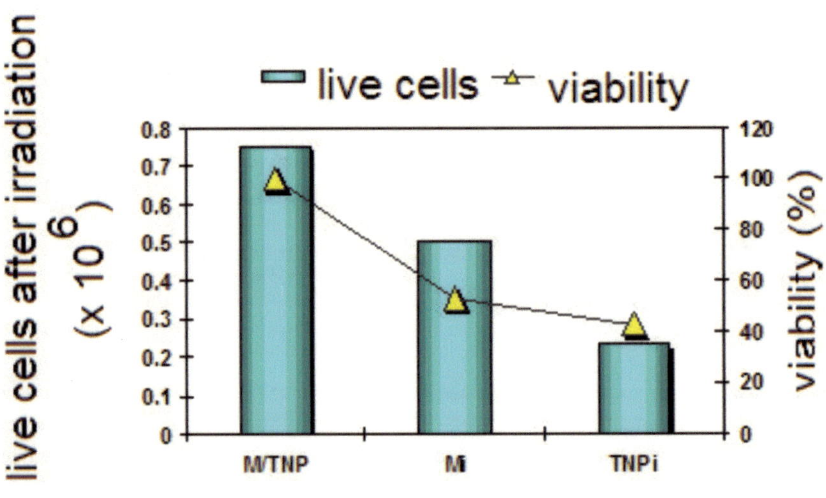


**Figure 15.** The viability of K562 loaded with TNP after the first hours after irradiation

At different time intervals after treatment, the cells were analyzed for caspase-3. At 4 h post-PDT, 100% of the cells displayed protease activities. After PDT, 30.6% of the cells showed total caspase activity. Thereafter, the fraction of cells with caspase activity increased to 48.6%.

When present in the cytoplasm, a number of caspases have been activated following PDT and responsible for the cleavage of multiple cellular proteins, DNA fragmentation, and cell death. Activation of procaspase-3 after PDT has been demonstrated in multiple experimental settings. The morphological manifestation of apoptosis ("execution" phase) can be ascribed as degra-

http://dx.doi.org/10.5772/60872

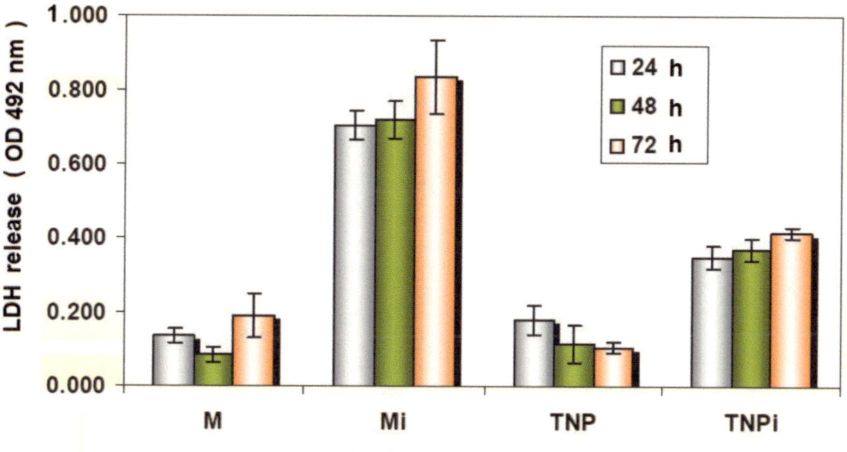


**Figure 16.** Time evolution of K562 cells viability after laser activation in the presence of TNP

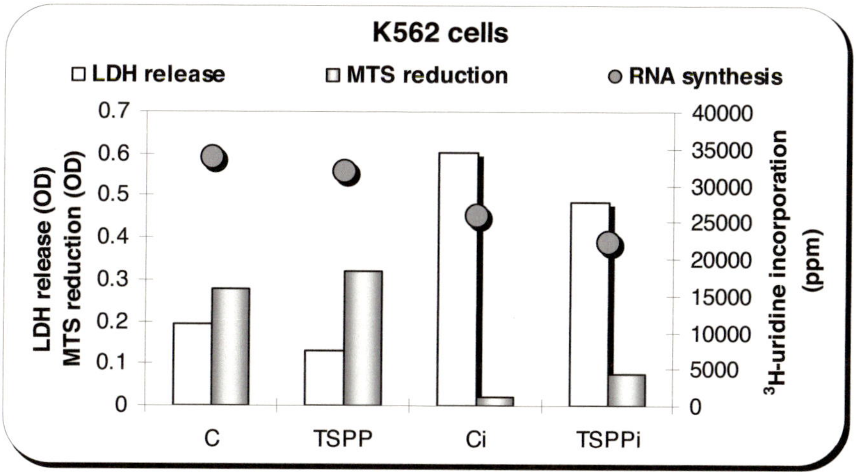


**Figure 17.** The viability and multiplication rate of K562 tumor cells at 24h post-irradiation. C = nonirradiated unloaded cells; TSPP = nonirradiated loaded cells; Ci = irradiated unloaded cells; TSPPi = irradiated loaded cells.

dation of various structural proteins and DNA. After loading K562 cells with TNP, and irradiation, a strong decrease has been obtained for caspase-3 activity (Figure 20).

Induced apoptosis of K562 cells loaded with 10 μg/ml TNP and activated with laser (TUNEL) has been shown in Figure 21, as results of TUNEL test. Many apoptotic cells characterized with brown nuclei can be seen in the TNP-PDT groups.

The PDT efficacy relies on the concerted action of sensitizer and light, with none of them alone inducing apoptosis.

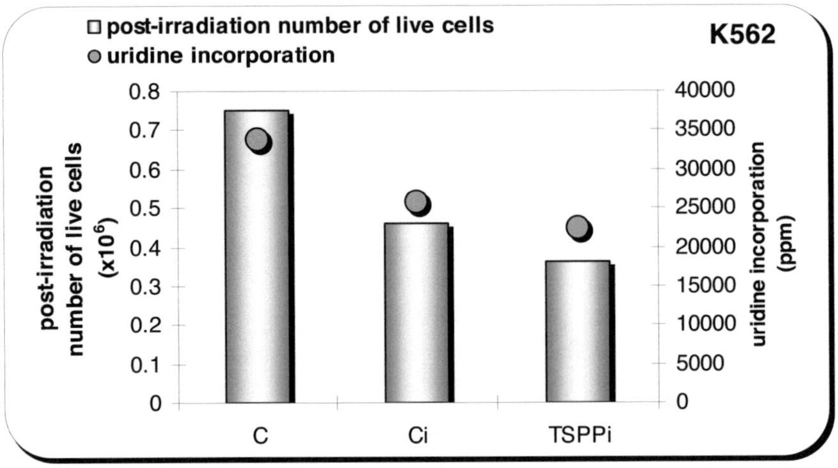


**Figure 18.** Correlation between uridine incorporation and number of live cells in 24 h post-irradiation. C = non-irradiated unloaded cells; Ci = irradiated unloaded cells; $H_2TSPP_i$ = irradiated loaded cells.

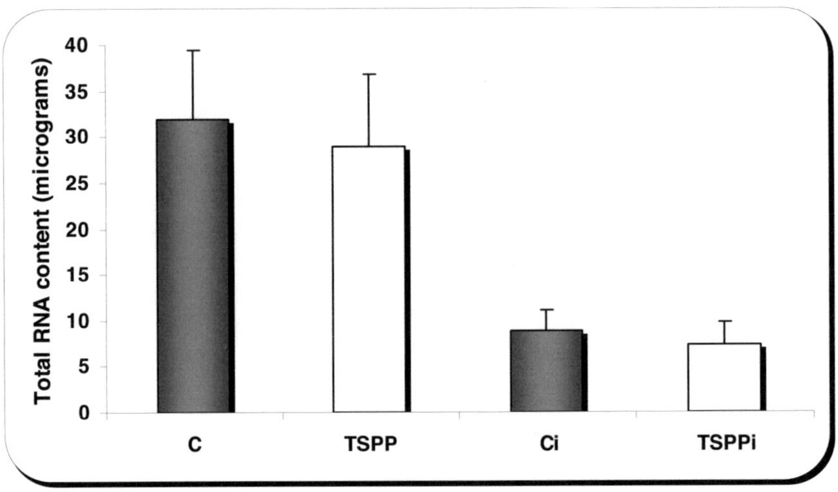


**Figure 19.** The amount of total cellular RNA isolated from the K562 cells. C = unloaded non-irradiated cells; Ci = irradiated unloaded cells; TSPP = non-irradiated loaded cells; TSPPi = irradiated loaded cells.

## 6. Natural products as antioxidants

An antioxidant is a molecule stable able to donate an electron to a free radical, neutralizing it, thus reducing its capacity to damage. The antioxidants delay or inhibit cellular damage mainly through their free radical scavenging property. A variety of dietary plants including grams, legumes, fruits, vegetables, tea, wine etc. contain antioxidants. The most important antioxidants seem to be nonenzymatic antioxidants derived from plant sources including vitamins (vitamin A, C, E, K), flavonoids (quercetin, catechin, epigallocatechin gallate, hesperidin,

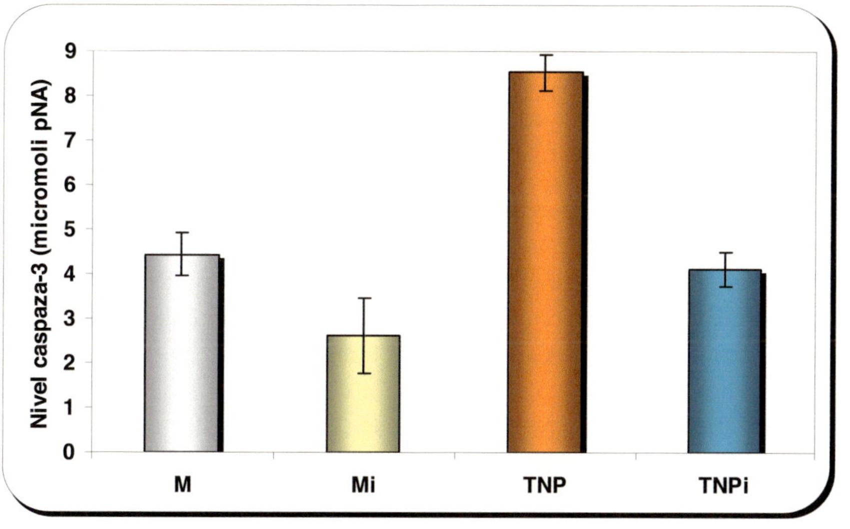


**Figure 20.** Caspase 3 activity in K562 cells loaded with 10 μg/ml TNP and activation by irradiation.

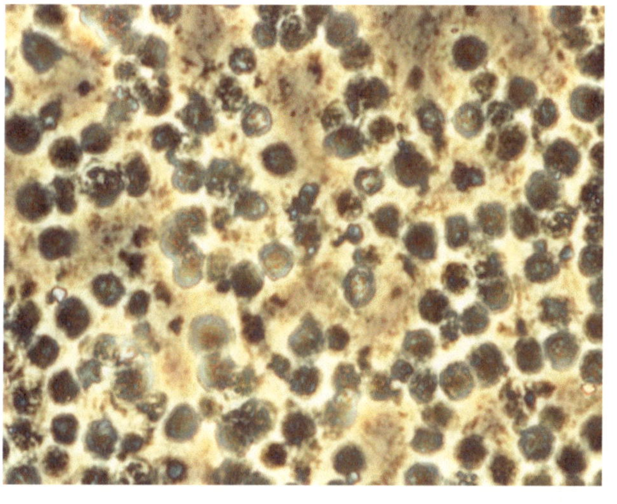

**Figure 21.** Induced apoptosis of K562 cells loaded with 10 μg/ml TNP and activated with laser (TUNEL). Results of TUNEL. Many apoptotic cells characterized with brown nuclei can be seen in the TNP-PDT groups.

hesperetin, diosmin, and many others), phenolic acids (cinnamic acid derivatives, curcumin, caffeine, catechins, gallic acid derivatives, salicylic acid derivatives, chlorogenic acid, resveratrol, folate, anthocyanins and tannins) [74–80].

Carotenoids, known as naturally fat-soluble pigments and synthesized by vegetal organisms, are sources of different colors [81]. They could be classified into carotenes (beta-carotene and lycopene) and xanthophylls (lutein and zeaxanthin).Their structure is shown in Figure 22. Beta-carotene and lycopene are widely regarded as being effective antioxidants, with small sizes (nm), specific absorption and fluorescence spectra, and are easy to detect (Figures 23–25). The synergic antioxidant effect of the mixture lycopene-beta-carotene-vitamin E on some cellular systems (in vitro and in vivo), has been reported [82–84]. Some antioxidants with health-protective effects (lycopene, beta-carotene, vitamin C, quercetin-glycosides, naringenin-chalcone, chlorogenic acid) are seen in tomato plants.

From ancient times, plants have been used intuitively for medicinal purposes. A large number of plants have been investigated and various species have been reported to exhibit antioxidant activity, including Marigold flower (*Calendula officinalis*), belonging to the Asteraceae family, which is a medicinal plant which contains oleanolic acid and other compounds, which present considerable potential health benefits, protective effects against the development of cancer, adverse effects of chemotherapy and radiation therapy, inhibition of existing tumor cells, anti-inflammatory activity, antioxidant activity, protective cardiovascular and antiviral effects [85]. The medicinal plants contain many ingredients with antioxidant capacity, as the pigments: antocyanins, chlorophyll, carotenoids, flavonoids, and so on [86]. Reactive oxygen species (ROS) comprise singlet oxygen (SO) and a range of oxidizing free radicals. The interaction of carotenoids with such species is important for the understanding of many important aspects of life such as photosynthesis, vision, various medical treatments from dermatology to cancer, as well as understanding possible deleterious reactions affecting man and also for commercial reasons, such as, investigations into the stability of carotenoids used as food dyes [87-90].

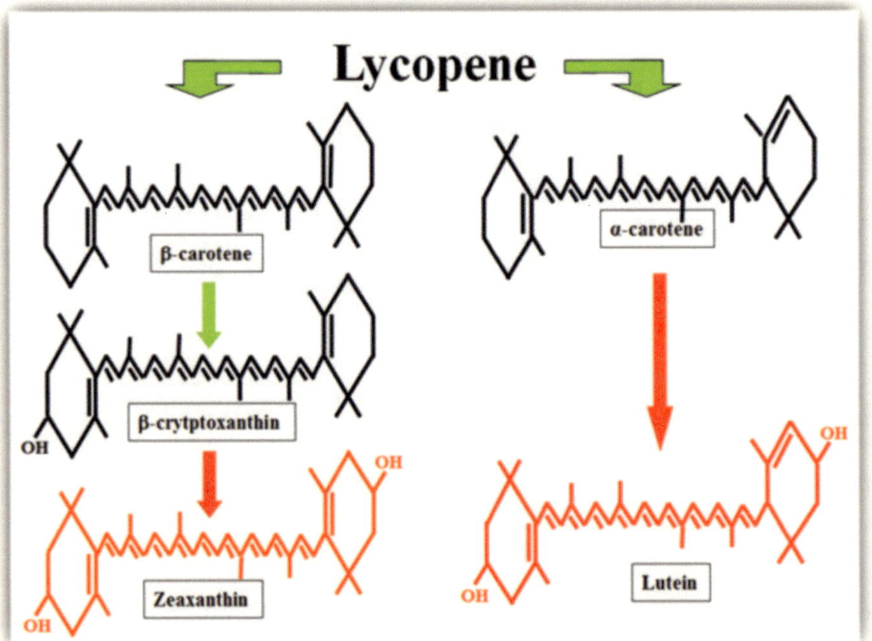


**Figure 22.** The transformation diagram of lycopene

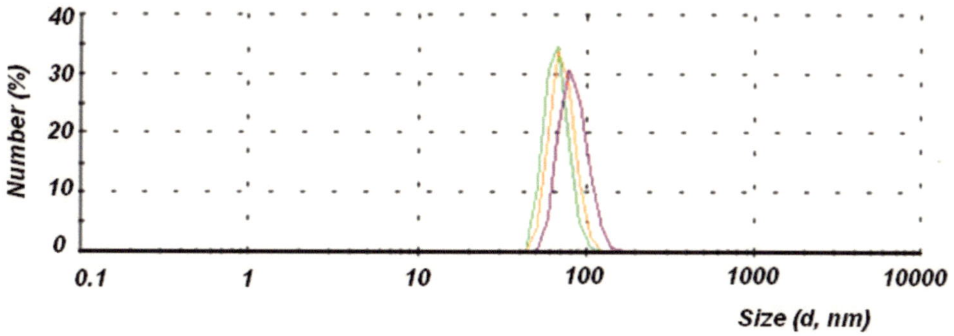


**Figure 23.** The DLS measurement of lycopene

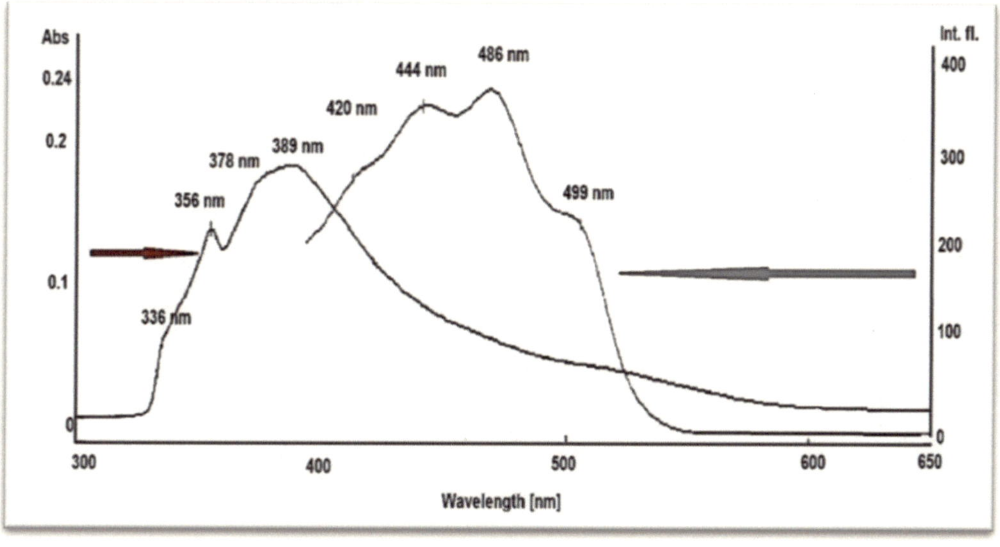


**Figure 24.** The absorption (left) and fluorescence (right) spectra of lycopene

Many antioxidants could be identified in tea leaves and fruits. For example, green tea, produced from the leaves of the plant (*Camellia sinensis*) contains polyphenols, which are potent antioxidants and, based on studies in preclinical models, have several photoprotective properties. The most active constituents are polyphenolic catechins, of which epigallocatechin-3-gallate is the most potent. Tea polyphenols have been shown to inhibit carcinogenesis in many animal models, and the significance of catechins, the main constituents of green tea, has been increasingly recognized to play a role in cancer prevention. Green tea contains some catechins, such as (–)-epigallocatechin-3-gallate (EGCG), epicatechin (EC), epicatechin -3-gallate (ECG), and epigallocatechin (EGC) (Figure 26), the first one being recognized as an efficient anticancer agent. Except the catechins, tea polyphenols exhibit carcinogenesis effect [91–101].

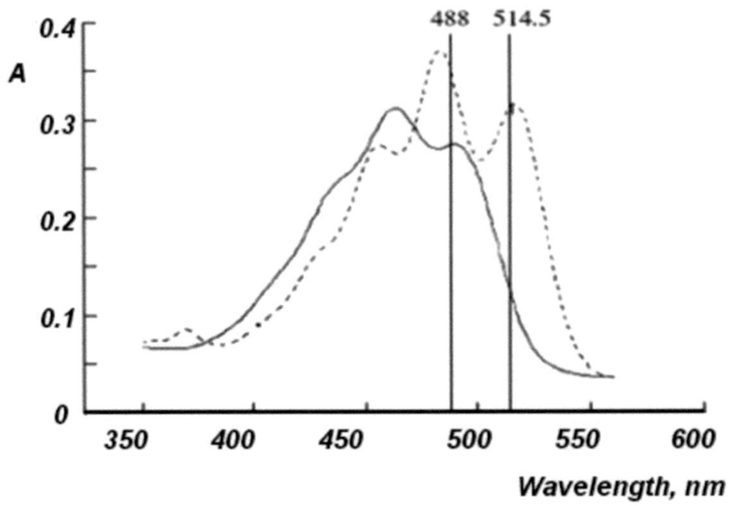


**Figure 25.** The absorption spectra of lycopene (- - -) and carotene (______)

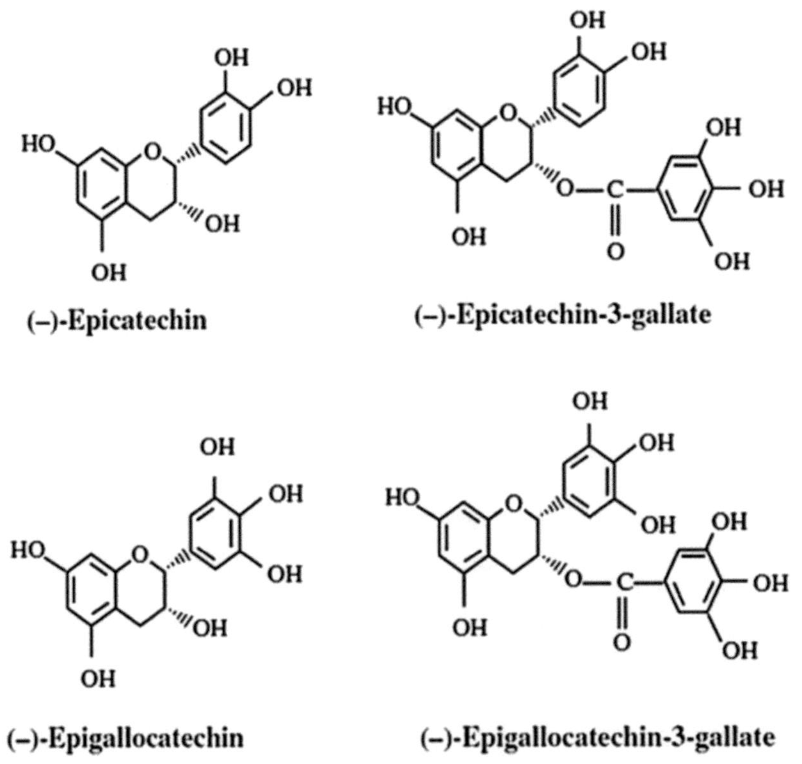


**Figure 26.** The chemical structure of the main epicatechin derivatives/polyphenols from green tea

http://dx.doi.org/10.5772/60872

The antioxidant activity (AA%) of the studied samples and their inhibitory effect against free radicals was evaluated using the DPPH method, by used the following formula:

$$AA\% = \left[A_{control} - A_{sample} / A_{control}\right] x\ 100$$

where: $A_{control}$ is the absorbance of a DPPH solution without sample, $A_{sample}$ is the absorbance of the sample mixed with DPPH solution.

Some citrus extracts are very important for their antioxidant activity such as hesperidin, hesperitin and diosmin (Figure 27). Among the flavonoids used in oral administration in chronic leg ulcer, hesperidin is a glucozid that is abundant in citrus fruits. Recently, formulating hesperidin in nanocrystals, has provided its dermatological application, assessing its antioxidant effect. In vitro studies have shown its clear antioxidant properties, and using them as nutrients has shown its vaso-protective action.

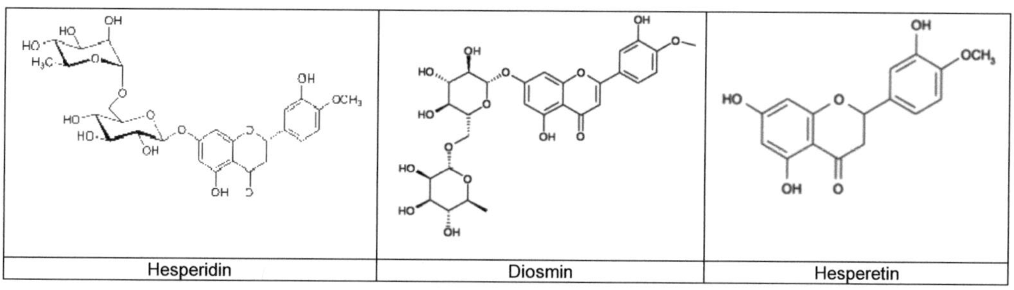


**Figure 27.** The structure of hesperidin, diosmin and hesperetin

Among mostly exploited flavonoids in chronic venous condition, hesperidin ($C_{28}H_{34}O_{15}$) [+/ −-3, 5, 7-trihydroxy-4'-methoxyflavanone 7-rhamnoglucoside] is a flavanone glycoside group of flavonoids found in large amounts in citrus fruits, grapefruit peels, lemon, oranges, blond grapefruit (*Citrus paradisi*), pummelo–blond grapefruit hybrid (*Citrus paradisi* var Jaffa Sweetie), or Chinese herbal medicine, with the highest concentration being found in the peel and the white parts of the fruit. Once formulated as nanocrystals, hesperidin also becomes dermally available and its antioxidant effect could be measured, a topic that is the subject of protected patents. In in vitro studies, this compound has a clear antioxidant action. In human nutrition, it contributes to maintenance of the integrity of blood vessels. Hesperidin has similar structure and properties with naringin, which is difficult to be separated from citrics [102]. In the same manner, diosmin is a hesperidin derivative, from the flavonoid family. As a synthetic drug, it is used for venous disease and for hemoroidal diseases [102]. Hesperetin, as a flavonoid from oranges and grapfruits, is a good protector of heart disease. It has antioxidant, anti-inflammatory, antiallergic and anticancer properties [103]. In spite of their structure differences, all of them show similar absorption spectra (Figure 28).

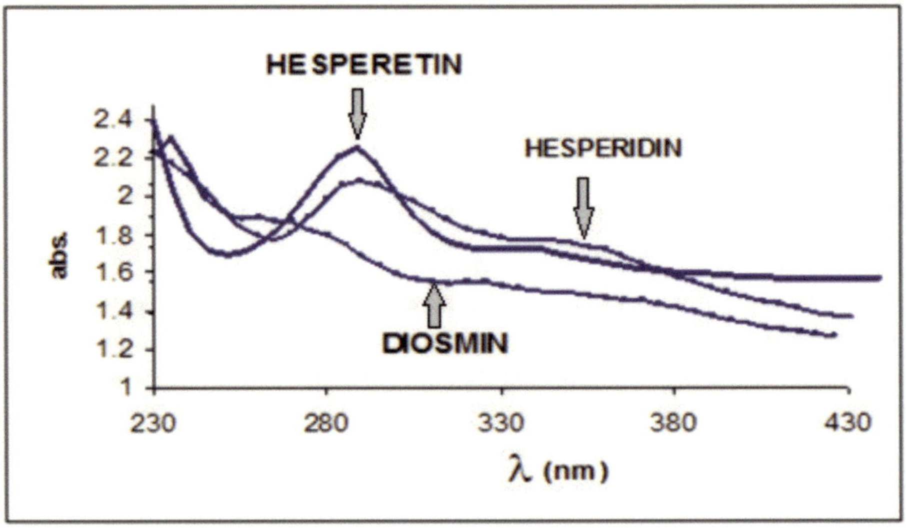


**Figure 28.** The absorption spectra of hesperidin, hesperetin and diosmin

For all of them, total flavonoids content (TFC), total poliphenols content (TPC) and antioxidant activity (AA%) have been calculated following literature methods (Table 5) [103–105].

| Sample | TFC (mg CE/L) | TPC (mg GAE/L) | AA (%) |
|---|---|---|---|
| Diosmin etalon in MeOH | 85.53 | 521.186 | 55.79 |
| Hesperidin etalon in MeOH | 43.1 | 466 | 59.31 |
| Hesperetin etalon in MeOH | 169.16 | 1814.406 | 73.34 |

**Table 5.** TFC, TPC and AA content

All the above measured parameters show very close values to the literature reports [106].

## 7. Conclusions

In spite of numerous advantages, the photodynamic therapy has a number of limitations. The light source should be close to the appropriate site in order to be effective before the diagnostic to be established. Persistent skin photosensitivity is observed some weeks after the treatment, which is considered as the main side effect. Also, this method is not possible without a light source and therefore entails high costs for the whole treatment. PDT acts both on ill cells and in a small manner on healthy cells. The intracellular $Ca^{2+}$ from the cells induce low levels of shear stress on them, without any morphological changes [112]. The reactive oxygen species are localized to the cancer cells selectively destroying them and not the surrounding normal

http://dx.doi.org/10.5772/60872

tissue. The presence of the lymphoid cells in tumor immunity has been demonstrated during PDT with various photosensitizers. The immunologic effects include the production of interleukin 1-beta, interleukin 2, tumor necrosis factor-alpha, and granulocyte colony-stimulating factor. PDT acts to induce oxidative stress by the generation of free radicals to damage DNA and proteins, and eventually cell death, by necrosis and apoptosis. The reactive oxygen species lead to the state called oxidative stress. The antioxidants offer protection against lipid oxidation, react and interfere with free radicals, reduce oxidative stress, and stop low-density lipoproteins from being oxidized. Also, they protect healthy tissues and lower the incidence of treatment-related side effects. In the context where many clinical studies with respect to the application of antioxidants as sensitizers are lacking, this chapter showed a systematic review by showing evidence of the antioxidant action in photochemotherapy and their comparison with synthetic sensitizers (porphyrins and phthalocyanines). Under such circumstances, PDT is extremely important for the treatment of different diseases: lung, bladder, and skin cancers (precancerous and even melanoma).

## Acknowledgements

We have received the financial support of the project PNII 185/2014 to submit this book chapter.

## Author details

Rodica-Mariana Ion[1,2*] and Ioana-Raluca Şuică-Bunghez[1]

*Address all correspondence to: rodica_ion2000@yahoo.co.uk

1 ICECHIM, Nanomedicine Research Group, Bucharest, Romania

2 Valahia University, Nanomaterials Center, Targoviste, Romania

## References

[1] Ion R.M. The use of phthalocyanines and related complexes in photodynamic therapy. In Nyokong T., Ahsen V. (eds.); Photosensitizers in Medicine, Environment and Security. Springer, 2012, pp. 315–349.

[2] Ion R.M. Porphyrins for tumor destruction in photodynamic therapy. Current Topics on Biophysics 2000;24(1) 30–42.

[3] Ion R.M. Photodynamic Nanomedicine Strategies in Cancer Therapy and Drug Delivery. In Advances in Bioengineering, Intech, 2015.

[4] Ion R.M. Photodynamic therapy (PDT): A photochemical concept with medical applications, Revue Roumaine De Chimie 2007;52(12):1093–1102.

[5] Clichici S., Filip A., Daicoviciu D., Ion R.M., Mocan T., Tatomir C. The dynamics of reactive oxygen species in photodynamic therapy with tetra sulfophenyl-porphyrin. Acta Physiologica Hungarica 2010;97(1):41–51.

[6] Pushpan S.K., Venkatraman S., Anand V.G., Sankar J., Parmeswaran D., Ganesan S., Chandrashekar T.K. Porphyrins in photodynamic therapy—a search for ideal photosensitizers. Current Medicinal Chemistry Anti-Cancer Agents 2002;2(2):187–207.

[7] R.M. Ion, Nanomedicine between laboratory and clinical applications, in: Nanostructuring and Nanocharacterization, in Micro and Nanoengineering Series (eds. M. Zaharescu, M. Ciurea, D.Dascalu), Ed. Academiei, Bucharest, Romania (2010):241–256.

[8] Frackowiak D., Planner A., Waszkowiak A., Boguta A., Ion R.M., Wiktorowicz K. Yield of phthalocyanines intersystem (singlet-triplet) crossing evaluated on the basis of time resolved photothermal method. Journal Photochemistry and Photobiology A: Chemistry 2001;141(2–3):101–108.

[9] Frackowiak D., Waszkowiak A., Manikowski H., Ion R.M., Cofta J., Wiktorowicz K. The interactions of phtalocyanines with stimulated and resting human peripheral blood mononuclear cells. Acta Biochimica Polonica 2001;48(1) 257–269.

[10] Frąckowiak D., Planner A., Ion R.M., Wiktorowicz K. Incorporation of dye in resting and stimulated leukocytes: in: Near-Infrared Dyes for High Technology Applications. NATO ASI SERIES, (eds. S. Daehne, U. Resch-Genger, O. S. Wolfbeis), Kluwer Academic Publishers, Dordrecht, (1998):87–114.

[11] R.M. Ion, A.A. Sorescu, A. Nuta. New metallo-sensitizers in photodynamic therapy, in: Proceedings in Advanced Research in Scientific Areas, EDIS-Publishing Institution of the University of Zilina, Slovakia (2014) 223–228.

[12] Ion RM. Porphyrins and photodynamic therapy of cancer, Bucharest; FMR; 2003.

[13] Ochsner M. Photophysical and photobiological processes in the photodynamic therapy of tumours. Journal Photochemistry Photobiology B 1997;39(1):1–18.

[14] Foote C.S. Definition of type I and type II photosensitized oxidation. Photochemistry and Photobiology 1991;54(5):659.

[15] Bergamini C.M., Gambetti S., Dondi A., Cervellati C. Oxygen, reactive oxygen species and tissue damage. Current Pharmaceutical Design 2004;10(14):1611–1626.

[16] Sharman W.M., Allen C.M., van Lier J.E. Role of activated oxygen species in photodynamic therapy. Methods Enzymol 2000;319:376–400.

[17] Halliwell B. Oxygen radicals: a commonsense look at their nature and medical importance. Medical Biology 1984;62(2):71–77.

[18] Fernandez J.M., Bilgin M.D., Grossweiner L.I. Singlet oxygen generation by photodynamic agents, Journal of Photochemistry and Photobiology B:Biology 1997;37(1-2): 131–140.

[19] Plaetzer K., Krammer B., Berlanda J., Berr F., Kiesslich T. Photophysics and photochemistry of photodynamic therapy: fundamental aspects. Lasers in Medical Science 2009;24(2):259–268.

[20] Bonnett R. Photosensitizers of the porphyrin and phthalocyanine series for photodynamic therapy. Chemical Society Reviews 1995;24:19–33.

[21] Boguta A., Wójcik A., Ion R.M., Wróbel, D. Photothermal methods as tools for investigation of weakly interacting non-fluorescent phthalocyanines. Journal of Photochemistry and Photobiology A: Chemistry, 2004;163(1-2):201–207.

[22] Ion R.M., Scarlat F., Gunaydin K.,.Niculescu V.I.R, Scarlat Fl. Radiation-induced degradation of DNA in photodynamic therapy of cancer. Studia Universitatis Babeş-Bolyai, Physica—Special Issue, 2001.

[23] Alarcon E., Edwards A.M., Aspee A., Moran F.E., Borsarelli C.D., Lissi E.A., Nilo D.G., Poblete H., Scaiano J.C.. Photophysics and photochemistry of dyes bound to human serum albumin are determined by the dye localization. Photochemical and Photobiological Sciences 2010;9(1):93–102.

[24] Ion R.M. Porphyrins for tumor destruction in photodynamic therapy. Current Topics on Biophysics 2000;24(1):30–42.

[25] Gunaydin K., Ion R.M., Scarlat F., Scarlat F.l., Niculescu V. I. R., Macau, C. (2004). Study of the fluorescence-quenching of Mg-TNP by anionic anthraquinones. Journal of Optoelectronics and Advanced Materials 2004;6(1):289–296.

[26] Davila J., Harriman A. Photoreactions of macrocyclic dyes bound to human serum albumin, Photochemistry and Photobiology 1990;51(1):9–19.

[27] Jasaitis A., Streckyte G., Rotomskis R. Spectroscopic studies of photosensitizer-human serum albumin complexes and their photostability, Proceedings of Photochemotherapy: Photodynamic Therapy and Other Modalities II, Vienna, 1996, SPIE, v.2924, 91–101.

[28] Ion R.M., Derivative UV-VIS spectrophotometry for porphyrins interactions in photodynamic therapy. Analytical Letters 2010;43(7):1277–1286.

[29] Chanon, M. Julliard M., Mehta G., Maiya B.G. Is $^1O_2$ Alone sufficient for DNA cleavage? Possible involvement of paramagnetic intermediates, Research on Chemical Intermediates 1999; 25(7):633–644.

[30] Vedaldi D., Caffieri S., Miolo G., Dall'Acqua F., Arslan P. Dark and photohemolysis of erythrocytes by furocoumarins. Zeitschrift für Naturforschung C. 1988; 43(11–12) 888–892.

[31] Ion R.M., Grigorescu M., Scarlat F., Niculescu V.I.R, Scarlat Fl., Porphyrins sensitization of DNA, Romain Journal of Physics 2001;46(9–10):637–646.

[32] Wang Q., Wang X., Yu Z., Yuan X., Jiao K. Spectroscopic and electrochemical studies on the binding mechanism of DNA with an anthraquinone biological dye, nuclear fast red, International Journal of Electrochemical Science 2011;6:5470–5481.

[33] Constantin C., Neagu M., Manda G., Ursaciuc C., Radu E., Ion R.M. Aspecte privind efectul terapiei fotodinamice cu porfirine de sinteza asupra continutului de acizi nucleici din celule tumorale, Proc. Al 35-lea simpozion de Imunologie, Timisoara, 2007, pp. 22.

[34] Ion R.M., Scarlat F., Niculescu V.I.R. Porphyrins as advanced materials in PDT. Romanian Journal of Physics 2003;48(1–2):339–346.

[35] Wróbel D., Hanyz I., Bartkowiak R., Ion R. M. Fluorescence and time-resolved delayed luminescence of porphyrins in organic solvents and polymer matrices. Journal of Fluorescence 1998;8(3):191–198.

[36] Lever A.B.P., The phthalocyanines, Advanced Inorganic Radiochemistry 1965;7:27–32.

[37] Stillman MJ., Nyokong T. In Phthalocyanines: Properties and Applications; New York; VCH; 1989.

[38] Frackowiak D., Planner A., Waszkowiak A., Boguta A., Ion R. M., Wiktorowicz K. Yield of intersystem (singlet-triplet) crossing in phthalocyanines evaluated on the basis of a time in resolved photothermal method. Journal of Photochemistry and Photobiology A: Chemistry 2001;141(2–3):101–108.

[39] Wróbel D., Goc J., Ion R. M. Photovoltaic and spectral properties of tetraphenyloporphyrin and metallotetraphenyloporphyrin dyes. Journal of Molecular Structure 1998;450(1–3):239–246.

[40] Ion R.M., Savi L., Savi G., Niculescu VIR. Photophysical parameters for EL-4 inactivation in PDT; Studia Universitatis Babes Bolyiai, Studia Physica 1 (2003).

[41] Ion R.M., Photochemical production and quenching of singlet oxygen by the porphyrins used in photodynamic therapy of cancer. Romanian Journal of Biophysics 1996;6(3–4):205–212.

[42] Kroemer G., Dallaporta B., Resche-Rigon M. The mitochondrial death/life regulator in apoptosis and necrosis. Annual Review of Physiology 1998;60:619–642.

[43] Kuwana, T., and Newmeyer, D.D. BCL-2-family proteins and the role of mitochondria in apoptosis. Current Opinion in Cell Biology. 2003;15(6);691–699.

[44] Kerr JF, Winterford CM, Harmon BV. Apoptosis. Its significance in cancer and cancer therapy. Cancer. 1994;73(8):2013–2026.

[45] Majno G, Joris I. Apoptosis, oncosis, and necrosis. An overview of cell death. The American Journal of Pathology 1995;146(1):3–15.

[46] Ion RM: Photodynamic therapy (PDT): a photochemical concept with medical applications. Revue Roumaine de Chemie. 2007;52(12):1093–1102.

[47] Pham-Huy L.A., He H., Pham-Huy C., Free Radicals, antioxidants in disease and health, International Journal of Biomedical Science 2008;4(2):89–96.

[48] Harman D. Role of free radicals in aging and disease. Annals of the New York Academy of Sciences 1992;673:126–141.

[49] Knight J. Free radicals: Their history and current status in aging and disease. Annals of Clinical & Laboratory Science 1998;28(6):331–346.

[50] Weishaupt. K.R., Gomer. C.J., Dougherty. T.J. Identification of singlet oxygen as the cytotoxic agent and photoinactivation of a murine tumor. Cancer Resources 1976;36(7PT1):2326–2329.

[51] Daicoviciu D., Filip A.G., Ion R.M., Clichici S., Decea N., Muresan A. Oxidative photodamage induced by photodynamic therapy with methoxyphenyl porphyrin derivatives in tumour-bearing rats. Folia Biologica 2011;57(1):12–19.

[52] Susan M., Baldea I., Senila S., Macovei V., Dreve S., Ion R. M., Cosgarea R. Photodamaging effects of porphyrins and chitosan on primary human keratinocytes and carcinoma cell cultures. International Journal of Dermatology 2011;50(3):280–286.

[53] Filip A.G., Clichici S., Daicoviciu D., Ion R.M., Tatomir C., Rogojan L., Opris I., Mocan T., Olteanu D., Muresan A. Possible in vivo mechanisms involved in photodynamic therapy using tetrapyrrolic macrocycles. Brazilian Journal of Medical and Biological Research 2011;44(1):53–61.

[54] Agarwal M.L., Larkin H., Zaidi S.I.A., Mukhtar H., Oleinick. N.L. Phospholipase activation triggers apoptosis in photosensitized mouse lymphoma cells. Cancer Resources 1993;53(24):5897–5902.

[55] Clichici S., Filip A., Daicoviciu D., Ion R.M., Mocan T., Tatomir C., Rogojan L, Olteanu D, Muresan A. The dynamics of reactive oxygen species in photodynamic therapy with tetra sulfophenyl-porphyrin. Acta Physiologica Hungarica, 2010;97(1):41–51.

[56] Matei C., Tampa M., Caruntu C., Ion R.M., Georgescu S.R., Dumitrascu G.R., Constantin C., Neagu M. Protein microarray for complex apoptosis monitoring of dysplastic oral keratinocytes in experimental photodynamic therapy, Biological Research 2014;47(1):33–42.

[57] Khaitan D., Dwarakanath B.S. Endogenous and induced oxidative stress in multi-cellular tumour spheroids: implications for improving tumour therapy, Indian Journal of Biochemistry & Biophysics 2009;46(1):16–24.

[58] Popescu T., Nenu I., Aldea M.D., Olteanu D., Gheban D., Tatomir C., Bolfa P., Muresan A., Ion R.M., Filip A.G. The effect of TSPP-mediated photodynamic therapy and Parecoxib in experimental tumours. Life Sciences 2014;117(2):75–83.

[59] Conti M, Moran PC, Levillain P. Improved fluorimetric determination of malondialdehyde. Clinical Chemistry. 1991;37(7):1273–1275.

[60] Ferrario A, Chantrain C, Von Tiehl K, Buckley S, Rucker N, Shalinsky D, Shimada H, DeClerck YA, Gomer J. The matrix metalloproteinase inhibitor Prinomastat enhances photodynamic therapy responsiveness in a mouse tumor model. Cancer Research 2004;64(7):2328–2332.

[61] Magi B, Ettorre A, Liberatori S, Bini L, Andreassi M, Frosali S, Neri P, Pallini V, Di Stefano A. Selectivity of protein carbonylation in the apoptotic response to oxidative stress associated with photodynamic therapy: a cell biochemical and proteomic investigation. Cell Death & Differentiation 2004;11(8):842–852.

[62] Reznick AZ, Packer L. Oxidative damage to proteins: spectrophotometric method for carbonyl assay. Methods in Enzymology. 1994;233:357–363.

[63] Saczko J, Kulbacka J, Chwilkowska A, Lugowski M, Banas T. Levels of lipid peroxidation in A549 cells after PDT in vitro. Rocz Akad Med Bialymst 2004;49(1):82–84.

[64] Mohammed AA, Ibrahim AA. Pathological roles of reactive oxygen species and their defence mechanism. Saudi Pharmaceutical Journal 2004;12:1–18.

[65] Nenu I., Popescu T., Aldea M.D., Craciun L., Olteanu D., Tatomir C., Bolfa P., Ion RM., Muresan A., Filip A.G. Metformin associated with photodynamic therapy—A novel oncological direction, Journal of Photochemistry and Photobiology B: Biology 2014;138: C80–91.

[66] Tampa M., Matei C., Popescu S., Georgescu S.R., Neagu M., Constantin, C., Ion, R.M. Zinc trisulphonated phthalocyanine used in photodynamic therapy of dysplastic oral keratinocytes, Revista de Chimie, 2013;64(6):639–645.

[67] Matei C., Tampa M., Ion R.M., Neagu M., Constantin C. Photodynamic properties of aluminium sulphonated phthalocyanines in human displazic oral keratinocytes experimental model, Digest Journal of Nanomaterials and Biostructures, 2012;7(4): 1535–1547.

[68] Constantin C., Neagu M., Manda G., Ion R.M., Iordachescu D., The effects of laser activation of 5, 10, 15, 20-tetra-sulphophenyl-porphyrin loaded in K562 cells and human normal mononuclear cells. Romanian Archives of Microbiology and Immunology 2004;63(3-4):159–168.

[69] Pop S.F., Ion R.M., Neagu M., Constantin C., Photodynamic therapy on B16 cells with tetrasulphonated porphyrin and different light sources. Journal of Materials Science and Engineering 2010;4(3):10–16.

[70] Neagu M., Ion R.M., Manda G., Constantin C., Radu E., Cristu Z., Antitumoral effect of calixarenes in experimental photodynamic therapy with K562 tumor cell line, Romanian Journal of Biochemistry 2010;47(1):17–35.

[71] Ion, R.M., Neagu M., Manda G., Constantin C., Calin M. Mechanisms in photodynamic therapy photosensitizers and cellular localization on K562 cells, In Progress in Biomedical Optics and Imaging - Proceedings of SPIE, 2007, 6632, 17–21 June, 2007, Munich, Germany.

[72] Neagu M., Manda G., Constantin C., Radu E., Ion, R.M. Synthetic porphyrins in experimental photodynamic therapy induce a different antitumoral effect. Journal of Porphyrins and Phthalocyanines, 2007;11(1): 58–65.

[73] Stoykova E., Nedkova K., Sabotinov O., Ion, R.M., Alexandrova R. In vitro cytotoxicity assessment of second-generation photosensitizers for photodynamic therapy. Journal of Optoelectronics and Advanced Materials, 2007;9(2):490–493.

[74] Aruoma OI, Methodological consideration for characterization for potential antioxidant actions of bioactive components in plants foods. Mutation Research 2003;(523–524):9–20.

[75] López-Varela S, González-Gross M, Marcos A. Functional foods and the immune system: A review. European Journal of Clinical Nutrition 2002;56(3):S29–33.

[76] Han R.M., Zhang J.P., Skibsted L.H., Reaction dynamics of flavonoids and carotenoids as antioxidants. Molecules 2012;17 2140–2160.

[77] Vidya AD, Devasagayam TP. Current status of Herbal drug in India: An overview. Journal of Clinical Biochemistry and Nutrition 2007;41(1):1–11.

[78] Furuta S, Nishiba Y, Suda I. Fluorometric assay for screening antioxidative activities of vegetables. Journal of Food Science. 1997;62(3):526–528.

[79] Wang H, Cao G, Prior RL. Total antioxidant capacity of fruits. Journal of Agricultural and Food Chemistry 1996;44(3):701–705.

[80] Rao A.L., Bharani M., Pallavi V. Role of antioxidants and free radicals in health and disease. Advances in Pharmacology and Toxicology 2006;7:29–38.

[81] Holloway DE, Yang M, Paganga G, Rice-Evans CA, Bramley PM, Isomerization of dietary lycopene during assimiltion and transport in plasma, Free Radical Research 2000;32(1):93–102.

[82] Tapas AR, Sakarkar DM, Kakde RB. Flavonoids as nutraceuticals: A review. Tropical Journal of Pharmaceutical Research. 2008;7(3):1089–1099.

[83] Baranska M, Schütze W, Schulz H., Determination of lycopene and beta-carotene content in tomato fruits and related products: Comparison of FT-Raman, ATR-IR, and NIR spectroscopy, Analytical Chemistry. 2006;78(24):8456–8461.

[84] Bunghez I.R., Raduly M., Doncea S., Aksahin I., Ion R.M., Lycopene determination in tomatoes by different spectral techniques (UV-VIS, FTIR and HPLC), Digest Journal of Nanomaterials and Biostructures, 2011;6(3):1349–1356.

[85] Gupta V.K., Sharma S.K., Plants as natural antioxidants, Natural Product Radiance, 2006;5(4):326–334.

[86] Stoica R., Velea S., Ilie L., Calugareanu M., Ghimis S.B., Ion, R.M., The influence of ethanol concentration on the total phenolics and antioxidant activity of *Scenedesmus Opoliensis* algal biomass extracts, Revista de Chimie, 2013;64(3):304–306.

[87] Bensasson RV, Land EJ, Truscott TG. Excited States and Free Radicals in Biology and Medicine, Oxford University Press, Oxford, 1993, p. 195.

[88] Cantrell A, McGarvey DJ, Truscott TG, Rancan F, Boehm F., Singlet oxygen quenching by dietary carotenoids in a model membrane environment. Archives of Biochemistry and Biophysics. 2003;412(1):47–54.

[89] Edge R, Truscott TG. Properties of carotenoid radicals and excited states and their potential role in biological systems; in Carotenoids: Physical, Chemical and Biological Functions and Properties. Landrum JT (ed), Boca Raton; CRC Press, 2010, pp. 283–304.

[90] Edge R, McGarvey D, Truscott TG. The carotenoids as antioxidants—A review. Journal of Photochemistry and Photobiology B: Biology, 1997;41(3):189–200.

[91] Stoica, R., Pop, S.F., Ion, R.M., Evaluation of natural polyphenols entrapped in calcium alginate beads prepared by the ionotropic gelation method Journal of Optoelectronics and Advanced Materials, 2013;15(7-8):893–898.

[92] Wang Z, Agarwal R, Bickers D, Mukhtar H. Protection against ultraviolet B radiation-induced photocarcinogenesis in hairless mice by green tea polyphenols. Carcinogenesis 1991;12(8):1527–1530.

[93] Lin JK, Lin CH, Ling YC, Lin-Shian SY, Juan IM. Survey of catechins, gallic acid and methylxantines in green, oolong, pu-erh and black teas. Journal of Agricultural Food Chemistry. 1998;46(9):3635–3642.

[94] Ahn WS, Huh SW, Bae SM, Lee IP, Lee JM, Namkoong SE, Kim CK, Sin JI. A major constituent of green tea, EGCG, inhibits the growth of human cervical cancer cell lines, CaSKI cells, through apoptosis, G1 arrest and regulation of gene expression. DNA Cell Biology 2003;22(3):217–224.

[95] Taniguchi S, Fujiki H, Kobayashi H, Go H, Miyado K, Sadano H, Shimokawa R. Effect of (−)-epigallocatechin gallate, the main constituent of green tea, on lung metastasis with mouse B16 melanoma cell lines, Cancer Letters, 1991; 65(1):51–54.

[96] Nakazato T, Ito K, Ikeda Y, Kizaki M. Green tea component, catechin, induces apoptosis of human malignant B cells via production of reactive oxygen species. Clinical Cancer Research 2005;11(16):6040–6049.

[97] Fierascu I., Bunghez I.R., Fierascu R., Ion, R.M., Dinu-Pîrvu C.E., Nuţa, D., Characterization and antioxidant activity of phytosynthesised silver nanoparticles using *Calendula officinalis* extract, Farmacia, 2013;62(1):129–136.

[98] Mittal A, Piyathilake C, Hara Y, et al.: Exceptionally high protection of photocarcinogenesis by topical application of ()-epigallocatechin-3-gallate in hydrophilic cream in SKH-1 hairless mouse model: Relationship to inhibition of UVB-induced global DNA hypomethylation. Neoplasia 2003;5(6):555–565.

[99] Katiyar S, Matsui M, Elmets C, Mukhtar H. Polyphenolic antioxidant (–)-epigallocatechin-3-gallate from green tea reduces UVB-induced inflammatory responses and infiltration ofleukocytes in human skin. Photochemistry and Photobiology 1999;69(2):148–153.

[100] Katiyar S, Afaq F, Perez A, Mukhtar H. Green tea polyphenol(–)-epigallocatechin-3-gallate treatment of human skin inhibits ultraviolet radiation-induced oxidative stress. Carcinogenesis 2001;22(2):287–294.

[101] Otsuka T, Ogo T, Eto T, Asano Y, Suganuma M, Niho Y. Growth inhibition of leukemic cells by(–)-apigallocatechin-3-gallate, the main constituent of green tea. Life Sciences. 1998;63(16):1397–1403.

[102] Srilatha D., Nasare M., Nagasandhya B., Prasad V., Diwan P. Development and Validation of UV Spectrophotometric Method for Simultaneous Estimation of Hesperidin and Diosmin in the Pharmaceutical Dosage Form. Hindawi Publishing Corporation, ISRN Spectroscopy 2013, ID 534830, 4 pp.

[103] Erlund I., Review of the flavonoids quercetin, hesperetin and naringenin. Dietary sources, bioactivities, bioavailability, and epidemiology. Nutrition Research 2004;24(10):851–874.

[104] Biju J., Sulaiman C.T., Satheesh G., Reddy V.R.K., Total phenolics and flavonoids in selected medicinal plants from Kerala. International Journal of Pharmacy and Pharmaceutical Sciences, 2014;6(1):406–408.

[105] Zhishen J, Mengcheng T, Jianming W., The determination of flavonoid contents in mulberry and their scavenging effects on Superoxide radicals, Food Chemistry, 1999;64(4):555–559.

[106] Jakobek L., Seruga M., Medvidović-Kosanović M., Novak I. Antioxidant activity and polyphenols of aronia in comparison to other berry species, Agriculturae Conspectus Scientificus, 2007;72(4):301–306.

[107] Mosquera O.M., Correra Y.M., Nino J. Antioxidant activity of plant extracts from Colombian flora. Brazilian Journal of Pharmacognosy 2009;19(2A):382–387.

[108] Gurushankar K., Gohulkumar M., Rajendra Prasad N., Krishnakumar N. Synthesis, characterization and in vitro anti-cancer evaluation of hesperetin-loaded nanoparticles in human oral carcinoma (KB) cells, Advances in Natural Sciences: Nanoscience and Nanotechnology 2014;5(1):015006(10pp).

[109] Tsai T.H, Chen Y.F., Determination of unbound hesperetin in rat blood and brain by microdialysis coupled to microbore liquid chromatography, Journal of Food and Drug Analysis, 2000;8(4):331–336.

[110] Tsimogiannis D., Samiotaki M., Panayotou G., Oreopoulou V., Characterization of flavonoid subgroups and hydroxy substitution by HPLC-MS/MS, Molecules 2007;12(3) 593–606.

[111] Kuntic V., Pejic N., Svetlana Micic S., Direct spectrophotometric determination of hesperidin in pharmaceutical preparations, Acta Chimica Slovenica, 2012;59(2) 436–441.

[112] Yin H., Zhang X., Pattrick N., Klauke N., Cordingley H. C., Haswell S. J., Cooper J. M., Influence of hydrodynamic conditions on quantitative cellular assays in microfluidic systems. Anal. Chem. 2007; 79(18):7139–7144.